The Inclusion of Environmental Education in Science Teacher Education

Alec M. Bodzin · Beth Shiner Klein
Starlin Weaver
Editors

The Inclusion of Environmental Education in Science Teacher Education

Springer

Editors
Alec M. Bodzin
Lehigh University
Bethlehem, PA
USA
amb4@lehigh.edu

Beth Shiner Klein
SUNY Cortland
NY
USA
Beth.Klein@cortland.edu

Starlin Weaver
Salisbury University
Salisbury, MD
USA
sdweaver@salisbury.edu

ISBN 978-90-481-9221-2 e-ISBN 978-90-481-9222-9
DOI 10.1007/978-90-481-9222-9
Springer Dordrecht Heidelberg London New York

Library of Congress Control Number: 2010932326

Printed on acid-free paper

Springer is part of Springer Science+Business Media (www.springer.com)

Preface

Imagine small groups of students scattered about a natural area adjacent to their school. One group is identifying local plant life using a field guide. They are choosing plants that will be part of their Nature Gallery that they will share with other students. Another group has a microphone connected to an iPod and is trying to record the sounds of cicadas for use in a podcast they are creating about how insects communicate. Finally, another group is using mobile technology to collect stream data. This data will be incorporated into a larger watershed management study that uses Google Earth tools. Ultimately, they are all engaged in place-based, meaningful, and active learning experiences.

As the students work, their teacher walks around the natural area observing, questioning, and advising the students on their projects. As she moves to the next group, she thinks to herself how lucky she was to have had rich pedagogical environmental education experiences during her preservice teacher preparation. These experiences prepared her to maximize learning for field trips in outdoor settings, supplied her with environmental education curriculum content, taught her that environmental education is interdisciplinary and that technology can be used appropriately to enhance outdoor explorations and investigations. This preparation really made the difference for her class. The students seem to grasp new concepts much more quickly when they are provided with active learning experiences and truly understand how many important science concepts, ideas, and learning approaches transcend discrete subject areas.

This vision summarizes the hope that the editors of this book and the Association for Science Teacher Education (ASTE) Environmental Education forum members have for the improvement of integrating environmental education (EE) into science teacher preparation. According to the NSF report, *Complex Environmental Systems: Synthesis for Earth, Life, and Society in the 21st Century* (Pfirman and AC-ERE 2003), in the coming decades, the public will be called upon more frequently to understand complex environmental issues, evaluate proposed environmental plans, and understand how individual decisions affect the environment at local to global scales. The report calls for raising the environmental literacy of the general public by providing quality environmental education and training. Support for environmental education (EE) in school curricula has been well established, with 95% of the general public supporting the teaching of EE in schools (NEETF/Roper 2001). However, there has been no significant progress in incorporating EE into K-12 school curricula (NEETF/Roper 2001; Ramsey et al. 1998).

Studies published during 1995–2005 and reviewed by the State Education and Environment Roundtable (SEER) have shown that schools that do manage to incorporate EE programs demonstrate significant growth in student achievement and improved student behavior. For example, the SEER publication *Closing the Achievement Gap: Using Environment as an Integrating Context for Learning* (Lieberman and Hoody 1998) indicated that students in schools that incorporate EE improved in standardized test scores; science, mathematics, and literacy achievement; problem-solving and critical thinking skills; and improved overall grade point average. However, in order for such learning benefits to occur, teachers must be willing and able to support the integration of EE in the K-12 curriculum.

One of the reasons that the K-12 EE integration is rare is that many preservice teacher education programs do not incorporate EE into their programs of study (Rakow 1985; McKeown-Ice 2000). In a national survey of 715 teacher education institutions, McKeown-Ice (2000) found that only half of the surveyed students in preservice programs received exposure to EE. Barriers to including EE in preservice teacher education programs include a lack of faculty knowledge about EE, the inflexible structure of preservice teacher education courses, lack of EE standards in teacher accreditation and certification requirements, and curriculum time constraints in the teacher education curriculum (Scott 1996; McKeown-Ice 2000).

While studies indicate that there are many benefits to incorporating EE into preservice teacher education, the current extent to which it is incorporated is low. When preservice EE preparation is implemented, the treatment of EE is often shallow (Heimlich et al. 2004; McKeown-Ice 2000; Lane et al. 1995). These studies noted that teachers were not confident to enter classrooms and implement EE pedagogical methodologies, including curricular teaching and learning strategies after completing their education training. To increase EE curriculum integration in our schools, teachers must be confident and willing to incorporate EE pedagogical practices and learning activities in their classrooms.

Environmental education, when taught in school settings, is predominantly integrated into a school's science curriculum. Often, it is taught within earth and environmental science topic areas pertaining to ecosystems and environmental issues such as energy, climate change, pollution, and natural resources. EE teaching and learning also commonly occurs in life sciences topic areas that include biodiversity, endangered species, and genetic engineering. These discipline-based topic areas are covered in most basal science textbooks curriculum programs that are marketed to schools for adoption (McComas 2002; Wilson 2000). Most states incorporate EE-related content within their state science standards. While a few states such as Pennsylvania (Pennsylvania Department of Education 2002) and Wisconsin (Wisconsin Department of Public Instruction 1998) have academic standards for EE, such states do not require an EE course for high-school graduation nor do they prescribe how EE should be taught at the local level. In most states, EE-related content is integrated into science curriculum standards (No Child Left Inside Coalition 2009) and is assessed to some degree on high-stakes state science testing. That said, there has been a recent national resurgence of EE through the proposed *No Child Left Inside Act of 2009* legislation to advance environmental education in US schools through the creation of state environmental literacy plans.

Through the integration of EE within the context of science teacher preparation, preservice and inservice teachers can become aware of teaching and learning strategies for cognitive, affective, and behavioral goals of EE and can help their students with becoming environmentally literate. This book focuses on the inclusion of EE content and pedagogy in science teacher preparation as it applies towards instructional practices in K-12 science classrooms, early childhood settings, and other learning environments. EE goals include developing citizens that are aware of and concerned about the environment and its associated problems. To achieve these goals, science teacher educators need to prepare preservice and inservice teachers with knowledge of EE understandings and essential skills to ensure that their future students will have opportunities to acquire knowledge, values, attitudes, and commitment to protect and improve the environment. In science teacher preparation, this involves the teaching and learning of pedagogical approaches for creative problem-solving skills, scientific and social literacy, ethical awareness and sensitivity for the relationship between humans and the environment, making informed decisions, and commitment to engage in responsible actions.

About This Book

The purpose of this book is to share knowledge and ideas about EE pedagogy in the context of science teacher preparation as it applies to teaching and learning in K-12 science classrooms and their associated learning environments. The chapters in this book share, examine, and discuss EE foundations and pedagogical principles through theoretical and practical applications as it primarily pertains to the preparation of preservice and inservice science teachers. This book is designed to inform science teacher educators about the historical and philosophical underpinnings of EE, current trends in EE as it pertains to science teacher education, and EE-specific pedagogical practices and content-pedagogical knowledge as it applies to science teacher education. The book includes a series of case studies that highlight the teaching and learning of EE content and concepts in science teacher education. Some chapters highlight EE exemplary practice with K-12 and early childhood students in traditional classroom settings in addition to nontraditional instructional settings such as outdoor and field-trip settings. In addition, this book describes innovative science teacher preparation programs that have found ways to address the barriers to EE integration that are inherent to many teacher preparation programs.

A few key themes permeate across the book chapters:

1. *Inquiry-based teaching and learning is an integral part of EE.* EE instructional methods such as role-playing simulations, naturalistic inquiry, and field investigations incorporate essential features of inquiry (National Research Council 2001).

2. *EE is multidisciplinary and provides many applications for the teaching and learning of science.* In addition to developing science content knowledge and science process skills, EE incorporates a multidisciplinary approach to learning that incorporates problem-solving, critical thinking, and literacy skills, that are each inherent to other core school disciplines.
3. *It is important to provide preservice and inservice teachers with professional development experiences in outdoor settings.* Preservice teachers have a lack of comfort in outdoor settings as a location for learning, and most having no experience with learning outside of a traditional classroom.
4. *EE is a collaborative endeavor.* It is not necessary to include EE into science teacher education in isolation. EE educators based in a variety of settings are available to collaborate in innovative ways with science teacher educators to promote science and environmental literacy.

The primary audience for this book is science teacher educators. That said, the chapters in the book will appeal to a wide audience including faculty in teacher preparation programs, classroom science teachers, and environmental educators who work with preservice teachers, inservice teachers, and K-12 and early childhood learners.

What's in the Book?

The book is organized into two main sections: (i) *Introduction to Environmental Education* and (ii) *Environmental Education Pedagogy.*

Introduction to Environmental Education discusses the historical and philosophical foundations of EE, how environmental science is different from EE, and current trends in EE as it pertains to science teacher education with a focus on inquiry-based teaching and learning, learning environments including early childhood settings, service learning, and ocean and aquatic sciences.

In the chapter "The History and Philosophy of Environmental Education," Carter and Simmons present the tumultuous history of EE and describe its relationship to other disciplines and fields of study. The chapter traces the historical and philosophical development of the EE field and relates EE as presently practiced to the mosaic of K-12 education with a focus on its relationship to science education. This chapter lays the foundation for further discussion of EE's place in the education of teachers of science for the twenty-first century.

In the chapter "Professional Preparation for Science Teachers in Environmental Education," McDonald and Dominguez discuss the importance of professional preparation for science teachers in EE. They describe reasons for why the professional preparation of including EE in teacher preparation programs has become a complex issue. The authors illustrate how the *National Science Education Standards* (NRC 1996) and the *Guidelines for the Initial Preparation of Environmental*

Educators (NAAEE 2000) provide guidance to how preservice teachers should be prepared to implement EE in K-12 classroom environments.

In the chapter "Approaches to Environmental Education," Winther, Sadler, and Saunders provide an overview of various teaching and learning approaches that exist within the field of EE. Like science education, environmental education is an interdisciplinary and complex field that offers a multitude of strategies for learning, dependent upon the variables of resources, time, space, curriculum, student characteristics, plus a full range of factors that can affect any kind of educational implementation.

In the chapter "Environmental Education Within the Early Childhood," Plevyak and Mayfield discuss incorporating EE into early childhood settings. They emphasize the importance of using EE as an integrated curricular context and provide many examples of how EE can be infused across the many different disciplinary areas taught in early childhood classrooms. Special emphasis is placed on acquiring knowledge about the environment, developing an environmental ethic, adapting EE activities for inclusion of students with special needs, and the use of assessment techniques.

In the chapter "Environmental Education Service-Learning in Science Teacher Education," Phillipson-Mower and Adams explore the history, theory, and use of service-learning as it relates to EE and teacher preparation. The authors describe instructional methods that engage both preservice and inservice teachers as well as classroom students in citizenship through decision-making, research, and community-building skills that meet the goals for both science and environmental education.

In the chapter "Beyond *Terra Firma*: Bringing Ocean and Aquatic Sciences to Environmental and Science Teacher Education," Payne and Zimmerman discuss the lack of ocean and aquatic science in environmental and science teacher education. The authors contend that such content is essential to global Earth systems science literacy. The chapter describes many resources designed to provide teacher educators and classroom teachers with tools to enhance the existing curriculum through the integration of ocean and aquatic sciences in their instruction.

The *Environmental Education Pedagogy* section is divided into three subsections that apply to science teacher preparation. These include (1) *Outdoor Learning and Place-Based Environments*, (2) *Instructional Strategies*, and (3) *Technology*.

Outdoor Learning and Place-Based Environments discuss teaching and learning in nontraditional learning environments. This section includes chapters on outdoor learning spaces, field-trip strategies, elementary teacher learning, and learning about local plant life. Three case studies are presented that include (1) EE classroom implementation of inservice teachers in an outdoor professional development program, (2) a service learning program, and (3) an urban education program.

In the chapter "Promoting the Use of Outdoor Learning Spaces by K-12 Inservice Science Teachers Through an Outdoor Professional Development Experience," Bloom, Holden, Sawey, and Weinburgh describe a summer professional development program designed to encourage inservice elementary and secondary school teachers to use outdoor learning spaces (OLSs) as part of their curriculum. As part of the professional development design, the authors identify the teachers' perceived and actual obstacles to integrating OLSs and then designed the professional development experiences to specifically address these. The authors provide recommendations for others who are considering developing professional development programs to promote the use of OLSs with K-12 teachers.

In the chapter "Integrating Environmental Education Field Trip Pedagogy into Science Teacher Preparation," Rebar and Enochs discuss how preservice science methods courses can make use of field trips to enhance EE integration into preservice teacher education programs. The authors describe a variety of research-based strategies for optimizing learning on field trips with secondary students. The authors provide practical implementation strategies and examples that illustrate that including field-trip pedagogy in the existing science methods courses may be accomplished without restructuring course objectives and without displacing other important materials to be covered.

In the chapter ""Eeew! There's Dew On My Toes": Common Characteristics of Preservice Elementary Teacher Learning in Environmental Education and Instructional Strategies for Science Teacher Educators," Hug describes important characteristics of preservice elementary school teachers that science teacher educators should understand in order to consider integrating EE activities into science methods course work. Chapter vignettes focus on inadequate content knowledge, ecophobia, avoidance of minor physical discomfort, and a need for highly structured learning environments. Instructional strategies and learning experiences are discussed to address these characteristics.

In the chapter "Name That Plant! Overcoming Plant Blindness and Developing a Sense of Place Using Science and Environmental Education," Frisch, Unwin, and Saunders describe "plant blindness," a phenomenon that attempts to explain why botanical education is often neglected in the implementation of school curricula, and why people have so much trouble "seeing" plants. The authors suggest that plants can and should be an integral part of life science education. They advocate place-based pedagogical practices that provide teachers and students the chance to learn about and explore plant life in their communities to enhance their environmental awareness and sense of place.

In the chapter "Place-Based Inquiry: Advancing Environmental Education in Science Teacher Preparation," Sarkar and Frazier describe a 3-year professional development project for inservice science teachers to implement place-based pedagogy and EE in science classrooms. The authors provide a framework for planning successful place-based investigations. Case studies are presented where teachers with their students engage in inquiry-based, place-based investigations for an extended period of time. The authors also discuss how place-based pedagogy is important for the preparation of science teachers and how such a strategy addresses a range of science concepts through deeper inquiry.

In the chapter "Summer Methods in Summer Camps: Teaching Projects WILD, WET, and Learning Tree at an Outdoor Environmental Education Center," Eick, Carrier, Perez, and Keasal describe an innovative partnership program in which elementary and secondary science preservice teachers at Auburn University teach

EE to summer camp children at the university's outdoor environmental education center as part of their first science methods course. The preservice teachers receive training in the use of *Project* curricula including Projects WILD, WET, and Learning Tree from a Cooperative Extension Specialist. The chapter describes the camp experiences and how the *Project* curricula are integrated into instruction.

In the chapter "Teachers Connecting Urban Students to Their Environment," Brown, Votaw, and Tretter describe the *Science Beyond the Classroom* program, a 10-day *Hands-on, Minds-on Summer Science Camp* led by preservice and inservice teachers for urban, low-SES middle-school students to learn about environmental science concepts through site visits to environmental community-based venues. The program is designed to nurture positive attitudes of urban students toward environmental science learning by increasing awareness of science in their community. As a result of participating in this innovative professional development program, K-12 inservice and preservice teachers gained an enduring awareness of the impact that they can have on the environment.

Instructional Strategies discusses specific instructional strategies for the inclusion of EE in science teacher professional development. This section includes chapters on instructional methods to elicit learner EE conceptions, use of concept mapping to promote EE knowledge and understandings, Science-Technology-Society role-playing simulations with environmental issues, problem-based learning methodologies, and collaborative activities between science teacher methods instructors and nonformal environmental educators.

In the chapter "Exploring Preservice Teachers' Mental Models of the Environment," Moseley, Desjean-Perrotta, and Crim describe the development of the Draw-An-Environment Test (DAET), a survey tool designed to uncover preservice teachers' mental models of the environment and a rubric, based on the NAAEE *Guidelines for the Preparation and Professional Development of Environmental Educators* for scoring the drawings produced in the DAET. The authors present their implementation findings from a sample of preservice teachers. The authors describe how their data findings influenced programmatic changes in their certification program.

In the chapter "Pedagogy, Environmental Education, and Context: Promoting Knowledge Through Concept Mapping," Austin and Schmidt describe how they used regional environmental questions to model concept mapping for content learning and collaborative learning with their secondary science methods students. They describe how their project used concept mapping activities to incorporate collaborative learning to develop a curriculum that promoted EE learning with a focus on the interdisciplinary nature of science, while integrating discipline-specific content standards.

In the chapter "Unraveling the Scientific, Social, Political and Economic Dimensions of Environmental Issues Through Role-Playing Simulations," MaKinster describes the implementation of a United States Senate Subcommittee hearing role-playing simulation on the use of Bt (*Bacillus thuringiensis*) genes in corn in a college-level interdisciplinary science course that is taken by many preservice teachers. The simulation incorporated a wide variety of teaching strategies and topics that are of current interest in science education including simulations, role-playing, driving questions, oral presentations, technology integration, portfolios, reflection, and concept mapping. The implementation findings demonstrated that environmental role-playing simulations can have a significant impact on students' understanding of how science is applied to environmental problems.

In the chapter "Exploring Environmental Education Through Eco-feminism: Narratives of Embodiment of Science," Spencer and Nichols discuss how care must be taken when teaching about the environment whether the underlying philosophical framework is one of sustainability, deep ecology, bioregionalism, or ecofeminism. The authors describe how problems-based learning as a pedagogical practice at first seems to be a logical way to learn about nature. However, the authors discuss problems with using this teaching strategy with EE topics from an ecofeminist perspective.

In the chapter "The Value of Nonformal Environmental Education-Based Professional Development in Preservice Science Teacher Preparation," Peffer and Bodzin describe the work of nonformal EE educators and discuss their potential role in science teacher preparation programs. Nonformal EE educators use a wide assortment of teaching methodologies in varying learning environments to encourage an environmentally literate citizenry. The benefits of collaborative relationships between nonformal EE educators and science teacher educators are discussed.

In the chapter "Using Environmental Education *Project* Curricula with Elementary Preservice Teachers," Schepige, Morrell, Smith-Walters, Sadler, Munck, and Rainboth describe the use of the *Project* curricula – Project WET, Project WILD, Project WILD Aquatic, and Project Learning Tree – as a means of introducing environmental education in preservice university courses. Four different case studies demonstrate diverse methods of integrating EE through the use of the *Project Guides* into preservice teacher coursework at four different universities. The authors describe how their instructional approaches strengthen elementary preservice teachers' science content knowledge, develop science process and inquiry skills, integrate literacy, and embed field work in educational settings.

Technology discusses the integration of technology to promote the teaching and learning of EE in science teacher preparation. This section includes chapters on instructional methods to incorporate geospatial technologies (including Google Earth and Geographic Information Systems), podcasts, and web-based inquiry activities in science teacher professional development coursework.

In the chapter "Situated Learning in Environmental Education: Using Geospatial Technologies with Preservice Secondary Teachers," Hagevik, Stubbs, and Whitaker describe how situated learning using Geospatial Information Technologies (GIT) in preservice teacher education courses can be used to study the environment. They describe how nature study and GIT were used in science teacher education courses on campus and through field experience courses in diverse natural locations. The courses promoted collaborative learning communities, where students became immersed in the natural world and were able to investigate their own investigative questions.

In the chapter "Using Podcasting to Address Nature-Deficit Disorder," Klein and Weaver discuss the issues associated with the digital native student population and their disconnection with the natural world. The authors describe two preservice teacher education podcast projects that integrate technology to encourage student connections with nature. In these projects, preservice teachers develop podcasts that are used as learning tools for outdoor field settings with elementary and secondary school students.

In the chapter "Integrating Web-Based Activities and Site-Based Experiences to Investigate Environmental Issues," Bodzin describes how an EE course at Lehigh University uses a hybrid approach of instruction using web-based activities and face-to-face site-based experiences to primarily focus on the study of environmental issues in the Lehigh River watershed. Course activities are presented that illustrate how technology can be used effectively to support EE teaching and learning with prospective and current science teachers. The chapter describes how course materials take advantage of easily available geospatial information technologies to foster spatial literacy in the curriculum and support learners with the ability to make use of data visualizations for analysis and interpretation when examining environmental issues such as sprawl and land use decision-making.

Bethlehem, PA — Alec M. Bodzin
Cortland, NY — Beth Shiner Klein
Salisbury, MD — Starlin Weaver

References

Heimlich, J. E., Braus, J., Olivolo, B., McKeown-Ice, R., & Barringer-Smith, L. (2004). Environmental education and preservice teacher preparation: A national study. *Journal of Environmental Education*, *35*(2), 17–21.

Lane, J., Wilke, R., Sivek, D., & Champeau, R. (1995). Strengths and weaknesses of teacher environmental education preparation in Wisconsin. *Journal of Environmental Education*, *27*(1), 36–45.

Lieberman, G. A., & Hoody, L. L. (1998). Closing the achievement gap: Using the environment as an integrating context for learning. San Diego, CA: State Education and Environment Roundtable (SEER).

McComas, W. F. (2002). The nature of the ideal environmental science curriculum: Advocates, textbooks, & conclusions (Part II of II). *The American Biology Teacher*, *65*(3), 171–178.

McKeown-Ice, R. (2000). Environmental education in the united states: A survey of preservice teacher education programs. *Journal of Environmental Education*, *32*(1), 4–11.

NAAEE. (2000). Guidelines for the Initial Preparation of Environmental Educators. 2nd edition. Rock Springer, GA: North American Association for Environmental Education.

National Environmental Education and Training Foundation (NEETF) and Roper Starch Worldwide. (2001). *Lessons from the environment: Why 95% of adult Americans endorse environmental education/The ninth annual national report card on environmental attitudes, knowledge, and behaviors*. Washington, DC: Author.

National Research Council (NRC). (1996). *National Science Education Standards*. Washington, DC: National Academy Press.

National Research Council. (2001). *Inquiry and the National education standards*. Washington, DC: National Academy Press.

No Child Left Inside Coalition. (2009). *Environmental literacy plans by state*. Retrieved August 25, 2009, from http://www.cbf.org/Page.aspx?pid=924

Pennsylvania Department of Education. (2002). *Academic standards for environment and ecology*. Harrisburg, PA: Author.

Pfirman, S., & The AC-ERE. (2003). *Complex environmental systems. Synthesis for earth, life, and society in the 21st century.* A report summarizing a 10-year outlook in environmental research and education for the National Science Foundation. Arlington, VA: National Science Foundation.

Rakow, S. J. (1985). A review of teacher inservice in environmental education: 1970–1980. *Journal of Environmental Education, 16*(4), 7–10.

Ramsey, J. M., Hungerford, H. R., & Volk, T. L. (1998). Environmental education in the K-12 curriculum: Finding a niche. In H. R. Hungerford, W. J. Bluhm, T. L. Volk, & J. M. Ramsey (Eds.), *Essential readings in environmental education* (pp. 111–124). Champagne, IL: Stipes Publishing.

Scott, W. A. H. (1996). Pre-service environmental teacher education: A critique of recent arguments about constraints, approaches and course design. *The International Journal of Environmental Education and Information, 15*, 307–318.

Wilson, A. H. (2000). *A content analysis of environmental education as presented in selected high school biology textbooks: 1910–1994*. Unpublished dissertation, University of Maryland, College Park, MD.

Wisconsin Department of Public Instruction. (1998). *Wisconsin's model academic standards for environmental education*. Milwaukee, WI: Author.

Acknowledgments

We wish to thank the leadership team of the Association for Science Teacher Education (ASTE) for their generous support and assistance with the development of this book. We especially thank Bill McComas, Janice Koch, and Jon Pedersen for sharing their valuable insights, wisdom, and knowledge to help us through the development and editorial process for this book. We express our appreciation to the members of the ASTE Environmental Education Forum, many of whom are chapter authors, for their feedback during the conceptualization phase of this book. We are extremely grateful to the authors of the chapters in this book for their timely contributions and innovative work with integrating EE into science teacher education.

We express our deepest gratitude to the reviewers of the book chapters who helped guide us with enlightening and thoughtful comments: Diana Beck, Lisa Blank, Charles Eick, Leigh Ann Haefner, Bill Harwood, Tom Howick, Joseph Kerski, Emilie Kudela, Thomas Kromer, Michael Kamen, David Martin, Patricia Morrell, Teddie-Phillipson Mower, Irene Plonczak, Robert Reinsvold, Leonie Rennie, Gail Tooker, Peter Veronisi, Deb Molina-Walters, Bryan Wee, and Brenda Weiser. Our reviewers' feedback has helped to make this book a most valuable contribution to the fields of both science teacher education and environmental education.

Contents

About the Authors

April Dean Adams is an Associate Professor of Science Education in the Department of Natural Sciences at Northeastern State University in Tahlequah, Oklahoma. She teaches undergraduate and graduate science content and science methods courses. She holds an Ed.D. in Curriculum and Instruction from the University of Houston and an M.S. in Biology and a B.S. in Physics from Purdue University. Her research interests include teacher education, inquiry-based instruction, and the nature of science. She is a member of the Association for Science Teacher Education, the National Association for Research in Science Teaching, the American Educational Research Association, the National Science Teacher Association, and the American Association of Physics Teachers.

Barbara Austin is an Assistant Professor of science education at Northern Arizona State University. She has done industrial research and development and is co-author on several patents for electrochromic devices. She has worked in the standardized testing industry, and as a secondary physics and mathematics teacher.

Mark A. Bloom, Ph.D., is an Assistant Professor of Science Education in the Andrews Institute of Mathematics, Science & Technology Education at Texas Christian University. He teaches elementary and secondary science methods courses as well as graduate courses in science education. Prior to teaching in the College of Education, he taught for over 10 years in Colleges of Science and Engineering (at TCU and Tyler Junior College). He holds a Bachelors Degree in Biology from Dallas Baptist University, a Masters Degree in Biology from Baylor University, and a Doctorate in Science Education from TCU. His research interests include nature of science, environmental education, and professional development for teachers.

Alec M. Bodzin is Associate Professor in the Teaching, Learning, and Technology program and Lehigh Environmental Initiative at Lehigh University. Dr. Bodzin's research involves the design of web-based inquiry learning environments; learning with spatial thinking tools including GIS, Google Earth, and remotely sensed images; design and implementation of inquiry-based environmental science curriculum; visual instructional technologies; and preservice teacher education and teacher professional development. He has co-developed 20 peer-reviewed instructional science and environmental education curricular projects including recently developed carbon cycle and remote sensing educational modules as part of a NASA Earth System Science Education grant. Dr. Bodzin is currently the Primary Investigator on the Toyota USA Foundation's *Web-enhanced Environmental Literacy and Inquiry Modules* (WELIM) project that will create, implement, and evaluate instructional modules for middle-school learners for energy, global climate change, and environmental issues using interdisciplinary environmental science instruction via geospatial information technologies and the web. He can be contacted at amb4@lehigh.edu.

Sherri L. Brown is an Associate Professor of science education at the University of Louisville where she has been since 2002. She earned her doctorate in science education from the University of Tennessee - Knoxville. Prior to that, she taught middle- and high-school Biology, Chemistry, and Physical Science in Tennessee.

Sarah Carrier teaches elementary science methods courses at North Carolina State University. Her research focus is on informal science instruction, primarily schoolyard science strategies. These strategies help students learn about environmental science concepts, providing inquiry opportunities in authentic settings. One of her goals is to help preservice teachers experience the potential for outdoor environmental science in their future classrooms. She is also involved in studies that explore the relationship between science and music, supporting interdisciplinary connections in the pursuit of developing science literacy.

Robert L. Carter serves on the graduate faculty in the Department of Teaching and Learning at Northern Illinois University at its main campus in DeKalb, Illinois. As part of the faculty in Outdoor and Environmental Education, he teaches graduate courses leading to the M.S. Ed. with a specialization in Environmental Education and a Certificate of Graduate Study in environmental education. Bob also teaches a senior-level undergraduate course required for all elementary education majors in which future teachers experience and practice outdoor education, environmental education, community-based education, and service learning. Coming from a background in resource interpretation, Bob has spent over 20 years teaching in and about the environment. He began his post secondary teaching career at Northern Illinois University's Lorado Taft Field Campus in Oregon, Illinois, where he became involved with the National Project for Excellence in Environmental Education (NPEEE) as a Project Assistant. He continues to provide workshops on NPEEE materials as need arises and time permits. Bob is presently a Consulting Editor for the *Journal of Environmental Education*. While becoming more immersed in academia, Bob has maintained strong contact with his roots in the field of resource interpretation as a regional representative of the Environmental Education Section of the National Association for Interpretation.

Courtney Crim received her doctorate in Curriculum and Instruction with an emphasis in Gifted Education from The University of Houston. She currently works at The University of Texas at San Antonio in the Department of Interdisciplinary Learning and Teaching in the areas of early childhood/elementary education. She has experience as both a teacher and as a trainer of teachers in the public school system. Her research interests focus on the connection between differentiation and professional development for both preservice and inservice teachers. Specifically, this line of research intersects the practice of differentiated instruction for university and elementary learning environments and provides a foundation for programmatic changes and the development of transformative leadership in the field of education.

Blanche Desjean-Perrotta is Associate Dean for Teacher Education and Associate Professor in the Department of Interdisciplinary Learning and Teaching at the University of Texas at San Antonio. Dr. Desjean-Perrotta's research focuses on early childhood teacher preparation, particularly in the areas of science and environmental education. Dr. Desjean-Perrotta has been a primary investigator for Eisenhower Grants for environmental education of classroom teachers, and she also conducts research in the public schools investigating the effectiveness of expeditionary learning models on young children's perceptions of the environment. Presently, Dr. Desjean-Perrotta is a co-investigator in a research project examining early childhood preservice teachers' perceptions of the environment and the impact of formal environmental education programs on their mental models. Dr. Desjean-Perrotta is actively involved in the North American Association of Environmental Education as co-chair of the Preservice Teacher Advisory Council.

Lynn A. Dominguez is an Assistant Professor in the Recreation, Parks, & Leisure Services Department at Central Michigan University in Mt. Pleasant, Michigan. She was an Interpreter in the Virginia State Park system for 5 years prior to completing her Ph.D. at Michigan State University. Lynn teaches Outdoor and Environmental Education Methods classes, Outdoor Recreation classes, and Environmental Interpretation classes at CMU. She has been conducting environmental education workshops for teachers since 1990. Her research interests include: the use of service learning for developing environmental empowerment; preservice teacher development in environmental education; nonformal educator development in environmental education; and constraint negotiation in outdoor recreation.

Charles J. Eick is Associate Professor in Elementary Education at Auburn University. His background is in Biology and Soil Science. He regularly teaches undergraduate and graduate curriculum and methods courses where he features a variety of EE curricula including Projects WILD, WET, and Learning Tree. Charles first became interested in environmental science as a middle grades student where his sixth-grade science teacher had him monitoring water quality and visiting with local company executives about their pollution control measures. Today, he uses a similar water-monitoring approach in his work with his teachers and the Alabama Water Watch Association.

Larry G. Enochs is currently Professor in the Department of Science and Mathematics Education at Oregon State University. His Doctorate in education is from Indiana University and also holds a Master of Science degree in earth science education from the University of Rochester. During the past 20 years, he has been involved in several teacher-enhancement efforts, as principal investigator, project evaluator, and advisory board member. In the area of science teacher development, he has published numerous articles on science teacher beliefs and earth science education. Prior to joining the faculty at Oregon State University, Dr. Enochs served as Director of the Center for Mathematics and Science Education Research at the University of Wisconsin - Milwaukee. Enochs also served 2 years as a Program Officer for the Teacher Enhancement and Research on Teaching and Learning Programs at the National Science Foundation. Prior to his NSF appointment, he was an Associate professor of science education at Kansas State University.

Richard Frazier is Associate Professor of Science Education at the University of Central Missouri. Dr. Frazier taught middle-school science for many years in the United States and abroad. His international experience includes teaching at schools in Sierra Leone, Saudi Arabia, and Singapore. As a returned Peace Corps volunteer from Sierra Leone, he still works with teachers there.

Jennifer Kreps Frisch has a B.S. in Biology, an M.S. in Environmental Education, and a Ph.D. in Biology Education. She has learned and taught in Florida, Thailand, New Mexico, Colorado, and Minnesota, and is currently an Assistant Professor of Biology Education at Kennesaw State University near Atlanta, GA. Her goal is to be able to name all of the plants in her yard by the time you are reading this in print.

Rita A. Hagevik, Ph.D., is an Assistant Professor of Science Education in the Department of Theory & Practice in Teacher Education at the University of Tennessee in Knoxville. She received her Ph.D. in Science Education and Forestry from North Carolina State University. Her teaching and research focuses on Geospatial Information Technologies, nature of science, conceptual change, and environmental and outdoor education. She has been teaching GIS integrated courses and developing curricula for GIS in environmental education for preservice and inservice teachers for over 10 years.

Molly Holden, M.S., P.G., is a professional geologist and environmental consultant, adjunct geology faculty at Tarrant Community College, and doctoral student in Science Education at Texas Christian University. She has nearly 20 years of experience in the environmental industry performing site cleanups and water resources investigations. She has been a part-time instructor in the Andrews Institute of Mathematics, Science, and Technology Education and the College of Science and Engineering at TCU, and currently teaches environmental health and safety for a private Fort Worth company, and introductory geology at the community college level. An active member of the ASTE, her research interests include teacher professional development and teacher efficacy pertaining to environmental issues and outdoor education.

J. William Hug is Assistant Professor and Director, Center for Excellence in Elementary Science & Mathematics Education at California University of Pennsylvania. His academic work focuses on elementary science teacher education, environmental education, and place-based education. Dr. Hug teaches undergraduate and graduate teacher education courses such as elementary science teaching methods, place-based and environmental education, and educational research methods. Dr. Hug's research focuses on preservice elementary teachers' understanding of environment, ecology, and natural history content; place-based curriculum design, implementation and evaluation; and citizen actions in the service of sustainability goals.

Doyle E. Keasal is an Environmental Education Specialist with the Alabama Cooperative Extension System at Auburn University, Auburn, Alabama. His primary focus is working with educators to develop school-based outdoor classrooms and supporting curriculum as a means to extend the traditional four-walls of the classroom to include the outdoors. In addition to helping educators develop outdoor classrooms, he conducts numerous EE professional development workshops for educators throughout the year and also makes a variety of presentations to children regarding our natural resources. Prior to his employment with the Alabama Cooperative Extension System, he taught at the elementary school level for 22 years where he worked to integrate EE into the curriculum through a multidisciplinary approach that also included the development of an outdoor classroom that contained a nature trail, butterfly garden, stream-side water-quality monitoring station, vegetable garden, and greenhouse.

Beth Shiner Klein is Professor in the Childhood/Early Childhood Education Department at SUNY Cortland. Dr. Klein teaches undergraduate elementary science methods and educational technology, and graduate courses in environmental studies and integrated math, science, and technology. Dr. Klein has published in the areas of environmental education, elementary preservice and inservice education, and educational technology. Dr. Klein has been involved in a number of environmental education grant projects that include curriculum development, using technology to connect environmental experiences, to students and teachers, and inservice and preservice teacher professional development. Dr. Klein is a pioneer on the use of collaborative technologies in her teaching and was on the SUNY Cortland pilot team to integrate the use of podcasting and iTunesU into her instruction. For the past 9 years, she has co-led a team of faculty in a model environmental themed learning community for preservice elementary teachers that includes outdoor and environmental learning experiences.

James G. MaKinster is an Associate Professor of Science Education and Environmental Studies at Hobart and William Smith Colleges in Geneva, NY. His teaching and scholarship focus on scientific inquiry, geospatial technologies, and teacher professional development within science and environmental education.

Amy Mayfield Field Service Associate Professor of Teacher Education, has served as a faculty member at the University of Cincinnati since 2002. Her interests include working with preservice teachers in their field experiences and teaching students how to use scientific inquiry in their teaching. She can be reached at Amy.Mayfield@uc.edu.

James T. McDonald is Associate Professor of science education in the Department of Teacher Education and Professional Development at Central Michigan University in Mt. Pleasant, Michigan. He was a classroom teacher at the elementary and middle level for 10 years prior to receiving his doctorate from Purdue University. He has been conducting environmental education workshops for teachers since 1992. His research interests include children's conceptions of earth science topics, science and service learning problem-based learning, and the pedagogical content knowledge of preservice science teachers.

Patricia D. Morrell is an Associate Professor in the School of Education at the University of Portland. She received a B.S. in Forest Biology from SUNY ESF and an M.S. and Ph.D. in Science Education from Oregon State University. She has focused much of her work on the preparation and professional development of preservice and inservice science teachers, K-16; as well as designing science curricula and experiences for elementary-aged students. She is active in environmental education consortia in Oregon.

Christine Moseley is Associate Professor of Interdisciplinary Studies in the Department of Interdisciplinary Learning and Teaching at the University of Texas at San Antonio. Dr. Moseley's teaching and research centers around the impact of the implementation of environmental education into teacher education courses and programs on the perceptions and beliefs that elementary education preservice teachers have regarding the environment. She is actively involved in the Texas Association of Environmental Education (Vice-president) and the North American Association of Environmental Education as co-chair of the Preservice Advisory Council.

Miriam Munck is an Associate Professor at Eastern Oregon University. She has been a science educator and science teacher educator for 25 years. She holds a doctorate degree from Boise State University; her undergraduate work is in Chemistry. Her work has centered on encouraging preservice teachers in science teaching, developing curriculum for K-8 students, and improving science teaching skills and content knowledge of inservice teachers. She belongs to an environmental education for preservice teacher's consortium.

Sherry E. Nichols is an Associate Professor of Science Education at the University of Alabama. She teaches elementary and secondary science teacher education courses. Her research has focused on understanding science teaching as a community-based practice. Her current work explores uses of graphical representations for science and new literacy learning, and children's engineering at a local public housing afterschool club. She has drawn on case-based pedagogy, feminist philosophies, visual ethnography, and narrative inquiry methodologies to inform her teaching and research.

Diana L. Payne serves as an Assistant Professor in Residence and the Education Coordinator for Connecticut Sea Grant. Her research focus is on the professional development of teachers in a variety of settings, with a particular interest in the marine and aquatic sciences. Recent research projects include investigating the effects of teacher research experiences in marine science laboratories and aboard research vessels on teacher instructional practices and on student attitudes toward science. She earned a Ph.D. in Educational Psychology from the University of Connecticut, and holds B.S and M.S. in Biology as well as secondary teaching certification (grades 7–12) in multiple subjects.

Tamara E. Peffer is a graduate student in Learning Sciences & Technology doctoral program at Lehigh University. Her research focus is the application of technology in environmental education efforts to improve both preservice teacher preparation and student environmental literacy. She has worked as an environmental educator in formal and nonformal settings through Nolde Forest Environmental Education Center, PA DCNR for over 14 years. She served as Watershed Education site coordinator for 7 years and developed a technology-integrated model for watershed education. Tamara also teaches introductory environmental science, biology, and social science technology as an adjunct instructor at Reading Area Community College.

Karni Perez has an undergraduate degree in German and Biology and a teaching credential. She has done graduate study in German, Biology, and Environmental Studies within the California State University System. She received a Master of Science in Sociology and a Graduate Interdisciplinary Minor in Environmental Studies from Auburn University in 1993. Karno has held a number of elementary teaching positions since 1976 and taught as part of the Environmental Volunteers, a volunteer EE organization in California. After receiving the M.S. in 1993, she worked in sociology of fisheries and forestry and as volunteer Education Director at the Southeastern Raptor Center. She has written an extension handbook on solid waste management, articles on various topics, and a book on the history of the Alabama catfish industry. Karni has volunteered in various capacities at the Forest Ecology Preserve and has been the Education Coordinator for school programs since 2002.

Teddie Phillipson-Mower is an instructor and Director of the UofL Center for Environmental Education in the Teaching and Learning Department of the College of Education and Human Development at the University of Louisville. She is a Ph.D. candidate in Science and Environmental Education at Indiana University Bloomington and hopes to defend in the spring of 2009. She holds an MAT in Biology from Indiana University, and a B.S. from Northern Michigan University in Biology, Chemistry, and Sociology. Teddie's research interests include environmental education, intellectual and ethical development, nature of science, and service learning. She is the Chair of the Four Year University Section of the National Association of Biology Teachers and a strand coordinator of Strand 14: Environmental Education of the National Association of Research in Science Teaching. In addition, she is also active in the North American Association of Environmental Education, the Kentucky Association for Environmental Education, and the American Education Research Association (Environmental Education SIG). She thinks everyone should have a rain garden.

Linda Plevyak, Associate Professor of Teacher Education, has served as a member of the University of Cincinnati faculty since 2000. Her research interests include a focus on environmental science in elementary education with emphasis on the development of children's knowledge, skills, and attitudes toward environmental education. She is also interested in scientific inquiry and how preservice teachers incorporate this concept into their teaching. Dr. Plevyak can be reached at Linda.plevyak@uc.edu.

Donna Rainboth is a Senior Instructor at Eastern Oregon University. She has been involved in science education for 20 years. Over the past 14 years, she has developed and coordinated numerous science curricula projects focused on science inquiry and environmental education.

Bryan M. Rebar is currently a doctoral candidate in Environmental Sciences with a specialty in environmental education at Oregon State University. Under the same program, Bryan completed his master's thesis with a study of changes in children's conception of nature following a residential environmental education program. Bryan has considerable career experience working with teachers, chaperones, and students as a naturalist, outdoor educator, adventure trip leader, outreach specialist, and residential science center program director.

Kim Cleary Sadler is an Associate Professor of Biology at Middle Tennessee State University in Murfreesboro. She received a B.S. and M.S. with a focus on Plant Ecology from Middle Tennessee State University and an Ed.D. in Curriculum and Instruction from Tennessee State University. Teaching primarily nonmajors biology content courses for general education and preservice teachers, she uses active learning strategies extensively in her classes. Her current research interests are related to informal science education practices in after-school programs and nature centers. She also serves as director for the Center for Cedar Glade Studies and works closely with schools regarding education about this unique ecological system.

Somnath Sarkar is a Professor of Chemistry at the University of Central Missouri. He has a Doctoral degree in chemistry with an emphasis in organic chemistry and chemical education. He has been offering professional development workshop in sciences for K-12 teachers for many years. He provides services for school improvement for predominantly minority schools. He teaches secondary science methods in addition to regular chemistry courses. In addition, he has taught courses such as "Risk and Benefits of Environmentalism," "Chemistry of Cooking," and "Chemistry of Art and Archaeology."

Gerry Saunders is a Professor of Education at Unity College in Maine. He has degrees from the Universities of Northern Colorado, Idaho, and Nebraska-Lincoln. He taught high-school biology, environmental science, and human physiology for 18 years. Currently, he is Director of Teacher Education and Coordinator of EE at Unity College. He also teaches courses in Science and EE and Biology.

April T. Sawey, M.Ed., is a doctoral student in Science Education at Texas Christian University. She has 12 years of experience as a science teacher and administrator and holds an undergraduate degree in Biology and a Master's degree in Curriculum and Instruction – both from Texas Wesleyan University. Her primary research interest involves teacher's conception of inquiry and the role of inquiry in the classroom.

Adele C. Schepige is a Professor of Science Education at Western Oregon University. Her doctorate is in Curriculum and Instruction from Portland State University. She has taught biology, earth, and physical science courses for both preservice and inservice elementary teachers for 20 years. She has been involved with a variety of environmental education consortia and is developing global climate change education professional development and curriculum materials for K-8 teachers and students.

Nina Schmidt is an Environmental Health Specialist with over 20 years of experience in the field of environmental technology. She has worked as an analytical chemist, regulator, community college instructor, and wastewater system designer. She recently completed a master's degree in sustainability education.

Bora Simmons serves as the founding director of the National Project for Excellence in Environmental Education. The Project has drawn on the insights of literally thousands of educators across the United States and around the world to craft guidelines for top-quality environmental education. After 20 years as a professor of environmental education at Northern Illinois University, Bora retired in 2007 and moved the Project to the Institute for a Sustainable Environment at the University of Oregon. Bora has been actively involved in environmental education research, evaluation, and professional development for over 30 years. She has taught courses, given presentations, and facilitated workshops throughout the United States and Canada as well as in Europe, Asia, Africa, Middle-East, and Latin America. She was chair of the NCATE environmental education standards writing committee. She served as president of NAAEE; serves on numerous steering committees and boards of directors, and is an executive editor of the *Journal of Environmental Education*. For her achievements, Bora received the Walter E. Jeske Award for Outstanding Contributions to Environmental Education, and the NAAEE Award for Outstanding Contributions to Research in Environmental Education.

Cindi Smith-Walters is a professor of Biology at Middle Tennessee State University and co-directs the MTSU Center for Environmental Education. She received a B.S. in Biology from East Central State University and both an M.S. in Curriculum and Instruction and Ph.D. in Environmental Science from Oklahoma State University. She is interested in formal and informal science teaching and learning and in 2007 received the Tennessee Environmental Education Association's Distinguished Service Award. When she isn't teaching or learning, you can find her outside playing or reading.

M.E. Spencer is a secondary science specialist with the Alabama Math, Science, and Technology Initiative and adjunct assistant professor at the University of Montevallo where she teaches environmental education for teachers among other graduate courses. Her research focuses on embodiment of science and science education, ecofeminist thought, and environmental education with particular attention to teachers and middle-school students throughout the state of Alabama.

Harriett S. Stubbs, Ph.D., is an Associate Professor Emerita, member of the Department of Mathematics, Science and Technology Education since 1988, and located in the Office of Professional Development at North Carolina State University. She is Director of the SCI-LINK/ GLOBE-NET Projects, author of books and articles, presenter of methodologies and strategies for professional development of educators and environmental topics of interest for teaching and learning. In the past 5 years, she has developed and coordinated international professional development experiences for educators in Brazil and for more than 25 years in the USA.

Thomas R. Tretter has been an Associate Professor of science education at the University of Louisville since 2004. He earned his doctorate in science education from the University of North Carolina - Chapel Hill. Prior to that, he taught high-school mathematics and science in North Carolina, South Carolina, Sudan, Africa, and Gabon, Africa.

Matthew M. Unwin is an Assistant Professor of Biology in the Department of Biology and Physics at Kennesaw State University. He holds a Ph.D. in Botany with research interests and experience in plant biodiversity, conservation, and systematics. At KSU, he teaches both introductory biology and advanced botany courses.

Nikki L. Votaw has been an Assistant Professor at Johnson Bible College since 2008. She earned her doctorate in science education from the University of Louisville. Prior to that, she taught elementary middle-school science and mathematics in Tennessee and Kentucky.

Starlin Weaver is a Professor of Science Education at Salisbury University. She teaches graduate and undergraduate courses in middle and secondary science education and classroom management. She also supervises science interns. Dr. Weaver serves as an NSTA program reviewer and NCATE Board of Examiner member. Her current scholarly interests include using technology to teach science methods, incorporating environmental education into science methods courses, and integration of content area reading into science methods courses.

Molly H. Weinburgh, Ph.D., is the William & Betty Adams Chair of Education and the Director of the Andrews Institute of Mathematics, Science, & Technology Education at Texas Christian University. She directs the Ph.D. in Science Education, teaches courses in science education, and teaches the Honors section of a nonmajors biology course. Her service to science education includes: President of the ASTE, ASTE Board of Directors, Strand Coordinator of Culture, Social and Gender Issues of the NARST, and Policy Committee of the SSMA. Her research interests include equity issues, inquiry science, and professional development for teachers.

Diane C. Whitaker is a doctoral student in science education and GIS at North Carolina State University. Her research interests include environmental chemistry and GIS technologies. She was recently recognized by the North Carolina Science Teachers Association and the Guilford County Business Advisory Board for outstanding science teaching as a high-school chemistry teacher in Greensboro, NC. She is currently teaching high-school Earth Science and developing and leading professional development for inservice teachers in GIS and related technologies.

Austin A. Winther is Assistant Professor in the Department of Graduate Education, Leadership, and Counseling at Rider University. He received his Ph.D. from Southern Illinois University at Carbondale. He taught in the Chicago Public Schools for 17 years. As science Department Chairperson at Paul Robeson High School, he helped implement the first use of ChemCom in the Chicago schools in 1987. He teaches classes in science education, social studies education, and EE. He is the Director of the Master of Arts in Teaching Program. He serves on the Rider University Energy and Sustainability Steering Committee. His research interests include both science and EE.

Timothy D. Zimmerman is Assistant Professor of Science Education at Rutgers University. Jointly appointed in the Department of Learning and Teaching and the Department of Marine and Coastal Sciences, his research focuses on teaching and learning of ocean sciences concepts. In particular, he seeks to understand the nature of learning across formal and informal science contexts and the role of science learning as it relates to environmental decision-making about marine environmental problems. Dr. Zimmerman is involved in the design and testing of internet-based classroom and mobile computing technologies as part of his research agenda. He received his Ph.D. from the University of California, Berkeley, in science education. In addition, he holds B.S. and M.S. in Marine Biology and has conducted field-based marine biology and oceanography research.

Contributors

Dr. April Dean Adams
Department of Natural Sciences, Northeastern State University, College of Science and Health Professions, 611 N. Grand, Tahlequah, OK 74464-2302, USA
adams001@nsuok.edu

Barbara Austin
Center for Science Teaching and Learning, Northern Arizona University, PO Box 5697, Flagstaff, AZ 86011, USA
Baa49@nau.edu

Mark A. Bloom, Ph.D.
Andrews Institute of Mathematics & Science Education, Texas Christian University, TCU, Box 297920, Fort Worth, TX 76129, USA
M.Bloom@tcu.edu

Alec M. Bodzin, Ph.D.
Lehigh University, College of Education, A113 Iacocca Hall, 111 Research Drive, Bethlehem, PA 18015, USA
amb4@lehigh.edu

Sherri L. Brown, Ph.D.
Johnson Bible College, Teacher Education, 7900 Johnson Drive, Knoxville, TN 37998, USA
s.brown@louisville.edu

Robert L. Carter, Ed.D., CIT
Department of Teaching and Learning, College of Education, Northern Illinois University, DeKalb, IL 60115, USA
carter@niu.edu

S. Carrier

Department of Curriculum and Teaching, Auburn University, AL 36849, USA

Dr. Courtney Crim
Education Department, Trinity University, One Trinity Place,
San Antonio, TX 78212-7200
courtney.crim@trinity.edu

Dr. Blanche Desjean-Perrotta
University of Texas at San Antonio, One UTSA Circle,
San Antonio, TX 78249, USA
Blanche.perrotta@utsa.edu

Lynn A. Dominguez, Ph.D.
Department of Recreation, Parks, and Leisure Services Administration,
Central Michigan University, 103 Finch, Mt. Pleasant, MI 48859, USA
domin1la@cmich.edu

Charles J. Eick, Ph.D.
Department of Curriculum and Teaching, Auburn University, AL 36849, USA
eickcha@auburn.edu

Dr. Larry G. Enochs
Science and Mathematics Education Department, Oregon State University,
237 Weniger Hall, Corvallis, OR 97331, USA
enochsl@onid.orst.edu

Dr. Richard Frazier
Department of Elementary and Early Childhood Education,
University of Central Missouri, Warrensburg, MO 64093, USA
frazier@ucmo.edu

Jennifer Kreps Frisch, Ph.D.
Department of Biology and Physics, Kennesaw State University, Kennesaw,
GA 30144, USA
jfrisch1@kennesaw.edu

Rita A. Hagevik
The University of Tennessee, A404 Bailey Education Complex, Knoxville,
TN 37996, USA
rhagevik@utk.edu

Molly Holden
Andrews Institute of Mathematics & Science Education, Texas Christian University,
TCU, Box 297920, Fort Worth,
TX 76129, USA
M.Holden@tcu.edu

J. William Hug, Ph.D.
Director, Center for Excellence in Elementary Science & Math Education,
California University of Pennsylvania, California, PA 15419, USA
hug@calu.edu

Doyle E. Keasal
Extension Specialist, Environmental Educator Discovering Our Heritage Coordinator, Alabama Cooperative Extension System, 625 Jennifer Drive, Auburn, Alabama 36830

Beth Shiner Klein, Ed.D.
School of Education SUNY Cortland, P.O. Box 2000, Cortland, NY 13045, USA
Beth.Klein@cortland.edu

James G. MaKinster, Ph.D.
Hobart & William Smith Colleges, Geneva, NY, 14456, USA
makinster@hws.edu

Amy Mayfield, M.Ed.
University of Cincinnati, One Edwards Center, Room 2150Q, Cincinnati, OH 45221-0002, USA

James T. McDonald, Ph.D.
Department of Teacher Education and Professional Development, Central Michigan University, EHS 134c, Mt. Pleasant, MI 48859, USA
jim.mcdonald@cmich.edu

Patricia D. Morrell
University of Portland/School of Education, 5000 N. Willamette Blvd, Portland, OR 97203, USA
morrell@up.edu

Dr. Christine Moseley
University of Texas at San Antonio, One UTSA Circle, San Antonio, TX 78249, USA
Christine.moseley@utsa.edu

Miriam Munck, Ed.D.
Eastern Oregon University, PO Box 100, Pendleton, OR 97801, USA
mmunck@eou.edu

Sherry E. Nichols, Ph.D.
University of Alabama, 230-E Graves Hall, Tuscaloosa, AL 35487-0232, USA
snichols@bamaed.ua.edu

Diana L. Payne, Ph.D.
Neag School of Education, Department of Educational Psychology, Connecticut Sea Grant, University of Connecticut at Avery Point, 1080 Shennecossett Road, Groton, CT 06340, USA
diana.payne@uconn.edu

Tamara Elizabeth Peffer
Lehigh University, College of Education, 120 Center Street,
Reading, PA 19606, USA
tep205@lehigh.edu

Karni Perez
Department of Curriculum and Teaching, Auburn University,
AL 36849, USA
k.r.perez@charter.net

Teddie Phillipson-Mower
292 CEHD, University of Louisville, Louisville, KY 40292, USA
t0phil01@louisville.edu

Linda Plevyak, Ph.D.
University of Cincinnati, One Edwards Center, Room 2150Q, Cincinnati,
OH 45221-0002, USA
Linda.plevyak@uc.edu

Donna Rainboth
College of Education, Eastern Oregon University,
One University Blvd, La Grande, OR 97850, USA
drainbot@eou.edu

Bryan M. Rebar
Oregon State University, 2046 Cordley Hall, Corvallis, OR 97331, USA
rebarb@onid.orst.edu

Kim Cleary Sadler
Biology/MTSU Center for the Study of Cedar Glades, Middle Tennessee
State University, PO Box 60, Murfreesboro, TN 37132, USA
ksadler@mtsu.edu

Dr. Somnath Sarkar
Department of Biochemistry Chemistry and Physic,
University of Central Missouri, Warrensburg, MO 64093, USA
Sarkar@ucmo.edu

Gerry W. Saunders, Ph.D.
Center for Experiential and Environmental Education at Unity College,
Unity, Maine 04988-9502, USA
gsaunders@unity.edu

April T. Sawey
Andrews Institute of Mathematics & Science Education,
Texas Christian University, TCU, Box 297920, Fort Worth, TX 76129, USA
A.T.SAWEY@tcu.edu

Adele C. Schepige
Division of Teacher Education, Western Oregon University,
345 N Monmouth Ave, Monmouth, OR 97361, USA
schepia@wou.edu

Nina Schmidt
Environmental Health Specialist,1665 N. Turquoise Dr., Flagstaff,
AZ 86001, USA
nlschmidt@inbox.com

Bora Simmons, Ph.D.
Institute for a Sustainable Environment, University of Oregon,
Eugene, OR 97403, USA
borasimmons@gmail.com

Cindi Smith-Walters, Ph.D.
Biology/MTSU Center for Environmental Education, Middle Tennessee State
University, PO Box 60, Murfreesboro, TN 37132, USA
csmithwa@mtsu.edu

M.E. Spencer, Ph.D.
University of Montevallo, 2970 Pelham Parkway, Pelham,
AL 35115, USA
spencerme@montevallo.edu

Harriett S. Stubbs
Department of Mathematics, Science, and Technology Education, College of
Education, & c/o GIS Program, College of Natural Resources, North Carolina
State University, Box 7106, Raleigh, NC 27695-7401, USA
h_stubbs@ncsu.edu

Thomas R. Tretter, Ed.D.
Johnson Bible College, Teacher Education,
7900 Johnson Drive, Knoxville, TN 37998, USA
tom.tretter@louisville.edu

Matthew M. Unwin, Ph.D.
Department of Biology and Physics, Kennesaw State University,
Kennesaw, GA 30144, USA
munwin@kennesaw.edu

Nikki L. Votaw
Johnson Bible College, Teacher Education,
7900 Johnson Drive, Knoxville, TN 37998, USA
nlvotaw01@louisville.edu

Starlin Weaver, Ph.D.
Seidel School of Education and Professional Studies, Department of Education Specialties, Salisbury University, Salisbury,
MD 21801, USA
sdweaver@salisbury.edu

Molly H. Weinburgh, Ph.D.
Andrews Institute of Mathematics & Science Education,
Texas Christian University, TCU, Box 297920, Fort Worth, TX 76129, USA
M.Weinburgh@tcu.edu

Diane C. Whitaker
CALS Center for Applied Aquatic Ecology, Education Specialist,
North Carolina State University, Box 7510, Raleigh, NC 27695, USA
dcwhitak@ncsu.edu

Austin A. Winther
Department of Graduate Education, Leadership, and Counseling, Rider University,
2083 Lawrenceville Road, Lawrenceville, NJ 08648-3099, USA
awinther@rider.edu

Timothy D. Zimmerman, Ph.D
University of Rutgers, 10 Seminary Place, New Brunsuick, NJ 08901-1183
timothy.zimmerman@gse.rutgers.edu

Abbreviations

AAAS	American Association for the Advancement of Science
BSP	Botanical Sense of Place
CEC	Council on Environmental Education
COSEE	Centers for Ocean Sciences Education Excellence
DAET	Draw-An-Environment Test
EBL	Environment-Based Learning
EE	Environmental Education
EIC	Environment as an Integrating Context for learning
GIS	Geographic Information Systems
GLOBE	Global Learning and Observations to Benefit the Environment
GPS	Global Positioning Systems
MOSS	Mapping Our School Site
NAAEE	North American Association for Environmental Education
NASA	National Aeronautics and Space Administration
NCATE	National Council for Accreditation of Teacher Education
NCLB	No Child Left Behind
NOAA	National Oceanic and Atmospheric Administration
NSTA	National Science Teachers Association
OLC	Outdoor Learning Center
OLS	Outdoor Learning Space
PBL	Problems-Based Learning
PCK	Pedagogical Content Knowledge
PD	Professional Development
PEAK	Public Educators Accelerating Kids
PLT	Project Learning Tree
SAVE	Strengthening Awareness and Valuing the Environment
SBC	Science Beyond the Classroom
SES	Socio-Economic Status

STES	Science-Technology-Environment-Society
STS	Science-Technology-Society
UNEP	United Nations Environment Program
UNESCO	United Nations Education, Scientific, and Cultural Organization
WREEC	Western Regional Environmental Education Council
WYS	United Nations Environment Program

Part I
Introduction To Environmental Education

The History and Philosophy of Environmental Education

Robert L. Carter and Bora Simmons

Stewart Udall, President John F. Kennedy's Secretary of the Interior, identifies a pair of events and the questions they evoked as the impetus for his landmark book of 1963 – *The Quiet Crisis*.

> One week last fall two events came to my attention which seemed to sum up the plight of modern man: the first was a press report which indicated that T.S. Eliot, the poet, was a victim of London's latest "killer fog" and lay gravely ill; the second was a call from a preservation-minded citizen of New Hampshire who informed me that Robert Frost's old farm—fixed for all time in memory by the poem "West-running Brook"—was now an auto junk yard.
>
> The coincidence of these two events raised questions in my mind: Is a society a success if it creates conditions that impair its finest minds and make a wasteland of its finest landscapes? What does material abundance avail if we create an environment in which man's highest and most specifically human attributes cannot be fulfilled?
>
> (Udall 1963, p. vii)

Those questions, and many others, are still being asked today and it is through the discipline of environmental education that we can provide answers and map the way to solutions. What follows is an exploration of the beginnings, the present, and the future of environmental education, its philosophical underpinnings, and its relationship to science teacher education.

A Brief History Lesson

Ask one scholar when the term *environmental education* (EE) first came into use and you will get one answer. Ask another and you will most likely get a different response. Over the years that EE has been a part of the educational vernacular

R.L. Carter (✉)
Department of Teaching and Learning, College of Education, Northern Illinois University, DeKalb, IL 60115, USA
e-mail: carter@niu.edu

B. Simmons
Institute for a Sustainable Environment, University of Oregon, Eugene, OR 97403, USA
e-mail: borasimmons@gmail.com

A.M. Bodzin et al. (eds.), *The Inclusion of Environmental Education in Science Teacher Education*, DOI 10.1007/978-90-481-9222-9_1,

there has been disagreement about the first use of the term. It is not the intent of the authors to settle the matter of when the name was first used but to shed some light on its development and its characteristics – the unique ones as well as those shared with other disciplines and fields, and to examine what EE means to teachers of science. For those interested in the etymology of the term, John Disinger's (1985) well-detailed treatment of that topic and EE's antecedents is highly recommended.

This chapter focuses on the history and development of EE in the USA. That story did not occur in isolation, however, so a context of world events is supplied as needed. The presentation here is primarily chronological, but in order to present as complete a picture as possible some temporal, as well as geographic, leapfrogging is occasionally necessary. EE has a rich and varied past, with its underlying philosophy informed by a range of source disciplines – a situation that often has given rise to confusion regarding EE's identity and application. In the following pages, we offer some context and sequence for that variety with the hope that it provides readers with a clearer picture of the rich background and educational power of EE.

Authors, Awakenings, and Achievements

From Emerson's *Nature* (1836), to Thoreau's *Walden* (1854), to George Perkins Marsh's *Man and Nature* (1864) one can trace the developing concerns regarding human interaction with nature expressed by the political and social commentators of a young and, in the view of many people, a still seemingly limitless USA. The dialog continued in the writing and public speaking of renowned naturalists and writers of the late nineteenth and early twentieth centuries such as John Muir (1838–1914), Enos Mills (1870–1922), Robert Marshall (1901–1939), and Aldo Leopold (1887–1948). But much of what was being written, discussed, and actually accomplished primarily took the forms of resource conservation and habitat preservation rather than the environmental quality, environmental awareness, and environmental literacy that are the central concerns of today (Gottlieb 1995; Stegner 1990).

A new focus on the state of the environment can be traced to the years immediately after the close of World War II although this attention did not coalesce into the modern environmental movement until the 1960s (Kline 2007). The postwar years saw a proliferation of efforts to reach international accords for the protection of the environment. The *Conference for the Establishment of the International Union for the Protection of Nature* (IUCN) convened at the Fontainebleau, Paris, France in October of 1948 and made its top priority the protection of nature and habitats. Subsequent conferences were scheduled as well in order to insure continued progress (UNESCO 1948). A flurry of related activities during this period set the stage for a burst of effort that would begin developing in earnest in the 1960s and spill into the 1970s with unprecedented energy.

Although the concept of EE as practiced today may arguably be traced back to at least 1948 and the IUCN Conference (Disinger 1985), it is certain that 1972 was a major turning point in EE internationally. The participants in the first *United Nations Conference on the Human Environment* in Stockholm, Sweden produced a declaration containing 26 principles. Principle 19 of the Stockholm Declaration specifically calls for "education in environmental matters, for the younger generation as well as adults" (UNEP 1972). Environmental quality was finally gaining some attention from the world at large, but in the USA a groundswell of awareness, concern, and effort was already well underway.

Authors

In June of 1948, just months before the first IUCN conference, Aldo Leopold, a pioneer in the modern conservation movement suffered an untimely death fighting a fire on a neighbor's farm (Meine 1988). His seminal work on the relationship between people and the environment would be published posthumously in 1949. *A Sand County Almanac* (Leopold 1949) became, and remains, the cornerstone of the American environmental movement and of modern environmental thinking and writing. It helped set the stage for later works that would move the country further toward the environmental awakenings of the 1960s and 1970s.

Leopold challenged the pursuit of affluence for its own sake. The wisdom of the pursuit of affluence at the cost of the environment began to be questioned in earnest in the 1950s. The success of John Kenneth Galbraith's *The Affluent Society*, published in 1958, was punctuated by the choking smog in California cities, and in his 1960 book, *The Waste Makers*, Vance Packard raised the alarm against pollution and sprawl (Rome 2003). But it took the works of a quiet, eloquent scientist, and an environmentally literate bureaucrat to really shake things up.

Two landmark books brought deepening environmental problems to the attention of the American public during the early 1960s. The 1962 publication of Rachel Carson's *Silent Spring* awakened readers to a situation that threatened the very fabric of the environment. Carson documented and reported that the arsenal of chemicals manufactured, and used with abandon, to "control" insect populations and weeds was having a deleterious effect far beyond the "pest control" for which it was intended. Hailed as a master work by the conservation movement and environmental groups both the book and its author were vilified by the chemical industry (Lytle 2007). But the alarm had been sounded and the American public began to become more acutely aware of a deteriorating environment as well as some of the underlying causes of that deterioration.

While the furor over *Silent Spring* continued, another book piled even more fuel on the fires of environmental controversy and awareness. Late in 1961, at the urging of author Wallace Stegner, Steward Udall, President John F. Kennedy's Secretary of the Interior began work on his own book (Finch 2008). *The Quiet Crisis* was published in November of 1963 and it provided the reading public with a view of the

American environmental legacy, both what had been lost, and what could yet be lost, due to a broad range of existing and imminent environmental threats. *Silent Spring* and *The Quiet Crisis* ushered in a decade of unprecedented environmental legislation and action from grassroots organizations to the Congress and the White House.

Awakenings

The Civil Rights Movement and the Vietnam War protests of the 1960s overshadowed most other events of that decade (Hall 2005; Reed 1986). But the protest culture of the 1960s was fertile ground for the growing concerns about environmental quality (Rome 2003). Much of that concern was reflected in a marked increase in environmentally focused legislation being passed and signed into law at a rate, and in a volume, that would only be exceeded during the 1970s. The Wilderness Act of 1964, the Species Conservation Act of 1966, and the Wild and Scenic River Act of 1968 signaled a concern for our relationship to the environment and what humans might be doing to it. The Solid Waste Disposal Act (1965) and the Clean Air Act of 1965 reflected national concerns over what postwar affluence was pumping into the environment in the form of waste and emissions. The momentum that was built through the literature and legislation of the 1960s culminated in three separate events that establish 1970 as a landmark year in things environmental.

On January 1, 1970 another new law came into effect. The National Environmental Policy Act of 1969 (NEPA) remains the environmental law of the USA today. The environmental concerns of the day were clearly reflected in NEPA's statement of purpose, which reads in part, "to promote efforts which will prevent or eliminate damage to the environment and biosphere and stimulate the health and welfare of man; to enrich the understanding of the ecological systems and natural resources important to the Nation" (42 U.S.C. § 4321). But it is not due to NEPA alone that 1970 is considered the benchmark year for environmental concern and efforts. That was the result of a much more publicly visible and far reaching event.

The protest movements for civil rights, against the Vietnam War, and for environmental quality spawned a flurry of populist actions during the 1960s and 1970s, many of which took the form of a generally passive activity known as the *sit-in*, in which large numbers of protesters would gather in a particular area with the intent of hampering normal operations simply by getting in the way. These sorts of actions became a popular tool of a range of activists and were all characterized by a large number of people gathering together for a specific purpose – usually a protest, but not always. According to Ling (2000), sit-ins began at segregated lunch counters during the civil rights movement but later variations included kneel-ins at churches, wade-ins at public pools, and stand-ins at ticket counters. Sit-ins became a popular form of protest on college campuses. One variation of the sit-in with a decidedly educational focus born out of antiwar protests was the *teach-in* (Hall 2005).

Gaylord Nelson, at the time a US Senator from Wisconsin, had for some time envisioned an environmental teach-in that would raise public awareness on critical environmental issues (Christofferson 2004). Denis Hays, a Harvard law student collaborated with Nelson in enlisting the aid of campus activists from across the country for an environmental teach-in that became known as Earth Day and on April 22, 1970 it involved an estimated 20 million people with participation by nearly 1,500 college campuses (Rome 2003).

NEPA and Earth Day were not the only landmark developments of 1970. A study conducted by the National Science Teachers Association (NSTA) in 1970 painted the picture of an educational landscape in desperate need of program and curriculum development. Among the schools and districts of the 50 states there existed only 54 programs with any EE element (National Science Teachers Association 1970). In an August 1970 address to Congress President Nixon stated:

> It is also vital that our entire society develop a new understanding and a new awareness of man's relation to his environment—what might be called "environmental literacy." This will require the development and teaching of environmental concepts at every point in the education process.
>
> (Nixon 1970, p. vii)

Nixon's comment and the NSTA study indicated that there was a gaping hole yet to be filled. Part of that need would be filled by legislation just over the horizon.

In October 1970, the Environmental Education Act became law. Provisions of the new law included the establishment of an Office of Environmental Education within the US Office of Education in the former Department of Health, Education, and Welfare and funding for states to implement EE within their K-12 systems through several means. A marked shortcoming of the act, however, was that it had a life span of only 5 years. Another shortcoming was the limited funding that accompanied its short life. Nonetheless, EE had finally made its way into federal law and was a part of the federal government's infrastructure.

Achievements

The decade of the 1970s is epitomized by prolific growth for EE. This was an era of exuberant capacity building for the field. The momentum of the legislation and activism of the 1960s continued to build on both the national and international levels. In 1971, a group of educators concerned about the development of EE materials formed the National Association for Environmental Education, which later was renamed the North American Association for Environmental Education (Disinger 2001). By this time the Journal of Environmental Education was already in print, having had its first issue published in the fall of 1969. In that inaugural issue William Stapp of the University of Michigan enumerated the societal necessity for EE and identified objectives of the nascent field (Stapp, et al. 1969). According to Hammerman (1979), before the end of the new decade there were EE coordinators within the school systems of all 50 states. Meanwhile, publishing houses around the country were rapidly producing EE materials (Minton 1980).

Although the federally funded programs initiated through the Environmental Education Act were limited in duration, nongovernmental organizations (NGOs) maintained a healthy pace of development and dissemination. The Western Regional Environmental Education Council (WREEC), later to become the Council on Environmental Education (CEE), was formed and subsequently spearheaded the development of a number of EE curriculum materials, beginning with the widely acclaimed and internationally recognized Project Learning Tree (Carter 2006).

A number of conferences were held throughout the decade, each addressing different aspects of the concern for and the development of EE. The topics of these conferences ranged from elementary and secondary education to higher education and addressed emerging issues in the field, culminating in the National Leadership Conference in Environmental Education in Washington, D.C. in 1978 (Stapp 1978). In the latter half of the decade, NGO support continued to blossom and expand as federal government support waned.

The Rest of the World Catches Up

The 1972 Stockholm conference may have set the stage for greater awareness of the need to advance EE internationally but two subsequent conferences still stand today as the seminal events for EE on the world stage. The International Workshop on Environmental Education, held in Belgrade, Yugoslavia in October of 1975 resulted in what became known as *The Belgrade Charter*. The Belgrade Charter built on the framework of Stockholm and described the goals, objectives, audiences, and guiding principles of EE and proposed what has become the most widely accepted definition of EE:

> Environmental education is a process aimed at developing a world population that is aware of and concerned about the total environment and its associated problems, and which has the knowledge, attitudes, motivations, commitments, and skills to work individually and collectively toward solutions of current problems and the prevention of new ones.
>
> (UNESCO-UNEP 1976, p. 2)

But the definitive codification of EE as an international enterprise ultimately came out of the world's first Intergovernmental Conference on Environmental Education held in Tbilisi, Georgia, USSR in October of 1977. The document now known as *The Tbilisi Declaration* was formulated during this conference and in many quarters remains the definitive statement on what EE is and ought to be. These goals provide the foundation for much of what has been done in the field since 1978:

(a) to foster clear awareness of, and concern about, economic, social, political and ecological interdependence in urban and rural areas;
(b) to provide every person with opportunities to acquire the knowledge, values, attitudes, commitment and skills needed to protect and improve the environment;

(c) to create new patterns of behaviour of individuals, groups and society as a whole towards the environment (UNESCO 1978, p. 26)

But while EE was gaining momentum internationally, the same could not be said of EE back here in the USA.

Rollercoaster Ride to the Twenty-First Century

The 1980s were not as kind to EE as the previous decade had been, at least within the US government. Under President Ronald Reagan, the federal purse strings known as the Omnibus Budget Reconciliation Act (OBRA) of 1981 eliminated nearly everything that had been established by the Nixon-era Environmental Education Act.

Another aspect of Reagan era politics was not only Reagan's apparent indifference to environmental quality and literacy, but the development of a decidedly anti-environmental movement dubbed variously as *brown-lash*, the *sagebrush rebellion*, or the *wise use* movement (Kline 2007). No matter the label, it amounted to a burgeoning effort by many of the consumptive, extractive, and pollution-producing businesses and industries to roll back environmental advances of the previous 20 years. Reagan, along with the pro-development appointees in his Cabinet managed to achieve many such rollbacks. Although, eventually, Congress began to balk at many of the changes the Reagan White House attempted (Kraft 2000). The advances of the environmentally heady decades of the 1960s and 1970s were now quickly receding into the past.

The years of the Reagan administration may be viewed as the beginning of a long downturn for EE but the election of George H.W. Bush to the presidency in 1988 marked the beginning of a politically turbulent era with regard to both the environment and education. Although a new National Environmental Education Act was signed into law by President Bush in 1990, the 4 years of the Bush administration and the succeeding 8 years of the Clinton administration saw gradual but substantial change in the federal government as the White House re-embraced environmental concerns while an increasingly conservative-dominated Congress went the other way (Warren 2003).

During this period EE itself came under fire. Described variously as incomplete at best or biased at worst, EE came under heavy attack from conservative think tanks that invariably had agendas as one-sided as those they ascribed to practitioners and proponents of EE (Holsman 2001). At the same time a new focus was being placed on the quality of EE materials and instruction.

By this time the academic standards movement driven by the 1983 publication of *A Nation at Risk* was well-developed (Resnick and Resnick 1983). An outgrowth of the standards movement was an initiative by the North American Association for Environmental Education (NAAEE) to develop standards for EE (Simmons 1995). As the idea grew and matured, it became the *National Project for Excellence in*

Environmental Education and today provides guidelines for the development and assessment of EE materials as well as benchmarks for practitioner and student knowledge on environmental topics (NAAEE 2004a, b, c). It could be inferred that the twentieth century drew to a close with little net gain for EE but a strong infrastructure had been established.

The first decade of the twenty-first century did not start off any better for EE than the previous century had ended with regard to support within the US government. The 2001 reauthorization of the Elementary and Secondary Education Act, commonly known as The No Child Left Behind Act, ignored EE while repeated attempts to reinstate the National Environmental Education Act languished and died in committee. But as all educators and scientists know, many things occur in cycles, and EE, as both a useful teaching tool and an engine of environmental literacy, is no exception.

The capacity building, curriculum development, and dialog that had taken place since the environmental flurry of the 1960s resulted in a rich knowledge base for EE grounded in both research and practice. Education in and about the environment remained a topic in educational circles regardless of governmental lethargy. Educators, authors, and researchers continued to promote, demonstrate, and document the benefits of involving children in the environment as a learning context. Most telling was the resurgence of interest in, and mounting evidence for, the benefits of interaction with the natural environment and developing problems due to the reduction or complete loss of that contact and the environmental price to be paid.

Rivkin (2000) commented on the essential need for especially young children to interact with and experience the environment through outdoor play spaces, and Chawla (2003) examined the relationship of environmental awareness to children's manipulation of the natural environment. Two researchers at the University of Illinois documented the positive effects of green play spaces on the symptoms of attention-deficit disorder (ADD) and attention-deficit/hyperactivity disorder (ADHD) in children noting that there was a "green advantage" in natural versus built play environments (Kuo and Taylor 2004).

The capstone of this era of research and publishing on the environment and environmental concerns came in 2005 with the publication of Richard Louv's *Last Child in the Woods: Saving Our Children From Nature-Deficit Disorder*. Louv's manifesto on the causes and consequences of a number of modern society's ills reawakened an interest in the outdoors, the environment, and EE, returning them to center stage. A national *No Child Left Inside* movement sprang up, spearheaded by the Chesapeake Bay Foundation, a not-for-profit organization dedicated to the cleanup and protection of the Chesapeake Bay. The upwelling of new support for education in and about the environment even reached the chambers of Congress and as of this writing, the US House of Representatives had passed, and sent on to the US Senate, the reauthorization of the National Environmental Education Act, alternatively named in this version as The No Child Left Inside Act (H.R. 3036 2008).

What's in a Name?

With such a broad base in time, geography, and intellectual underpinnings it is not surprising that EE has, for many, been difficult to define or even conceptualize. Nonetheless, EE is a discrete discipline with identifiable roots and unique characteristics. EE as practiced today taps into knowledge generated by a wide range of source disciplines and EE practitioners transmit that knowledge through sound pedagogical principles (Archie 2003). A closer examination of the pedigree and practice of EE can shed some light on why defining or conceptualizing it seems to be such an intractable situation.

Predecessor Disciplines

Disinger (1985) identifies three antecedents to EE: nature study, conservation education, and outdoor education. Nature study gained prominence in the USA during the late eighteenth and early nineteenth centuries. The writing and public speaking of John Muir and Enos Mills popularized wild nature as a source of recreation, replenishment, and solace throughout the early 1900s (Nash 1989; Drummond 1995). The Cornell University biologist, Liberty Hyde Bailey perpetuated that growth well into the first half of the twentieth century (Hammerman et al. 2001). His student and protégé, Anna Botsford Comstock became the first female faculty member at Cornell University and her 1911 publication, *Handbook of Nature Study* remains a valuable teaching resource (Chase 1985). Conservation education extended the ideas of enjoyment, relaxation, and health embodied in nature study while emphasizing the need to conserve natural resources so that both nonconsumptive and extractive pursuits could be maintained in perpetuity. Conservation, as proposed by Aldo Leopold, espoused sensible resource consumption balanced with maintaining habitat quality, even to the point of leaving some wilderness intact for its own sake (Lorbiecke 1996).

As conservation education began to grow and develop, the Dust Bowl stamped an indelible exclamation point on the need for just such a discipline. The problems predicted by Leopold and Marsh came to pass in a swift and highly visible manner. On April 14, 1935, in Washington, D.C., Hugh Bennett, director of the US Soil Erosion Service spoke to Congress about the need to end destructive farming and ranching practices. As if on cue, the chamber was blackened by a cloud of soil that had blown in from the Great Plains states, a distance of 2,000 miles. Bennett's point had been made more powerfully than any words could express. Less than 2 weeks after that episode Congress passed the bill creating the Soil Conservation Service (Lookingbill 2001). Conservation, and the education for its need, had finally become a cause célèbre in the USA. Conservation education steadily gained momentum throughout the middle of the twentieth century and remains a robust part of the educational mosaic today (Swan 1975; Roth 2008).

Whereas nature study and conservation education are generally considered to be content areas, outdoor education is more often viewed as a teaching method that draws from both nature study and conservation education (Disinger 1985). Outdoor education's underlying philosophy can be traced back to John Amos Comenius (1592–1670) and his emphasis on sensory learning (Hammerman 1980). In the years immediately following World War II, outdoor education combined elements of nature study and conservation education with what at the time was known as *school camping*. The links between school camping and outdoor education were further developed throughout the postwar years as outdoor education became a more common aspect of the regular school experience (Sharp and Partridge 1947).

Outdoor education, conservation education, and nature study, remain active fields of endeavor that continue to contribute to the knowledge base of EE while benefiting from EE's own products and practitioners. The links between these varied fields of practice are both permanent and mutually beneficial.

Contributing Disciplines

EE taps into a broad range of source disciplines for its content. Science, mathematics, language arts, social science, politics, and philosophy make up just a part of the mix. It also draws from a broad base for its pedagogy. As previously noted, its historical roots can be found in nature study, conservation education, and outdoor education, but, at its best, EE also draws from a deep well of pedagogical best practice (Archie 2003).

A major contributor to the EE knowledge base is environmental science. But in recent years educators have often had difficulty distinguishing environmental science from EE. In daily practice they often blend almost seamlessly, while theoretically and conceptually they remain very different. Part of the issue is the variability found in definitions of these terms. A major contributing factor may be the broad topical net cast by educational materials produced for, and used in, environmental science courses. As a case in point, in the preface to their most recent text, Raven, Berg, and Hassenzahl state: "[It] integrates important information from many different fields, such as biology, geography, chemistry, geology, physics, economics, sociology, natural resources management, law, and politics." They go on to state: "[B]ecause environmental science is an interdisciplinary field, this book is appropriate for use in environmental science courses offered in a variety of departments, including (but not limited to) biology, geology, geography, and agriculture" (2008, p. vii). While the authors are not claiming that their multidisciplinary text on the environment is, itself, environmental science, a net cast so widely can certainly contribute to confusion. Nonetheless, the essential characteristics of EE and environmental science are fairly straightforward and distinct.

Environmental science is the engine of data collection and knowledge creation, while EE is the vehicle for dissemination and application of that knowledge with environmental literacy as the ultimate goal. In a position paper on EE adopted by the National Science Teachers Association, that organization's Board of Directors recognizes and emphasizes the nature of EE, noting that "environmental education [is] a way to instill environmental literacy in our nation's pre-K-16 students" (National Science Teachers Association 2003, p. 1).

There can be no argument that EE and environmental science are very closely intertwined and interdependent, but to say that they are one and the same is to say that science and education are the same.

The Focus on Environmental Literacy

At the heart of environmental education is developing an environmentally literate citizenry, and environmental literacy requires knowledge and skills that both build upon and go beyond the environmental sciences. Although there are many different definitions and descriptions of environmental literacy, the National Project for Excellence in Environmental Education has identified four key elements of environmental literacy (NAAEE 2004b). First, environmental literacy depends on a willingness and ability to ask questions about the surrounding world, speculate and hypothesize, seek and evaluate information, and develop answers to questions. Second, environmental literacy is contingent upon understanding environmental processes and systems, including human systems. Third, the environmentally literate citizen is able to identify, investigate, and formulate potential solutions to environmental issues. Finally, students are motivated, and understand that what they do as individuals and in groups makes a difference in their world.

Since environmental education begins close to home, it encourages learners to understand and forge connections with the environment in their own neighborhoods and communities. It is through these connections that students gain the knowledge and skills that help them make sound decisions. Recent variations on this theme are *environment-based education* and *place-based education* (Broda 2007). Ultimately, the goal of environmental education is a democratic society in which environmentally literate citizens participate actively. The challenge, of course, is to develop an education program that fosters environmental literacy. Environmental literacy depends on skills and knowledge drawn from the sciences, social sciences, and humanities. This vision of environmental literacy is also reflected in the newly adopted National Council for the Accreditation of Teacher Education (NCATE) Standards for the Initial Preparation of Environmental Educators, wherein teachers of environmental education are expected to be environmentally literate themselves (NAAEE 2007).

Environmental Education in the Post-NCLB Classroom

A commentary by Alston Chase in the November 1988 issue of Outside Magazine focused on the roots of a problem still being addressed today. In a brief but eloquent and readable article he proposed that many of our continuing, and developing, environmental problems were either caused by or exacerbated by, what he termed, "academic tunnel vision," the means by which practitioners in a multitude of disciplines and higher education advance through increasingly narrowly focused specialization entirely within their one, specific field. Chase noted that despite that dominant paradigm, true advances and breakthroughs often occurred on the cusps between disciplines where influences and knowledge from other fields provided a richer environment for innovation and development (Chase 1988). In essence, knowledge may be acquired through narrowly defined study, but applying that knowledge well often requires a more holistic approach.

EE, as envisioned and practiced today, is the embodiment of that holistic approach. As a content area it is a gathering place, a collecting jar, of knowledge and data, derived from a range of source disciplines in the sciences, the humanities, and the arts. As a teaching method it emphasizes the best of what current pedagogical knowledge has to offer and guides the pursuit of hands-on, minds-on learning toward the development of an environmentally literate citizenry. Simply stated, "[e]nvironmental education is good education." (NAAEE 2004a, p. 1)

References

Archie, M. (2003). *Advancing education through environmental literacy*. Alexandria, VA: Association for Supervision and Curriculum Development.

Broda, H. (2007). *Schoolyard-enhanced learning: Using the outdoors as an instructional tool, K-8*. Portland, ME: Stenhouse.

Carson, R., Darling, L., & Darling, L. (1962). *Silent spring*. Boston: Houghton Mifflin.

Carter, R. (2006). *Listening to the soloists in the choir: A study of the life experience of exemplary K-12 environmental educators*. Doctoral dissertation, Northern Illinois University, DeKalb, IL.

Chase, C. (1985). A chronology of environmental education: Please throw stones at this author. *Nature Study, 38*(2–3), 21–22, 25.

Chase, A. (1988, November). Scientific breakdown: The cultural weakness behind our ecological failures. *Outside*, pp. 45–47.

Chawla, L. (2003). Bonding with the natural world: The roots of environmental awareness. *NAMTA Journal, 28*(1), 133–154.

Christofferson, B. (2004). *The man from Clear Lake: Earth Day founder Gaylord Nelson*. Madison, WI: University of Wisconsin Press.

Disinger, J. F. (1985). What research says: Environmental education's definitional problem. *School Science and Mathematics, 85*(1), 59–68.

Disinger, J. (2001). K-12 education and the environment: Perspectives, expectations, and practice. *The Journal of Environmental Education, 33*(1), 4–11.

Drummond, A. (1995). *Enos Mills: Citizen of nature*. Boulder, CO: University of Colorado Press.

Environmental Education Act. (1970). P.L. No. 91-516, 84 Stat. 1312.

Finch, L. B. (2008). *Legacies of Camelot: Stewart and Lee Udall, American culture, and the arts*. Norman, OK: University of Oklahoma Press.

Gottlieb, R. (1995). Beyond NEPA and Earth Day: Reconstructing the past and envisioning a future for environmentalism. *Environmental History Review, 19*(4), 1–14.

H.R. 3036, 110th Cong. § 2 (2008).

Hall, S. (2005). *Peace and freedom: The civil rights and antiwar movements of the 1960s*. Philadelphia: University of Pennsylvania Press.

Hammerman, E. (1979). A Delphi formulation of environmental education objectives (Doctoral dissertation, Northern Illinois University, DeKalb, IL, 1979). *Dissertation Abstracts International, 40*, 2447.

Hammerman, W. (1980). *Fifty years of resident outdoor education 1930-1980: Its impact on American education*. Martinsville, IN: The American Camping Association.

Hammerman, D. R., Hammerman, W. M., & Hammerman, E. L. (2001). *Teaching in the outdoors* (5th ed.). Danville, IL: Interstate Publishers.

Holsman, R. (2001). The politics of environmental education. *The Journal of Environmental Education, 32*(2), 4–7.

Kline, B. (2007). *First along the river: A brief history of the U.S. environmental movement* (3rd ed.). Lanham, MD: Rowman & Littlefield.

Kraft, M. (2000). U.S. environmental policy and politics: From the 1960s to the 1990s. In O. Graham (Ed.), *Environmental politics and policy, 1960s – 1990s*. University Park, PA: The Pennsylvania State University.

Kuo, F., & Taylor, A. (2004). A potential natural treatment for Attention-Deficit/Hyperactivity Disorder: Evidence from a national study. *American Journal of Public Health, 94*(9), 1580–1586.

Leopold, A. (1949). *A sand county almanac and sketches here and there*. New York: Oxford University Press.

Ling, P. (2000). Racism for lunch. *History Today, 50*(2), 36–38.

Lookingbill, B. (2001). *Dust bowl, USA*. Athens, OH: University of Ohio Press.

Lorbiecke, M. (1996). *Aldo Leopold: A fierce green fire*. Helena, MT: Falcon Publishing.

Louv, R. (2005). *Last child in the woods: Saving our children from nature-deficit disorder*. Chapel Hill, NC: Algonquin Books.

Lytle, M. H. (2007). *The gentle subversive: Rachel Carson, Silent spring, and the rise of the environmental movement*. New York: Oxford University Press.

Meine, C. (1988). *Aldo Leopold: His life and work*. Madison, WI: University of Wisconsin Press.

Minton, T. G. (1980). The history of the nature study movement and its role in the development of environmental education. *Dissertation Abstracts International, 41*, 967. (UMI No. 8019480)

NAAEE. (2004a). *Environmental education materials: Guidelines for excellence*. Washington, DC: Author.

NAAEE. (2004b). *Excellence in environmental education: Guidelines for learning (K–12)*. Washington, DC: Author.

NAAEE. (2004c). *Guidelines for the preparation and professional development of environmental educators*. Washington, DC: Author.

NAAEE. (2007). *Standards for the initial preparation of environmental educators*. Washington, DC: Author.

Nash, R. (1989). *The rights of nature: A history of environmental ethics*. Madison, WI: The University of Wisconsin Press.

National Environmental Policy Act of 1969, 42 U.S.C. § 4321 (2004).

National Science Teachers Association. (1970). *Programs in environmental education*. Washington, DC: Author.

National Science Teachers Association. (2003). *NSTA position statement: Environmental education*. Retrieved July 15, 2008 from, http://www.nsta.org/about/positions/environmental.aspx

Nixon, R. (1970). President's message to the Congress of the United States. In *Environmental quality, the first annual report of the Council on Environmental Quality, together with the President's message to Congress*. Washington, DC: U.S. Government Printing Office. (ERIC Document Reproduction Service No. ED 062 109)

Omnibus Budget Reconciliation Act of 1981. P. L. No. 97–35, 95 Stat. 357.
Raven, P., Berg, L., & Hassenzahl, D. (2008). *Environment* (6th ed.). Hoboken, NJ: Wiley.
Reed, A. L. (1986). *Race, politics, and culture: Critical essays on the radicalism of the 1960s*. Contributions in Afro-American and African Studies, No. 95. Westport, CT: Greenwood Press.
Resnick, D., & Resnick, L. (1983). Improving educational standards in American schools. *Phi Delta Kappan, 65*(3), 178–180.
Rivkin, M. (2000). *Outdoor experiences for young children*. (ERIC Digest No. ED448 013)
Rome, A. (2003). Give Earth a chance: The Environmental movement and the sixties. *Journal of American History, 90*(2), 525–554.
Roth, C. (2008). Paul F-Brandwein lecture 2006: Conservation education for the 21st century and beyond. *Journal of Science Education and Technology, 17*(3), 211–216.
Sharp, L. B., & Partridge, E. (Eds.). (1947). Camping and outdoor education. *The Bulletin of the National Association of Secondary-School Principals, 31*(147).
Simmons, D. (1995). *The NAAEE standards project: Papers on the development of environmental education standards*. Troy, OH: North American Association for Environmental Education.
Stapp, W. (Ed.). (1978). *From ought to action in environmental education*. A report of the National Leadership Conference on Environmental Education, Washington, DC, March 28–30, 1978. Ohio State University Information Resource Center for Science, Mathematics, and Environmental Education. (ERIC No. ED 159 046)
Stapp, W., Havlick, S., Bennett, D., Bryan, W., Jr., Fulton, J., MacGregor, J., et al. (1969). The concept of environmental education. *The Journal of Environmental Education, 1*(1), 30–31.
Stegner, W. (1990). It all began with conservation. *Smithsonian, 21*(1), 35–43.
Swan, M. (1975). Forerunners of environmental education. In N. McInnis & D. Albrecht (Eds.), *What makes education environmental?* Medford, NJ: Plexus.
Udall, S. L. (1963). *The quiet crisis*. New York: Holt, Rinehart and Winston.
UNEP. (1972). *Stockholm declaration on the human environment*. United Nations Conference on the Human Environment, Stockholm, Sweden, 1972. New York: United Nations Environment Programme.
UNESCO. (1948). *Proceedings of the conference for the establishment of the International Union for the Protection of Nature*. Paris: Author.
UNESCO. (1978). *Final report, Intergovernmental Conference on Environmental Education*, organized by UNESCO in cooperation with UNEP, Tbilisi, USSR, 14–26 October 1977. Paris: Author.
UNESCO-UNEP. (1976). The Belgrade Charter. *Connect: UNESCO-UNEP Environmental Education Newsletter, 1*(1), 1–2.
Warren, L. (Ed.). (2003). *American environmental history*. Malden, MA: Blackwell.

Professional Preparation for Science Teachers in Environmental Education

James T. McDonald and Lynn A. Dominguez

Introduction

The ability of a new science teacher to incorporate and teach environmental concepts in their classroom requires content knowledge but also the skills of *how* to teach the concepts. The use and integration of environmental education (EE) pedagogy into teacher preparation programs assists with the skills of "how to teach." EE resources developed to assist current teachers in integrating environmental concepts into their science classrooms are also appropriate for the professional preparation of preservice teachers. These resources may include national EE *Project* curricula such as Project WILD (2000), Project Learning Tree (2006), and Project WET (1995). Additional resources are available from the North American Alliance for Environmental Education (NAAEE), the Environmental Protection Agency (EPA), the National Wildlife Foundation (NWF), the Leopold Education Project (LEP), the World Wildlife Fund (WWF), and many others. However, simply having resources available is not enough; EE methods and skills need to be purposefully integrated into preservice teacher preparation programs. Thoughtful integration of concepts, ideas, pedagogy, and skills for EE should reach beyond subject area barriers to include the rich scope of knowledge included in many areas of study. The purpose of this chapter is to introduce: background literature about preservice teacher preparation in EE; challenges of EE preparation; National Science Education Standards (NRC 1996) for professional development and the Guidelines for the Initial Preparation of Environmental Educators (NAAEE 2000); development of knowledge, skills, and pedagogy for EE; mentoring preservice teachers; and a section on recommendations.

J.T. McDonald (✉)
Department of Teacher Education and Professional Development, Central Michigan University, EHS 134C, Mt. Pleasant, MI 48859, USA
e-mail: jim.mcdonald@cmich.edu

L.A. Dominguez
Department of Recreation, Parks, and Leisure Servises Administration, Central Michigan University, 103 Finch, Mt. Pleasant, MI 48859, USA
e-mail: domin1la@cmich.edu

A.M. Bodzin et al. (eds.), *The Inclusion of Environmental Education in Science Teacher Education*, DOI 10.1007/978-90-481-9222-9_2,

The process of "doing" EE is intended to be interdisciplinary and supplemental throughout the K-12 curriculum, not confined to one subject area. Unfortunately, most higher education institutions are not set up to work across departments and subjects areas. Consequently, preservice teachers may learn content knowledge in one specific subject and teaching methods from different departments. The state certification process will usually only recognize certification in a specific subject area. Because of the interdisciplinary nature of EE it tends not to fit neatly with one subject area. How then do we manage to adequately prepare future teachers in EE?

To teach quality EE, there is a definite need for appropriate pedagogies to deliver effective EE in schools. There is also a need for teacher education institutions to address the specific teacher competencies for EE and provide adequate training and preparation for beginning teachers. A wide range of essential skills is required by individuals or groups to effectively participate in dealing with environmental issues. This is one reason that teachers, teacher educators, and education administrators have not adequately addressed the reality of exploring effective pedagogies for delivering EE within the context of the current *National Science Education Standards* (NRC 1996).

Theoretical Framework

The notion that knowledge is constructed from the interaction of prior and new experience is the basis of contemporary approaches to experiential education. John Dewey (1966, p. 140) suggested that to "learn from experience" was to use prior knowledge and present experience to develop connections between things in order to move forward. Dewey (1966) stated that the "nature of experience can be understood only by noting that it includes an active and a passive element peculiarly combined. On the one hand, experience is *trying*" (p. 139) – a meaning which is made explicit in the connected term experiment. "On the passive, it is *undergoing*" meaning that when we experience something, "we act upon it, we do something with it; then we suffer or undergo the consequences" (Dewey 1966, p. 139).

Dewey made the point that if experience involved being affected by something, then it embraced the concept of thinking about our actions. If our actions shift and become a change in behavior, then Dewey believed that true learning had occurred. However, it is the connection of what Dewey calls the passive and active elements of experience that determines the true value of an experience. Therefore, he argued that the activity by itself "does not constitute experience" (1966, p. 139).

Multiple research studies in EE have demonstrated the truth of Dewey's words. Research into the effectiveness of EE has demonstrated that simply having knowledge of an issue does not result in behavioral change (Hungerford and Volk 1990). Instead, for students to accept responsibility for the environment they need to take ownership for issues and feel empowered to do something about those issues (Hungerford 1996). Although EE programs are common to nonformal educational settings, many studies have also shown the need for the inclusion of EE in formal education venues.

Since the early 1970s numerous studies have been completed concerning the need for EE to be infused throughout schools and around the world. The frequently cited Belgrade Charter (UNESCO-UNEP 1976) provides a goal statement that is the generally accepted definition of EE. This statement goes beyond cognitive knowledge about the environment and encompasses education whereby learners develop responsible environmental behaviors. The Tiblisi Declaration (UNESCO 1978) was adopted in 1978 and provided three objectives for EE that build upon the Belgrade Charter (Disinger and Howe 1990; Stone 1989). A comprehensive framework for the delivery of EE was conceptualized and included elements of informal and formal educational settings. Early on, competencies needed by both formal and nonformal teachers for EE instruction were identified. These included the ability to select, utilize, and implement EE curricular programs; an understanding of the goals of EE; the ability to infuse EE into the curriculum; knowledge in environmental issues and concepts; the ability to investigate and evaluate environmental issues; and the knowledge and skill in taking environmental action (Stone 1989; Wilke et al. 1987).

EE is perceived by many teacher educators in the world (NIER 1993, 1996) as not part of the mainstream school curriculum. It is not a specific learning area in its own right, and therefore has low status. The lack of policy guidelines and a national framework for effective teaching and learning of EE in the USA (prior to 2002) has resulted in ad hoc delivery that was superficial and primarily focused on delivering information about the state of the environment (Powers 2004). It was often characterized by gathering information *about* local and global issues and presenting hypothetical solutions. There are few examples in formal education of quality EE programs that require individuals to reflect on their own behavior and explore the range of skills that produce solutions resulting in a change of attitudes and values, either "*for*" or "*with*" the environment. In most cases, the teaching and learning strategies are inappropriate and reflect a lack of commitment to EE by administrators (Plevyak et al. 2001).

However, the willingness of inservice teachers to engage in EE has been well-documented (Disinger and Howe 1990; Simmons 1998; Stone 1989). Unfortunately, these same teachers identify many barriers to the implementation of EE methods in their classrooms. These include a lack of content and pedagogical knowledge, a lack of skills in taking children into outdoor settings, an overcrowded curriculum, a lack of perceived preparation time, a lack of adequate resources, and a lack of personal commitment to EE (Kim and Fortner 2006; Samuel 1993; Simmons 1998; Stone 1989). In an effort to increase the impact of EE professional development and employ the "multiplier effect" – whereby teachers are taught and the knowledge is multiplied by their students being taught – preparing preservice teachers in EE may be the answer (Powers 2004).

The North Carolina Department of Public Instruction's Division of Science Education (1973) stated that preservice teachers represent "the most effective, long-range means of diffusing EE throughout the general curriculum" (p. 2).

Consequently, centering the focus of preparation for EE on the professional development of preservice teachers, in particular the development of new science teachers, would be effective (Heimlich et al. 2004; McKeown-Ice 2000; Stone 1989; Westing 1993).

The challenges of preparing preservice teachers to bring EE methods and environmental issues into the classroom are well-known (Heimlich et al. 2004; McKeown-Ice 2000; Powers 2004). These begin with the concept of EE itself as interdisciplinary and supplemental to a wide range of school subject areas. In the USA, new teachers are prepared to teach within specific disciplines. If a preservice teacher identifies that they are preparing to be a biology, social studies, or mathematics teacher everyone can conceptualize a basic framework of what they will be prepared to teach. Identifying oneself as an EE teacher is a conflict in terms. Will you teach an "environmental education course"? Will you be certified by your state to teach "environmental education"? Even many teachers certified for elementary education have specialty areas such as science, reading, language arts, or mathematics in which they are certified to teach. The concept of EE as a *method* of teaching is foreign to many State Departments of Education along with the issues of interdisciplinary and cross-curricular methodologies (Disinger and Howe 1990).

Currently, preservice teacher education programs have been tasked with including numerous general and professional education courses in their preparation programs. Finding room for EE is difficult, especially when its interdisciplinary nature is taken into account. Consequently, very few universities have any type of required EE coursework or fieldwork. In general, preservice teachers have very limited access to EE content or teaching methods in their course work (McKeown-Ice 2000; Mastrilli 2005). If EE is included in a teacher preparation program, science or social studies methods classes are usually used rather than an interdisciplinary approach (Plevyak et al. 2001).

Key research findings for the preparation of effective teachers have found that both the knowledge of the subject being taught and the "knowledge and skill in *how to teach* that subject" (NCATE 2006, p. 4) are critical to classroom success. Managing students in an outdoor classroom setting, using methods designed to increase their environmental awareness, and being able to successfully critique and evaluate appropriate EE resources are all skills related to teaching about the environment. Preparing preservice teachers to successfully use EE methods and resources requires exposure to appropriate teaching materials that are non-biased and based on science.

The best methods are based on inquiry techniques that allow preservice teachers to be learners, encourage active engagement, and model appropriate teaching methods for use in their own classrooms (Bell et al. 2003). The advantage of incorporating professional development early and often into teacher preparation programs is to expose preservice teachers to best professional practices early in their career. As they enter the classroom, new teachers are then better equipped and have resources ready for curriculum planning, development, and use (Van Petegem et al. 2005).

Preservice Science Teachers as Environmental Educators

Two documents have been developed to guide the professional development of science teachers. These documents are also important to guide the professional preparation of preservice teachers in both science and EE. The *National Science Education Standards* (NSES) (NRC 1996) contains a section on professional development for inservice teachers and provides a description of professional development:

> Professional development for teachers should be analogous to professional development for other professionals. Becoming an effective science teacher is a continuous process that stretches from preservice experiences in undergraduate years to the end of a professional career. Science has a rapidly changing knowledge base and expanding relevance to societal issues, and teachers will need ongoing opportunities to build their understanding and ability. Teachers also must have opportunities to develop understanding of how students with diverse interest, abilities, and experiences make sense of scientific ideas and what a teacher does to support and guide all students. And teachers require the opportunity to study and engage in research on science teaching and learning, and to share with colleagues what they have learned (p. 55).

The second document, *The Guidelines for the Initial Preparation of Environmental Educators* (*The Guidelines*) (NAAEE 2000) was developed by the North American Association for Environmental Education (NAAEE). *The Guidelines* contains "a set of recommendations about the basic knowledge and abilities educators need to provide high-quality environmental education" (NAAEE 2000, p. 1). Emphasized within *The Guidelines* is the need for all teachers to pursue ongoing professional development opportunities. Also important is the development of relationships with mentors and advisors who can model teaching methods, provide different ideas about environmental issues, and assist in expanding the skill base of beginning teachers. These professional development frameworks from these two documents (see Fig. 1) will guide our discussion of professional preparation for preservice teachers in the field of EE.

There is some agreement between these two documents when it comes to how teachers should obtain professional development in EE. Standard A of the NSES (NRC 1996) states that professional development should be a lifelong process. This aligns with guideline 3.3 of *The Guidelines* (NAAEE 2000) that states that professional development should be ongoing. It takes time to learn all of the EE curricula that have been developed for use in the USA. Therefore, it is critical to prepare preservice teachers for the need to pursue professional development in new and emerging EE projects, programs, and curricula. These projects in EE continue to be developed by a variety of agencies. Ultimately, it becomes the teacher's responsibility to determine which EE resources best fit their curriculum.

Professional development needs to go beyond familiarity with the various EE *Project* curricula and resources. Familiarity with the content of the curricula does not mean that preservice teachers or experienced teachers possess the proper skills to create positive learning environments for their students in science or EE.

National Science Education Standards for Professional Development	**Guidelines for the Initial Preparation of Environmental Educators: Professional Responsibilities of Environmental Educator**
A. Professional development for a teacher of science is a continuous, lifelong process.	3.1 Exemplary environmental education practice.
B. The traditional distinctions between "targets," "sources," and "supporters" of teacher development activities are artificial.	3.2 Emphasis on education, not advocacy.
C. The conventional view of professional development for teachers needs to shift from technical training for specific skills to opportunities for intellectual professional growth.	3.3 Ongoing learning and professional development.
D. The process of transforming schools requires that professional development opportunities be clearly and appropriately connected to teachers' work in the context of the school.	

Fig. 1 The standards for professional development from the *National Science Education Standards* (NRC 1996) and the professional responsibilities section of *The Guidelines for Initial Preparation of Environmental Educators* (NAAEE 2000)

Standards C and D of the NSES (NRC 1996) pertain to the need to go beyond acquiring knowledge about a variety of activities to use in the classroom. The "intellectual professional growth" that it talks about is in accordance with *The Guidelines* (NAAEE 2000) standard 3.2 to emphasize education and not advocacy.

The emphasis on opportunities for professional growth states that professional development should be continuous. In addition, professional development should be connected to the context of school as an area related to acquiring pedagogical content knowledge and relevant science and EE content knowledge by preservice teachers and practicing teachers.

Pedagogical Content Knowledge

Shulman (1986, 1987) developed a framework for teacher education by introducing the term "pedagogical content knowledge" (PCK). Rather than considering the knowledge of teaching from the perspective of either content or pedagogy, Shulman

(1986, 1987) believed that elements should be combined of these two knowledge domains. Various scholars have further developed conceptualizations of PCK (e.g., Appleton 2003; Gess-Newsome 1999; Loughran et al. 2006; Van Driel et al. 1998). PCK has become a way of understanding the complex relationship between teaching and content through the use of specific teaching approaches. Understanding of this relationship is developed through an integrated process rooted in classroom practice (Van Driel et al. 1998). Preservice teachers who have only an awareness of the science content studied in the academic discipline are not necessarily prepared with the understandings needed to teach that content. Their academic knowledge must be transformed into instructional activities appropriate for classroom instruction. Encouraging preservice teachers to reflect on their own teaching may well allow them to develop insights into their thinking about science subject matter, science teaching, and their own professional development. However, in so doing, there is a need to explicate the knowledge used for teaching and to establish ways of thinking about science teaching beyond the accumulation of pedagogical strategies.

The NSTA Standards for Science Teacher Preparation (NSTA 2003) help address issues of PCK and are based on a review of the professional science education literature and on the goals set forth in the National Science Education Standards. These standards outline the knowledge that teachers ought to have about specific content in four areas of scientific study: biology, chemistry, earth sciences, and physics. Within each of these domains, numerous objectives describe the most important ideas teachers ought to understand and demonstrate throughout their preservice experiences.

A Call for Strong Science Content Knowledge

Since the science education movement began in the 1960s, the study of student misconceptions about scientific phenomena has been prolific in the literature. Students develop these misconceptions as a result of either personal experience, from other people, or through the media (Ausubel 1978; Driver et al. 1985). Driver (1985) reported that different people have different misconceptions in different areas of science. In teacher education, it is critical to evaluate the conceptions of preservice teachers. If they have misconceptions, it is likely they will pass the incorrect content on to their future students. The result of persistent wrong conceptions about scientific phenomena is an ill-informed citizenry and a reduced possibility of appropriate preventive actions by these citizens against future problems (Boyes, Chamber, & Stanisstreet 1995).

This is a cascading effect that has not been widely addressed. For example, an analysis of survey data indicated that many high-school preservice teachers possess an array of misconceptions about the causes and effects of the greenhouse effect, ozone depletion, (Boyes, Chamber, & Stanisstreet 1995) and acid rain (Khalid 2003). The problem grows more complex due to mismatched concept and student developmental levels. Inaccuracies in textbooks, incorrect information provided by instructors, and student memorization of prior concepts without meaningful understanding of the basic concepts compound the problem. Ultimately, a lineage of confused science concepts – and confused students – is created (Westbrook and Marek 1992). Both preservice and practicing teachers need to know the potential misconceptions that students can have in examining EE concepts and topics.

Call for Quality EE Teacher Preparation

Research points to the urgent need for teacher education to embrace teaching and learning approaches that support and complement the aims and objectives of EE. Beginning teachers must also be assisted with reflecting on their own teaching practices. The global significance of this issue is illustrated by the importance placed on teacher education for EE in the 1975 International Belgrade workshop (UNESCO-UNEP 1976), the ministerial-level Tbilisi Conference of 1977 (UNESCO 1978), the 1987 Moscow Congress (UNESCO-UNEP 1990), the Brundtland Report (WCED 1987), the "Earth Summit" in 1992 (UNESCO-UNEP 1992) and the UNESCO-ACEID conference on "Environmental Education in Teacher Education in Asia and the Pacific" (NIER 1993).

More recently, an Asia-Pacific UNESCO-ACEID project "Learning for a Sustainable Environment" (UNESCO 1997) addressed the issue of quality teacher education for EE by implementing a region-wide action research project that focused on enhancing professional practice (Fien et al. 1997). Independent researchers have also emphasized the need to develop pedagogical practices in preservice teacher education programs for EE (Fien 1993; Hart 1990; Robottom 1987c; Stapp et al. 1980; Tilbury 1995).

Robottom (1987a, b, c, d) posits that teacher education for EE involves a dual pedagogical challenge. The first challenge is the need to address the social change objectives of EE that seek to transform the "business as usual" mind-set to ecological sustainable approaches. The second challenge addresses the need for professional preparation experiences in EE that assist preservice teachers to become critically reflective practitioners.

Robottom (1987a) has proposed five principles in order to address his dual challenge. Teacher preparation in EE should:

- Be participatory and practice-based
- Be enquiry-based
- Involve ideological critique
- Be community-based
- Be collaborative

These are the approaches that have guided a mentoring program in EE developed at Central Michigan University.

Mentoring Preservice Science Teachers

In order to develop a notion of PCK and how to know when to use appropriate teaching techniques in EE, a mentoring program has been developed for several preservice teachers to help them become exemplary science teachers. In this program, preservice teachers work as "facilitators in training" with a mentor faculty member to assist with professional development workshops. These experiences provide intensive, hands-on opportunities for skill development. Preservice teachers first take part in four EE *Project* curriculum workshops: Project Wild (2000), Aquatic Wild (2001), Water Education for Teachers (Project WET) (1995), and Project Learning Tree (PLT) (2006).

In Michigan, coordinators for the various EE *Project* curricula report a predominance of preservice teacher participants in the majority of workshops (D. Elshoff, K. Fischer, and J. Vail, personal communication, 2008). At Central Michigan University, we have found this participation trend to hold true in each of the two annual workshops in EE that we offer. Of the participants 95% are our own preservice teachers or students studying for careers in outdoor education. In order to develop project facilitators for tomorrow, we are "growing" our own facilitators for the four EE *Projects*.

After students have completed the four EE *Project* workshops as participants, interested preservice teachers can become facilitators in training. This addresses Robottom's (1987b) principle that teacher education should be participatory and practice-based. If preservice teachers have been recent participants in EE classes, they can apply that experience to becoming an effective facilitator when they have to lead their own workshop.

Two faculty mentors, a science educator and an outdoor educator, work with the selected preservice teachers as they plan and implement their own EE professional development. Preservice teachers learn to be an EE facilitator and combine that experience with their previous participant experience to plan effective EE professional development. The experience is collaborative since the preservice teachers work and team with the mentors.

The preservice teachers must plan three EE professional development experiences before they become facilitators. After each is planned and implemented, reflective meetings are held with mentors to evaluate the effectiveness of the experience. Evaluations from participants are reviewed and a discussion ensues as to how the next professional development session can be enhanced. The preservice teachers also reflect on how they can become better facilitators, and by association, better science teachers or outdoor educators.

Preservice teachers also work to mentor local alternative high-school students who then provide EE programs for elementary school children. This process of teaching others helps students to develop their own PCK. This local collaboration addresses Robottom's (1987b) principle that teacher preparation in EE needs to be community-based.

Preservice teachers have been encouraged to present EE activities at state and national conferences. This assists them in practicing their teaching skills and connects them to professional networks. Many of the preservice teacher mentees that we have worked with at Central Michigan University have presented at area conferences of the National Science Teachers Association (NSTA) and at the Michigan Science Teacher Association annual conference. Many of these preservice teachers are members of the NSTA preservice chapter NSTA-CMU. Working with this organization has given us access to a committed group of students who wish to actively engage in professional development activities. Many of the above preservice-teachers-turned-EE-facilitators are among the mentees that have presented EE activities at conferences.

The above example addresses several of Robottom's (1987a) principles. The presentations that the preservice teacher mentees gave at the science education conferences were participatory and practice-based. The activities presented were from the four main EE *Project* curricula mentioned above. Preservice teachers worked in collaborative teams to present in a share-a-thon format with activities that were enquiry-based.

Recommendations and Conclusion

Our experiences with the professional preparation of preservice teachers in EE have led to the following recommendations.

1. Faculty involved in the preparation of preservice teachers in EE should be familiar with and incorporate the *NAAEE Guidelines for the Initial Preparation of Environmental Educators* (NAAEE 2000) in their methods courses. *The Guidelines* (NAAEE 2000) provides science methods instructors with a framework with which to integrate EE into a science methods course. The instructional strategies, methods, and skills applied during EE activities are particularly applicable to science processes.
2. EE should be a part of science methods courses in teacher preparation programs since many of these programs do not have a separate EE methods course. Although the basic premise of EE calls for an interdisciplinary approach, the reality of resource availability for preservice preparation may mandate a single subject approach. Therefore, the most reasonable alternative would be the use of science methods courses for EE preparation delivery.
3. Examples of EE curriculum (for example, *Project* curricula) should be introduced to preservice teachers as a way to promote the use of science process skills, the integration of EE with science and other subjects, and as a way to introduce hands-on, minds-on instructional strategies that encourage active learning. Student participation in the EE *Project* curriculum workshops provides preservice teachers with a toolbox of exemplary activities appropriate for immediate use in their science classroom. At the elementary level, the EE activities

promote integration with science and other subjects. These activities contain multiple opportunities for teachers to develop and enhance their students' process skills.

4. EE can be used as a way to promote environmental literacy for preservice teachers and an interdisciplinary, multicultural, and global approach to instruction. Environmental perspectives can also be used along with indigenous ways of knowing to present global issues. Using an EE perspective in science promotes global thinking because of its interdisciplinary approach. A global approach is needed for preservice teachers to be prepared to educate the next generation of global citizens.
5. Global and local issues related to the environment are a way for preservice teachers to think critically, while examining both sides of an issue, and how they can be involved in providing a solution for a problem. Environmental issues need to be presented in a balanced, nonbiased manner so that students can see both sides of an issue. Examples of issues that need to be presented in a balanced manner include the harvesting of the Redwood groves in northern California, whaling, the rainforest, energy use and policy, global warming, and endangered species.
6. EE can serve as a way to introduce the use of appropriate technologies to preservice teachers. Outdoor environmental data collection experience can provide preservice teachers with opportunities to learn how to use technologies such as GPS and probeware for environmental investigations. The use of these technological tools can promote higher-order thinking skills for students and allow them to apply science to their own lives.

In conclusion, preservice teachers need opportunities to practice instructional strategies for EE. Instructional strategies as outlined by NAAEE (2000) include hands-on observation and discovery in the environment, inquiry, cooperative learning, service learning, problem-based learning, and other methods. Preservice teachers must also develop their own environmental awareness and an attitude toward environmental responsibility and stewardship to be effective environmental educators. Preparation in EE methods and strategies at the preservice level gives new teachers a variety of resources, skills, and knowledge to assist them with implementation of environmental lessons in their science classrooms.

References

Appleton, K. (2003). How do beginning primary school teachers cope with science? Toward an understanding of science teaching practice. *Research in Science Education, 33*, 1–25.

Ausubel, D. (1978). In defense of advance organizers: A reply to the critics. *Review of Educational Research, 48*, 251–257.

Bell, C., Shepardson, D., Harbor, J., Klagges, H., Burgess, W., Meyer, J., et al. (2003). Enhancing teachers' knowledge and use of inquiry through environmental science education. *Journal of Science Teacher Education, 14*(1), 49–71.

Boyes, E., Chamber, W., & Stanisstreet, M. (1995). Trainee primary teachers' ideas about the ozone layer. *Environmental Educational Research, 1*(2), 133–145.

Dewey, J. (1966). *Democracy and education*. Toronto, Canada: Collier-Macmillan Canada.

Disinger, J. F., & Howe, R. W. (1990). *Trends and issues related to the preparation of teachers for environmental education*. (ERIC Document Reproduction Service No. ED335233)

Driver, R., Guesne, E., & Tiberghien, A. (1985). *Children's ideas and the learning of science. In Children's Ideas in Science*. Philadelphia: Open University Press.

Fien, J. (1993). *Education for the environment: Critical curriculum theorizing and environmental education*. Geelong, Australia: Deakin University Press.

Fien, J., Heck, D., & Ferreira, J. (1997). *Learning for a sustainable environment: A professional development guide for teacher educators*. Bangkok, Thailand/Brisbane, Australia: UNESCO-Asia Pacific Centre of Innovation for Development//Griffith University.

Gess-Newsome, J. (1999). Pedagogical content knowledge: An introduction and orientation. In J. Gess-Newsome & N. G. Lederman (Eds.), *Examining pedagogical content knowledge* (pp. 3–17). Boston: Kluwer.

Hart, P. (1990). Environmental education in Canada: Contemporary issues and future possibilities. *Australian Journal of Environmental Education, 6*, 45–65.

Heimlich, J. E., Braus, J., Olivolo, B., McKeown-Ice, R., & Barringer-Smith, L. (2004). Environmental education and preservice teacher preparation: A national study. *The Journal of Environmental Education, 35*(2), 17–21.

Hungerford, H. R. (1996). The development of responsible environmental citizenship: A critical challenge. *Journal of Interpretation Research, 9*(1), 25–37.

Hungerford, H. R., & Volk, T. L. (1990). Changing learner behavior through environmental education. *Journal of Environmental Education, 21*(3), 8–21.

Khalid, T. (2003). Pre-service high school teachers' perceptions of three environmental phenomena. *Environmental Educational Research, 9*(1), 10–21.

Kim, C., & Fortner, R. W. (2006). Issue-specific barriers to addressing environmental issues in the classroom: An exploratory study. *The Journal of Environmental Education, 37*(3), 15–22.

Loughran, J. J., Mulhall, P., & Berry, A. (2006). *Understanding and developing science teachers' pedagogical content knowledge*. Rotterdam, The Netherlands: Sense Publishers.

Mastrilli, T. (2005). Environmental education in Pennsylvania's elementary teacher education programs: A statewide report. *The Journal of Environmental Education, 36*(3), 22–30.

McKeown-Ice, R. (2000). Environmental education in the United States: A survey of preservice teacher education programs. *The Journal of Environmental Education, 32*(1), 4–11.

National Council for Accreditation of Teacher Education. (2006). *What makes a teacher effective?* Washington, DC: NCATE.

National Research Council. (1996). *National science education standards*. Washington, DC: National Academy Press.

National Science Teachers Association. (2003). *Standards for science teacher preparation*. Washington, DC: Author.

NIER. (1993). *Environmental education in teacher education in Asia and the Pacific*. Tokyo: National Institute for Educational Research.

NIER. (1996). *Environmental education in teacher education in Asia and the Pacific*. Tokyo: National Institute for Educational Research.

North American Alliance for Environmental Education. (2000). *The guidelines for the initial preparation of environmental educators*. Washington, DC: Author.

North Carolina Department of Public Instruction's Division of Science Education. (1973). *Environmental education: Preservice preparation of teachers*. Raleigh, NC: State Department of Public Instruction.

Plevyak, L. H., Bendixen-Noe, M., Henderson, J., Roth, R. E., & Wilke, R. (2001). Level of teacher preparation and implementation of EE: Mandated and non-mandated EE teacher preparation states. *The Journal of Environmental Education, 32*(2), 28–36.

Powers, A. L. (2004). Teacher preparation for environmental education: Faculty perspectives on the infusion of environmental education preservice methods courses. *The Journal of Environmental Education, 35*(3), 3–11.

Project Learning Tree. (2006). *PreK-8 environmental education activity guide*. Washington, DC: American Forest Foundation.

Project WET. (1995). *Curriculum and activity guide*. Bozeman, MT: Montana State University.

Project WILD. (2000). *K-12 curriculum & activity guide*. Houston, TX: Council for Environmental Education.

Project WILD Aquatic. (2001). *K-12 curriculum & activity guide*. Houston, TX: Council for Environmental Education.

Robottom, I. (1987a). Contestation and consensus in environmental education. *Curriculum Perspectives, 7*(1), 23–27.

Robottom, I. (1987b). Towards enquiry-based professional development in environmental education. In I. Robottom (Ed.), *Environment education: Practice and possibility*. Geelong, Australia: Deakin University Press.

Robottom, I. (1987c). The dual challenge for professional development in environmental education. In A. Greenall (Ed.), *Environmental education past, present and future*. Proceedings of the Third National Environmental Education Seminar and Workshop (pp. 72–83). Canberra, Australia: AGPS.

Robottom, I. (1987d). Two paradigms of professional development in environmental education. *The Environmentalist, 7*(4), 291–298.

Samuel, H. R. (1993). Impediments to implementing environmental education. *The Journal of Environmental Education, 25*(1), 26–29.

Shulman, L. S. (1986). Those who understand: Knowledge growth in teaching. *Educational Research, 15*(2), 4–14.

Shulman, L. (1987). Knowledge and teaching: Foundations of the new reform. *Harvard Educational Review, 57*(1), 1–22.

Simmons, D. (1998). Using natural settings for environmental education: Perceived benefits and barriers. *The Journal of Environmental Education, 29*(3), 23–31.

Stapp, W., Caduto, M., Mann, L., & Nowak, P. (1980). Analysis of pre-service environmental education of teachers in Europe and an instructional model for furthering this education. *Journal of Environmental Education, 12*(2), 3–10.

Stone, J. M. (1989). Preparing teachers to become involved as environmental educators. *Contemporary Education, 60*(3), 159–162.

Tilbury, D. (1995). Environmental education for sustainability: Defining the new focus of environmental education in the 1990s. *Environmental Education Research, 1*(2), 2–8.

UNESCO. (1978). The Tbilisi Declaration: Final report of Intergovernmental Conference on Environmental Education organized by UNESCO in cooperation with UNEP, Tibilisi, USSR, 14–26 October 1977. *Connect, 3*(1), 1–5.

UNESCO. (1997). *Education for a sustainable future: A transdisciplinary vision for concerted action*. International Conference Thessaloniki, Greece.

UNESCO-UNEP. (1976). The Belgrade Charter. *Connect: UNESCO-UNEP Environmental Education Newsletter, 1*(1), 1–2.

UNESCO-UNEP. (1990). Environmentally educated teachers: The priority of priorities? *Environmental Education Newsletter*, pp. 1–3.

UNESCO-UNEP. (1992). UNCED: The Earth summit. *Connect, 17*(2), 4–6.

Van Driel, J. H., Verloop, N., & de Vos, W. (1998). Developing science teachers' pedagogical content knowledge. *Journal of Research in Science Teaching, 35*, 673–695.

Van Petegem, P. V., Blieck, A., Imbrecht, I., & Van Hout, T. (2005). Implementing environmental education in pre-service teacher training. *Environmental Education Research, 11*(2), 161–171.

Westbrook, S. L., & Marek, E. A. (1992). A cross-age study of student understanding of the concept of homeostasis. *Journal of Research in Science Teaching, 29*, 51–61.

Westing, A. H. (1993). The global need for environmental education. *Environment, 35*(7), 4–5, 45.

Wilke, R. J., Peyton, R. B., & Hungerford, H. R. (1987). *Strategies for the training of teachers in environmental education.* UNESCO-UNEP International Environmental Education Programme, Environmental Education Series 25. UNESCO, Paris: Division of Science, Technical, and Environmental Education.

World Commission on Environment and Development. (1987). *The Brundtland Report. In our common future: Report of the World Commission on Environment and Development.* Retrieved on July 1, 2009, from http://www.worldinbalance.net/agreements/1987-brundtland.php

Approaches to Environmental Education

Austin A. Winther, Kim Cleary Sadler, and Gerry Saunders

As discussed in the introductory chapter on the history and philosophy of environmental education (EE), the EE field is quite broad. It includes teaching learners of all ages to appreciate and value nature and to understand natural systems and how they work. EE also helps people develop understandings of how humans interact with their environment, both natural and man-made, and how to take responsible, democratic action to preserve and protect the environment. With such a wide range of goals, it is evident that there is no one "right" pedagogical approach for EE; there are many approaches. The approach that will work best in any given case depends on numerous factors including, but not limited to: student interest, maturity, age, and abilities; the curricular goals being addressed; and the instructional setting including resources and the time available for instruction. It also depends on the knowledge, skills, interests, and educational philosophy of the instructor or instructors involved. Educators who work in nature centers, aquaria, and zoos, for example, can provide learners with experiences that classroom teachers cannot provide. Conversely, classroom teachers who work with students over a period of a semester or a year can do different things with students than educators who only see students for a few hours or a few days.

This chapter is intended to provide educators who are relatively new to EE a summary of some of the approaches that have been found to be effective. This chapter is not meant to be either prescriptive or exhaustive, but rather provides an

A.A. Winther (✉)
Department of Graduate Education, Leadership, and Counseling, Rider University, 2083 Lawrenceville Road, Lawrenceville, NJ 08648-3099, USA
e-mail: awinther@rider.edu

K.C. Sadler
Biology/MTSU Center for the Study of Cedar Glades, Middle Tennessee State University, PO Box 60, Murfreesboro, TN 37132, USA
e-mail: ksadler@mtsu.edu

G. Saunders
Director of Teacher Education, Center for Experiential and Environmental Education, Unity College, 90 Quaker Hill Road, Unity, ME 04988-9502, USA
e-mail: gsaunders@unity.edu

A.M. Bodzin et al. (eds.), *The Inclusion of Environmental Education in Science Teacher Education*, DOI 10.1007/978-90-481-9222-9_3,

introduction to the diversity of ways in which science and EE intersect. In the cases of approaches that have centers or organizations associated with them, we have provided contact information for those who wish more information about a particular approach.

Environmental Education and the Schoolyard

The use of the school grounds as a core part of the curriculum has a rich and diverse history. In modern times, use of school grounds can be traced to the nature study movement of the late 1800s and early 1900s (Dewey 1959; Tolley 1994). Today, using schoolyards as extensions of the classroom can take many forms and be included at all grade levels. The list below provides a small sampling of the diverse ways in which the school campus can be used for promoting both science education and EE.

- Sixth-grade students in a mid-coast Maine middle school design exploratory activities to be used by the school's fourth-grade students as they investigate the working forest adjacent to the school.
- Third-grade students in an eastern Nebraska elementary school design, plant, and schedule maintenance for a butterfly garden on the school grounds.
- Students in a Colorado middle school learn about scale, map reading, and the use of a compass as they navigate a school ground orienteering course.
- Maryland middle school students raise sea grasses as part of a Chesapeake Bay restoration project.
- Elementary students plan and maintain an outdoor laboratory in the schoolyard to study native plants and animals in Tennessee (Sadler et al. 2006).
- Students at various schools in California grow vegetables through the School Garden Program (California Department of Education 2009) for the school cafeteria.

For science educators, using the campus provides opportunities to explore nature in context rather than isolation. For example, often the north and south sides of the school will show variations in plant growth. Sidewalks will create heat sinks, creating microclimates and convection currents. While studying these microclimates, students become aware of nature or become sensitive to natural processes and establish a connection to place. Knowledge of ecological processes, sensitivity to the environment, and connection to place are fundamental knowledge and attitudes necessary for developing the concern needed for personal involvement (Marcinkowski 1988; Peterson and Hungerford 1981; Sivek and Hungerford 1989).

Work by Cronin-Jones (2000) demonstrated the effectiveness of using the schoolyard in promoting understanding of selected environmental science topics when compared with classroom only instruction. Martin (2003) found evidence of students developing positive environmental attitudes and changed behaviors as a result of schoolyard-based instruction. For example, if the students focus on the awareness of texture through feeling different barks or leaves during a sharing walk with a blindfolded partner, the student is not only developing a sensitivity to the his/her surroundings, but also developing observation and inference skills that are basic science process skills. The awareness of how these skills interplay in helping the students become aware of the richness of their learning is an essential part of achieving the full wealth of the outdoor learning experience (National Research Council 2009).

Examination of common applications of schoolyard EE shows that this approach is most effective at developing the foundation-level goals of EE: awareness and sensitivity to the local environment, gaining experience and basic understanding of the environment, and developing a feeling and concern for the environment (UNESCO 1978). Students can spend significant time getting to know their home system in a variety of contexts, developing a deep connection and knowledge of the area. The China Elementary and Middle School in Maine provides a specific example of how this can be done. Students in grades 1–8 use the community forest adjacent to the school grounds in a variety of subject areas. Primary-grade students are taught to use all their senses to make observations. As students progress through the grades, they use the forest for science explorations and investigations, literature activities, and mathematics applications. In the process, they develop a more complex understanding of forest processes and the interactions of forest management practices and wildlife habitat. These experiences support content learning objectives in addition to promoting understandings and appreciation of ecological systems.

Broda (2007) describes four major benefits of using the schoolyard: (1) providing concrete experiences to clarify abstract concepts; (2) providing motivation for the reluctant learner; (3) adding variety to teaching and learning; and (4) helping increase student achievement. Taken collectively, these benefits provide an important rationale for taking students outside and are consistent with the goals of both science and EE.

A sample of programs available to provide support and training for teachers interested in using their schoolyards as learning sites can be found in Appendix 1.

Place-based Education

In many ways, place-based education is an extension of the schoolyard. Rather than being limited by the schoolyard, the community becomes the classroom. Students develop knowledge and skills in the context by using the resources of the local community.

As discussed in the chapter "The History and Philosophy of Environmental Education," the framework for the goals of EE are based on the Belgrade Charter (UNESCO-UNEP 1976), Tbilisi Declaration (UNESCO 1978), and the curriculum framework for EE described by Hungerford, Peyton, and Wilke (1980). These documents emphasized the ultimate goal of EE as developing environmentally literate citizens, with the sensitivity, awareness, skills, and willingness to become knowledgeable and responsible citizens. Place-based education tries to achieve this goal through utilizing the needs and resources of the local community. Students are immersed in the local environment, history, problems, and resources to provide an educational experience that involves students by putting learning in a local context and emphasizing solving real problems (Center for Place-based Education 2008).

The curricula of two Maine middle schools provide examples of place-based education in action. At one middle school, students study where their food comes from by raising melons, corn, salad greens, and other vegetables that are then sold at a local farmer's market and a roadside produce stand. Other students design the packaging for seeds from plants that have been selectively bred; the seeds are also sold at a local food co-op. In another environmental project, students conducted an energy audit of their school, determining the sources of the school's energy. They then wrote a grant to obtain solar panels for the school to reduce its carbon footprint and environmental impact on local rivers. Students at another Maine middle school learn about forest ecology, the economics of forestry, and issues of forest management by studying a working community forest adjacent to the school.

These students actively learn about important concepts and skills as described in the Maine State Standards (Maine Department of Education 2007). Their learning occurs within personal contexts of their daily lives. In addition, they are actively involved in solving local problems – an important feature of place-based education.

People's sense of place, how they are connected to where they live, affects how they learn, and the decisions they make about land use and personal consumption. The Antioch New England Institute (Center for Place-based Education 2008) describes place-based education as a focus on learning within the local community of the student; they further identify that the goal of place-based education is to develop citizens who are active in their communities and take care of the environment. This goal may be considered a contemporary view of the Belgrade Charter (UNESCO-UNEP 1976) and the curriculum framework for EE described by Hungerford, Peyton, and Wilke (1980). Similar to schoolyard education, place-based education has its roots in Dewey's work that called for the curriculum of schools to connect the classroom with the world of the child (Dewey 1959).

In common with EE, place-based education encompasses more than science. It seeks to engage students in the entire community including its history, culture, and ecology. The underlying philosophy is that by developing knowledge and skills in a local context, with first-hand application, students will be better able to understand more abstract and distant phenomena. It also recognizes that for students to be engaged, knowledgeable, and skilled citizens, they must have a connection to a place with which to be engaged.

Unlike many EE programs, place-based education takes a very holistic perspective of place, including history, literature, and the arts, as well as local ecology and environmental problems and issues. The *Foxfire* cultural anthologies originally developed by students at Rabun Gap-Nacoochee School and edited by Eliot Wigginton (1971–2004) are one of the best known examples of place-based education.

Greg Smith (2002) describes five thematic patterns of place-based education that have been used in a variety of settings. These patterns include cultural studies, nature studies, real-world problem-solving, internships and entrepreneurial opportunities, and induction into community processes. When taken collectively, these themes align to the Tbilisi goals (UNESCO 1978) and curriculum framework for EE described by Hungerford, Peyton, and Wilke (1980). While no single student project may achieve the goal of developing people who are "environmentally knowledgeable and, above all, skilled and dedicated citizens who are willing to work, individually and collectively, toward achieving and/or maintaining a dynamic equilibrium between quality of life and quality of the environment" (Hungerford et al. 1980, p. 42), place-based education offers a potent tool to achieving this goal.

Population Connection and the EE *Project* Curricula (PLT [Project Learning Tree], WILD [Wildlife in Learning Design], and WET [Water Education for Teachers])

In addition to the schoolyard and place-based education, another approach to EE is through a thematic curriculum integration perspective. Collectively, Population Connection and the *Project* activities are resources disseminated by organizations with a long history of providing quality materials that combine environmental and science education. Population Connection – formerly Zero Population Growth or ZPG (Population Connection 2009), Project Learning Tree – PLT (American Forest Foundation 2008), Project Wildlife in Learning Design – WILD (Council for Environmental Education 2008a, b), and Project Water Education for Teachers – WET (The Watercourse and Council for Environmental Education 2007) are EE curriculum integration approaches that have had a major impact on both formal and nonformal education practices. While they are not curriculum in the classic definition, they are thematic programs that provide training and educator activity guides that contain instructionally sound activities to help educators integrate EE into their existing curriculum. Although Population Connection is not usually thought of as one of the *Project* curriculum, it is one of the older EE curriculum enhancement materials, first published in 1975, that provides age-appropriate content and learning activities about population growth and environmental impact. Project Learning Tree was instituted in 1976 by educators and resource managers to develop an educationally sound and unbiased program for teachers and their students. Project Learning Tree was established by a partnership between the American Forest Institute (now called the American Forest Foundation) and the Western Regional EE Council (WREEC); WREEC has since evolved into the Council for EE (CEE). Project WILD was developed in 1979 and became available in 1983 through the Western Regional EE Council (WREEC). Project WET was established in 1984 by the North Dakota State Water Commission and is currently supported by the Project WET International Foundation and CEE.

Each of the *Project* curriculum guides presents a unique perspective through which the goals of EE can be achieved. For example, Population Connection promotes sustainability of the Earth's resources through stabilization of the world's population. Project Learning Tree uses the forest as a means to enhance critical and creative thinking about the environment, develop informed decision-making abilities about environmental issues, and have students respond appropriately. Project WILD provides wildlife-based EE to assist all age learners with abilities to make informed decisions, and take constructive action in a responsible manner toward wildlife and the environment. Project WET provides educator-ready materials to support awareness, appreciation, knowledge, and stewardship toward water conservation practices. Collectively, these perspectives represent a consistent approach to environmental education. Each guide uses something of high intrinsic interest to students, such as wildlife or the forest, to provide motivating learning contexts for students. These topics provide a coherent focus for activities that help achieve the goals of EE, while learning essential science knowledge and skills.

Although each *Project* curriculum offers a different content area through which EE can be examined, the framework for dissemination is through a "train the trainer" model. Workshop facilitators receive comprehensive training to become certified to teach others. As workshops are conducted, new trainers are recruited and the program outreach expands. Usually, a state coordinator oversees the facilitation of a specific program in that respective state. Both Population Connection and the *Project* curricula include comprehensive curriculum activity guides that provide essential details necessary to teach each activity (objectives, background content information, materials, requisite skills, implementation procedures, assessments, curricular extensions), in addition to correlations to national and state learning standards. Distribution of activity guides and supplementary materials is only possible by attending a workshop conducted by a trained facilitator. Each activity guide has been through an intensive field-testing and evaluation process. The revision of each activity guide is an on-going process with later additions that include additional resources, and in the case of Population Connection, activities that are distributed on CD-ROM and on the Internet.

The conceptual frameworks used for activity development and organization of the activity guides are the goals of EE. In its conceptual framework, PLT uses themes ranging from diversity to patterns of change, to develop and organize its instructional materials. Project WILD's conceptual framework includes ecological knowledge, social and political knowledge, and sustaining fish and wildlife resources. The conceptual framework for Project WET includes the physical and chemical properties of water, connecting Earth systems, water as a natural resource, water management, and water existing with social and cultural contexts. These frameworks closely align to the goals of EE presented by the Tbilisi Declaration (UNESCO 1978) and the recommendations for curriculum development presented by Hungerford, Peyton, and Wilke (1980). Learning activities include developing sensory and environmental awareness, ecological knowledge, and the development of responsible citizen action skills. In addition, each activity is clearly identified with corresponding content objectives that support developing understandings of fundamental EE and science concepts. In addition to science, activities support learning across the curriculum, including art, mathematics, literacy, and civics. Analysis of the depth and breadth of the *Projects*' activity lessons is beyond the scope of this chapter. However, the abiding theme throughout each curriculum guide is not to tell students what to think, but to teach them how to think about the environment. They provide effective instructional tools for bridging science and EE in diverse learning settings.

Science and EE in Nature Centers, Zoos, and Museums

Another approach to EE that is further discussed in Chapters "Summer Methods in Summer Camps: Teaching Projects WILD, WET, and Learning Tree at an Outdoor Environmental Education Center," "Exploring Preservice Teachers Mental Models of the Environment," and "The Value of Nonformal Environmental Education-Based Professional Development in Preservice Science Teacher Preparation" is the important role that nonformal EE centers including nature centers, zoos, and museums play in regard to science and environmental education. The American Association of Museums (2009) has defined nature centers as facilities with an educational, scientific, and cultural purpose with a trained professional staff. Open to the public, centers manage and interpret the landscape and native organisms to promote an understanding of nature and natural processes. Nature centers can also be associated with conducting frequent EE programs and activities for the public (NSYF 1990). By definition, this also includes botanical gardens or arboretums, zoos, natural history museums, environmental learning centers, and national or state park nature interpretation centers. Although these designed settings sound more like destinations than an approach to EE, these facilities serve as outdoor learning labs that provide students with opportunities to interact with and generate awareness about the local environment. Primarily directed toward informal learning environments, students also have opportunities to gain knowledge through self-guided, exploratory, and assessment-free experiences. Since our attitudes and knowledge about the environment is shaped by multiple experiences over time, nature centers, zoos, and natural history museums serve as another venue for learning about the environment in a multisensory manner, beyond textbooks, papers, and pencils. The smells of the forest, the sounds of birds calling and the leaves moving in the wind, the sensation of a cricket crawling on your hand, or the texture of rough bark on the tree engage our most primitive senses. Learning in this context is not only meaningful but also inspiring. Such experiences can positively shift patterns of student behaviors because they are rooted in emotional experiences. Drawing on the goals of EE, rural and urban nature centers serve as institutions that advance learner awareness, appreciation, and action in regard to the environment.

Nonformal learning centers are potentially powerful approaches for connecting science and EE. In particular, nature centers have the advantage of being located in places that promote people's connection to the environment through voluntary experiences in natural settings. This positive affective environment can be effectively combined with interpretive signage, and structured programming to develop people's sensitivity to the environment as well as knowledge of earth systems and environmental problems. There are many outstanding informal learning facilities in this country, too many to be named or discussed in this chapter. However, all are unified as an approach to learning about science and the environment that is based on the reality that people can choose to visit these spaces and take ownership of their learning (National Research Council 2009).

Investigating and Evaluating Environmental Issues and Actions

A major goal of EE is to produce environmentally responsible action on the part of students. A goal of science education is to teach science process skills to students. Teaching students how to investigate environmental issues and to develop citizenship action plans align to both these goals. The instructional approach, *Investigating and Evaluating Environmental Issues* (Hungerford et al. 2003), teaches students to analyze environmental issues in terms of the science behind the problem or issue as well as the social dimensions surrounding the issue. This approach is designed to meet the goals of the Tbilisi Declaration (UNESCO 1978) and the four levels of EE identified by Hungerford, Payton, and Wilke (1980).

Goal Level I: The Ecological Foundations Level. Instruction at this level seeks to provide learners with sufficient ecological knowledge to permit him/her to eventually make ecologically sound decisions with respect to environmental issues.

Goal Level II: The Conceptual Awareness Level. This level of instruction seeks to guide the development of a conceptual awareness of how individual and collective actions may influence the relationship between quality of life and the quality of the environment and, also, how these actions result in environmental issues, which may be resolved through investigation, evaluation, values clarification, decision-making, and finally citizenship action.

Goal Level III: The Investigation and Evaluation Level. Education at this level provides for the development of the knowledge and skills necessary to permit learners to investigate environmental issues and evaluate alternative solutions for solving these issues. Similarly, values are clarified with respect to issues and alternative solutions.

Goal Level IV: Action Skills Level-Training and Application. Education at this level seeks to guide the development of those skills necessary for learners to take positive environmental action for the purpose of achieving and/or maintaining a dynamic equilibrium between the quality of life and the quality of the environment (Hungerford and Volk 1990, p. 13).

The Issue Analysis

In the first phase of this approach, referred to as issue analysis, students are taught to identify environmental problems and issues. A problem is any situation in which something valuable is at risk. An issue arises when two or more parties, called players (or stakeholders), disagree about the solution to a problem. In an environmental problem or issue, some part of the environment is at risk, but so may be jobs, homes, health, cultural or recreational resources, or other things of value. Issues may arise when two players have different knowledge or different values about an issue. Value descriptors are provided to help students identify the values of the players (Winther et al. 1994). To practice these skills, students read articles from a variety of media sources (newspapers, magazines, broadcast media, and web-based sources). Cooperative learning is also encouraged throughout the skill development activities. During this phase, students are required to examine all aspects of an issue, not just the ones they agree. This allows them to develop a fuller understanding of the issue and to consider several points of view.

Issue Investigation

One of the strongest features of the Investigating and Evaluating Environmental Issues approach from the standpoint of science education is that the students carry out an original scientific investigation of a particular environmental issue. This may be done in either of two ways. Younger students, less capable of independent work, can carry out a class case study, whereas older students can conduct small group or individual investigations. In all implementations of the issue investigations, all key elements are retained. Issue investigations meet all the criteria listed by the National Science Teachers Association for a Science-Technology-Society program (National Science Teachers Association 1990).

Having conducted an issue analysis, the students then formulate research questions on the issue in their community, based on guidelines provided to them (Winther et al. 1994). The students collect data in their communities to attempt to answer their research questions. The data collected may take a variety of forms including natural science data (such as acid rain deposition or stream sedimentation), or social science data (such as knowledge, beliefs, and/or values of a sample population).

After data is collected, the students must tabulate and interpret their findings. This involves the construction of data tables and graphs to understand and report their data. It also includes formulating conclusions and making inferences from their data. Students are expected to report their results to their class and often they also report their results to the larger community.

Citizenship Action

After students complete their issue investigation, they take a position on the issue based on their analyzed scientific information and their own values. The final phase of issue analysis is the formulation of an action plan on the basis of their background research and their original research findings. With older students, it is not necessary that everyone within a group reach consensus on their issue. It is usually the case that there is a unanimous decision, but by now the students understand that well-informed citizens may take different positions based on different values. With students capable of independent work, these action plans are usually formulated on the basis of a second issue investigation conducted individually or in pairs. If a case study approach is used, then the class can formulate an action plan under the guidance of the teacher. In either case, students are guided to consider individual as well as group actions. Students are encouraged, but not required, to carry out their action plans. Four types of actions are considered: persuasion, consumerism, political action, and eco-management. Action plans are formulated on the basis of 14 action criteria provided to the students (Winther, et al. 1994).

Action Analysis Criteria

A set of 14 critical questions are important to ask before proceeding with a citizen action. The questions are:

1. Is there sufficient evidence to warrant action on this issue?
2. Are there alternative actions available for use? What are they?
3. Is the action chosen the most effective one available?
4. Are there legal consequences of this action? If so, what are they?
5. Will there be social consequences of this action? If so, what are they?
6. Will there be economic consequences of this action? If so, what are they?
7. What are the ecological consequences of this action?
8. Do my personal values support this action?
9. Do I understand the beliefs and values of others involved in this issue?
10. Do I understand the procedures necessary to take this action?
11. Do I have the skills needed to complete this action?
12. Do I have the courage to take this action?
13. Do I have the time needed to complete this action?
14. Do I have all the other resources needed to make this action effective?

An important aspect of the issue investigation approach is that students do not have to reach consensus on an issue. They may adopt positions that are not the same as other students or the instructor. Many times, students support positions different from those around them because they hold different values. The only requirement is that they are able to support their position with sound argument and evidence.

Another aspect of the decision-making process is consideration of consequences prior to action. This is particularly important when working with adolescents or people new to civic engagement.

The National Science Education Standards (National Research Council 1996) identify seven criteria for understanding scientific investigation: identifying questions, conducting investigations, using technology and mathematics in investigation and communication, formulating and revising explanations, recognizing alternatives, communicating and defending results, and understanding scientific inquiry. The *Investigating and Evaluating Environmental Issues* approach incorporates each of these criteria. The chapter "Unraveling the Scientific, Social, Political, and Economic Dimensions of Environmental Issues Through Role-Playing Simulations" describes an example of implementing this approach with preservice teachers.

Action Research and Environmental Education

The heart of action research is to have students engage in trying to solve a real-world environmental problem or address issues in their community. There are many different approaches that are used, but each includes having students plan and implement a solution to a problem they have identified. Action research begins with students becoming more familiar with their community and identifying problems or issues that affect it. This can occur through field trips, walking trips, and/or from learning from community leaders and experts (Hammond 1994; Stapp and Wals 1994; Wals et al. 1990). According to Stapp and Wals (1994), it is important to engage the school administration and faculty before engaging students with action research. Faculty and administrators select a class to implement the action research process (Wals et al. 1990). In some cases, students may be selected on the basis of their demonstrated leadership ability (Hammond 1994).

In action research, there is emphasis on group processes, consensus building, and team building. This is accomplished by teaching groups skills and through ongoing discussions about real-life issues in the community. A residential camping experience may also occur; although such an experience is not essential for action research, it has at least two advantages. First, such an experience promotes team-building capacity. Second, it provides students with intimate contact with nature that many students may not otherwise experience. Action research requires certain skills to carry out a project. These skills include identifying and researching issues, effective communication, managing conflicts, understanding alternatives, defining objectives, working with other stakeholders, determining likely appropriate actions, and evaluating outcomes. Students must also develop persistence in pursuing their goals.

Collaboratively, a class develops a list of problems and issues facing their community. They then work at prioritizing the list in order to identify a specific problem to approach. It is important that the students, and not adult teachers or mentors, define the problem. Next, students research the problem and gather background information to become knowledgeable about the problem. This can be very time-consuming and involves considerable effort. Care must be taken to define the scope of the problem in such a way that it is neither trivial nor so large that it cannot be addressed within a reasonable time frame.

After defining the problem and conducting research, a class develops a citizenship action plan they can implement in order to achieve their objectives. They must consider a range of alternative actions and, working as a group, narrow down the list of possible actions to a list of actions all agree to carry out. A class must be persistent in implementing their action plan and evaluating the results of their actions. Action plans should not be quick fixes but should recognize that significant changes usually take place over the long term and require a developing series of actions. Using McTaggert's Action Research Spiral (Stapp and Wals 1994; Wals et al. 1990) the action research model is formalized, where there is a recurring spiral of planning, implementing, and evaluating actions. The sequence is repeated with appropriate modifications until the students reach the objectives they have defined for themselves. Throughout the process, the students consider constraints, opportunities, and possibilities. There is considerable discussion and negotiation among the student teams.

The scope of problems students undertake through action research can vary considerably. Hammond (1994) has identified three levels the action research project might achieve:

- Level I: Carrying out actions that result primarily in an end product. The outcome is distinguished by a product or completion of the project within the school year or possibly longer.
- Level II: Carrying out actions that result in ongoing environmental processes. These actions are not a single completed outcome, but rather are operational processes that are sustained in some form of multiyear perpetuity. This level requires all requisite skills for a Level I project, in addition to organizational skills to design and implement a system to sustain the project as an ongoing endeavor.
- Level III: Carrying out actions that result in some level of policy change. This requires the most complex set of skills. It is directed at changing or creating a new public policy. Students engaged in this endeavor need all the skills necessary in the other levels as well as sophisticated skills in lobbying, mass media governmental processes, and positive efficacy within the democratic system.

Examples of student citizen action projects in school settings include:

- Middle school student citizen students on the small Hawaiian island of Molokai became concerned about the negative impact of discarded plastic bottles on the island. They began a campaign to get a bottle bill passed in their state. There is now a state law in Hawaii requiring a five cent deposit on all plastic beverage bottles in the state.
- A group of elementary students became concerned that the reusable plates in their school cafeteria were replaced by disposable Styrofoam plates. They wrote

a letter to the principal stating their concerns. Nothing was done. One day, the students collected all the disposal plates from the cafeteria, put them in plastic bags, and left the bags in the principal's office. The next school year, disposable plates were not used in the cafeteria.

- A group of high-school students investigated the impact of vegetation removal along the bank of a stream near their school on the water quality of the stream. They decided to work to reestablish native vegetation on the stream bank. This project resulted in an improvement of the water quality in the stream.
- A group of fifth graders became concerned about the proposed citing of a low-level nuclear waste disposal facility near their community. They organized a campaign against the facility and the proposal was dropped. These projects were completed using the The Investigating and Evaluating Environmental Issues and Actions model (Hungerford et al. 2003).

The scope of the project students undertake depends on many factors including student skill level, student interest, time available, and available resources. The use of action research in a school classroom entails several important principles: respect for democratic principles and the need for citizenship participation of students of all ages. This is what Stapp and his colleagues refer to as "praxis, the union of reflection and action," (1994, p. 56) and real-life community problem-solving. In fact, Stapp and his colleagues refer to their model as "Action Research and Community Problem Solving" (Stapp and Wals 1994, p. 55).

The structure of action research is relatively open and it is recursive in the way it works. There is a strong resemblance between action research and the action agenda of a community organization. Hammond identifies three important outcomes that can be expected for the students who participate in action research: "… students tend to become bonded to (1) natural systems though direct experience and expanded knowledge of how nature works; (2) democracy though personal empowerment and expanded sense of locus of control; and, (3) their community by developing a 'sense of place' or connectedness to a physical place and cultural context" (1994, p. 47).

The STS (Science-Technology-Society) Approach to Science Education and Environmental Education

The Science-Technology-Society (STS) approach to science education was first published in the Project Synthesis report (Harms and Yager 1981) and has been an integral component for science education reform documents including *Project 2061* (American Association for the Advancement of Science 1989) and the *National Science Education Standards* (National Research Council 1996). STS is much broader in scope than EE. It may include any topic pertaining to the social or personal application of science and technology. These include science and environmental applications related to diet, health, safety, agriculture, transportation, construction, communication, national defense, and public policy. STS may also focus on appropriate technology for a specific place and purpose, and include social and natural impacts. Learning with an STS approach has the potential to make an individual's life richer, safer, and healthier. It may also help students become better citizens by providing deeper understandings of the science behind critical social issues such as climate change. The National Science Teachers Association (NSTA) STS position statement lists characteristics common to STS programs. These include:

- Student identification of problems with local interest and impact
- The use of local resources (human and material) to locate information that can be used in problem resolution
- The active involvement of students in seeking information that can be applied to solve real-life problems
- The extension of learning going beyond the class period, the classroom, the school
- A focus on the impact of science and technology on individual students
- A view that science content is more than concepts, which exist for students to master on tests
- An emphasis on process skills, which students can use in their own problem resolution
- An emphasis on career awareness, especially careers related to science and technology
- Opportunities for students to experience citizenship roles as they attempt to resolve issues they have identified
- Identification of ways that science and technology are likely to impact the future
- Some autonomy in the learning process (as individual issues are identified) (National Science Teachers Association 1990)

Just as there is a wide range of purposes for EE, from understanding how nature works to resolving environmental issues, so too are there a wide range of goals in the STS approach to science education. One purpose is to make the learning of science concepts richer and deeper by providing more context and application for scientific concepts. This permits learners to develop a more elaborated conceptual network than is possible with a more traditional approach of lectures followed by verifying laboratories that may limit student attention to science concepts isolated from social phenomenon. The chemistry curriculum, *ChemCom: Chemistry in the Community* (American Chemical Society 2002) is an example of a basal curriculum program that uses the STS approach to learning. The curriculum provides learners with a rich array of concepts concerning science-related social issues with the intent of making chemistry more meaningful. Many of the STS topics in *ChemCom* (American Chemical Society 2002) have EE connections including topics related to air, water quality, and energy resources (including carbon-based and alternative fuels).

ChemCom's (American Chemical Society 2002) first unit concerns water. The text provides a scenario that describes how the citizens of Riverwood experience a fish kill in their local river. The actions taken in response to the fish kill lead to water rationing and the cancellation of a fishing tournament that is a major source of revenue for the town. To understand the reasons for the fish kill and develop a solution to this problem, students learn about the chemistry of water, mixtures, solutions, solubility, water purification, water supply, and water consumption. The unit includes a great deal of chemistry, but in the context of larger social issues. This curriculum approach is designed to make chemistry more interesting and more meaningful for students who have not yet developed an interest in the subject.

Biology: A Community Context (Leonard and Penick 2003) and *EarthComm: Earth System Science in the Community* (It's About Time 2002) are both high-school science curricula that use an STS approach to learning. *Biology: A Community Context* (Leonard and Penick 2003) begins each unit with an initial inquiry into a social issue that leads students to develop knowledge and skills needed to understand the issue. The unit ends with a "congress" where students present what they have learned in addition to alternative solutions to the associated problems. Several of the units primarily focus on environmental issues including waste disposal, biogeochemical cycles, and human populations. The curricular approach is consistent with the primary goals of EE – building student awareness, knowledge, and ability to analyze environmental issues and problems. *EarthComm* (It's About Time 2002) uses a similar format, framing curricular content in EE areas including energy, mineral, and water resources. In the curriculum, content knowledge is developed through science understandings associated with resource use and its associated problems using personal contexts.

In addition to basal textbook curricula, STS approaches to learning have been used to develop EE curricular enhancement units. *Acid, Acid Everywhere: A Problem Based Unit* (The College of William and Mary Center for Gifted Education 1997) uses a problem-based approach to help students understand key science concepts pertaining to acid rain. *Investigating and Evaluating STS Issues and Solutions* (Hungerford et al. 1990) uses a more learner-centered STS approach in which students select their own environmental issues to analyze, evaluate, and take action. These curricular materials have often been implemented in secondary science classrooms in conjunction with more traditional science teaching practices (Winther et al. 2002).

STS science materials should not be viewed as "watered-down" science. It can be argued that *ChemCom* (American Chemical Society 2002), for example, contains more complex organic chemistry and biochemistry than are contained in most high-school chemistry books. Aikenhead (2003) cites 11 studies that show that STS-based science courses produce higher achievement scores on traditional tests than non-STS-based courses. Studies have found that middle school students taught with STS approaches applied science concepts better, developed more positive attitudes towards science, showed more creativity, and were better able to use science in real-life situations than students learning with a non-STS approach (Yager and Akcay 2008; Yager et al. 2006).

Many EE issues lend themselves inherently to STS topics. There are STS topics that do not pertain directly to the environment, but many STS issues do have environmental ramifications. Some of these issues include applications of technological development and implementation, medical technology, national security issues, health, and safety.

Conclusion

EE is a discipline in its own right. It has a long history and a well-defined structure. At the same time, it can be an important tool to be used when learning about science. It provides students the ability to integrate scientific knowledge and to apply it in important and meaningful ways. In addition, EE is a powerful pedagogical framework for future teachers. It allows them to integrate science, social studies, mathematics, language arts, and the arts into curricular contexts.

The goals of EE and science education naturally overlap in many areas. A wide variety of EE approaches can be used to enhance learning. Some of these, such as using the schoolyard and place-based education strategies, focus on developing understandings of where we live and connecting and applying these places to important science topics. The EE thematic resource *Project* curricula approaches use intrinsically interesting topics, such as forests or wildlife, to stimulate student interest and form a unifying theme for learning about core science content and the environment while developing citizenship skills. Nature centers and other related facilities offer informal and engaging approaches to EE that provide personal connections to the environment. The STS approach also emphasizes this approach to learning by promoting knowledge in personal contexts that develop contextual understandings around specific problems. Action research and issues investigations are very powerful approaches for developing actively involved citizens able to use their content knowledge to address local, regional, and national environmental problems.

There are many ways to achieve the goals of developing responsible, actively involved, scientifically literate citizens, who are able to balance the needs of the environment and humans. We can achieve this goal by consciously combining science and EE. It is up to the individual educator to select and apply a particular approach that will be most effective in specific settings.

Appendix 1: Representative Programs Providing Training and Support for Educational Use of Schoolyards

There are several programs available to provide support and training for teachers interested in using their schoolyards as learning sites. Some examples include:

- The National Wildlife Federation Schoolyard Habitat Program – http://www.nwf.org/schoolyard/. This program provides encouragement, rationale, and examples of how to develop school grounds as wildlife habitats.
- Many states have a "Homes for Wildlife" program. These are usually administered through a state's Fish and Game Department or Department of Conservation or Natural Resources. New Hampshire has developed an excellent guide that has been adapted by many states for their specific ecosystem (Wyzga 1995).
- Project Learning Tree's Greenworks! grant program can provide grants for environment-related service learning projects. Information available at: http://www.plt.org/cms/pages/21_22_18.html

- The Maryland Association for Environmental and Outdoor Education – http://www.maeoe.org/habitat/. This organization provides support and training for teachers interested in schoolyard habitat projects.
- The U.S. Fish and Wildlife Service – http://www.fws.gov/chesapeakebay/schoolyd.htm. This government organization coordinates schoolyard habitat programs with state agencies and school districts to promote the development of school grounds as wildlife habitats. They provide resources, training, and funding.

References

Aikenhead, G. (2003, August). *Review of research on humanistic perspectives in science curricula*. Paper presented at the European Science Education Research Association (ESERA) 2003 Conference, Noordwijkerhout, Netherlands.

American Association for the Advancement of Science. (1989). *Project 2061. Science for all Americans*. Washington, DC: American Association for the Advancement of Science.

American Association of Museums. (2009). *What is a museum?* Retrieved June 16, 2009, from http://www.aam-us.org/aboutmuseums/whatis.cfm

American Chemical Society. (2002). *ChemCom: Chemistry in the community*. New York: W. H. Freeman.

American Forest Foundation. (2008). *Project Learning Tree: Pre-K environmental education activity guide*. Washington, DC: American Forest Foundation.

Broda, H. W. (2007). *Schoolyard enhanced learning: Using the outdoors as an instructional tool K – 8*. Portland, ME: Stenhouse Publishers.

California Department of Education. (2009). *A garden in every school*. Retrieved July 1, 2009, from http://www.cde.ca.gov/Ls/nu/he/garden.asp

Center for Place-Based Education. (2008). *What is place-based education?* Retrieved October 5, 2009, from http://www.anei.org/pages/99_place_based_education.cfm

College of William and Mary Center for Gifted Education. (1997). *Acid, acid everywhere: A problem-based unit*. Dubuque, IA: Kendall/Hunt.

Council for Environmental Education. (2008a). *Project WILD: K-12 curriculum and activity guide*. Houston, TX: Council for Environmental Education.

Council for Environmental Education. (2008b). *Project WILD Aquatic K-12 curriculum & activity guide*. Houston, TX: Council for Environmental Education.

Cronin-Jones, L. L. (2000). The effectiveness of schoolyards as sites for elementary science instruction. *School Science and Mathematics, 100*(4), 203–211.

Dewey, J. (1959). School and society. In M. Dworkin (Ed.), *Dewey on education*. New York: Teachers College Press.

Hammond, W. F. (1994). Action within schools. In L. V. Bardwell, M. C. Monroe, & M. T. Tudor (Eds.), *Environmental problem solving: Theory, practice and possibilities in Environmental Education* (pp. 40–48). Troy, OH: North American Association for Environmental Education.

Harms, N. C., & Yager, R. E. (Eds.). (1981). *What research says to the science teacher* (Vol. 3). Washington, DC: Nation Science Teachers Association.

Hungerford, H. R., Litherland, R. A., Volk, T. L., Ramsey, J. M., & Peyton, R. B. (2003). *Investigating and evaluating environmental issues and actions: Skill development modules: A curriculum development project designed to teach students how to investigate and evaluate science-related social issues*. Champaign, IL: Stipes Publishing.

Hungerford, H. R., Peyton, R. B., & Wilke, R. J. (1980). Goals for curriculum development in Environmental Education. *Journal of Environmental Education, 11*(3), 42–47.

Hungerford, H. R., Volk, T. L, & Ramsey, J. M. (1990). *Science-Technology-Society: Investigating and evaluating STS issues and solutions. Champaign*, IL: Stipes Publishing.

Hungerford, H. R., & Volk, T. L. (1990). Changing learner behavior through environmental education. *Journal of Environmental Education, 21*(3), 8–21.

It's About Time. (2002). *EarthComm earth system science in the community*. Armonk, NY: It's About Time.

Leonard, W. H., & Penick, J. E. (2003). *Biology: A community context*. Columbus, OH: Glencoe/McGraw-Hill.

Maine Department of Education. (2007). *Maine learning results: Parameters for essential instruction*. Retrieved February 17, 2009, from http://www.maine.gov/education/lres/pei/index/html

Marcinkowski, T. J. (1988). An analysis of correlates and predictors of responsible environmental behavior (Doctoral dissertation, Southern Illinois University at Carbondale, 1988). *Proquest Dissertations*, AAT 8903716.

Martin, S. C. (2003). The influence of outdoor schoolyard experiences on students' environmental knowledge, attitudes, behaviors, and comfort levels. *Journal of Elementary Science Education, 15*(2), 15–23.

National Research Council. (1996). *National science education standards*. Washington, DC: National Academy Press.

National Research Council. (2009). *Learning science in informal environments: People, places, and pursuits*. Washington, DC: National Academies Press.

National Science Teachers Association. (1990). *NSTA position statement on Science/Technology/Society: A new effort for providing appropriate science for all*. Retrieved July 1, 2009, from http://www.nsta.org/about/positions/sts.aspx

Natural Science for Youth Foundation. (1990). *Natural science centers: Directory and NSYF blue ribbon reports from the Natural Science for Youth Foundation*. (ERIC Document Reproduction Service No. ED319 619)

Peterson, N. J., & Hungerford, H. R. (1981). Developmental variables affecting environmental sensitivity in professional Environmental Education. In A. B. Sacks, L. A. Iozzi, & R. J. Wilke (Eds.), *Current issues in environmental education and environmental studies* (pp. 111–113). Columbus, OH: ERIC Clearinghouse for Science and Mathematics Education.

Population Connection. (2009). *Teaching materials and tools*. Retrieved February 19, 2009, from http://www.populationeducation.org/

Sadler, K. C., Smith-Walters, C., Ring, T., & Lasater, M. (2006). Thinking outside the box: No child left inside at Campus School. In *Exemplary science in grades PreK–4: Standards-based success stories*. Arlington, VA: NSTA Press.

Sivek, D. J., & Hungerford, H. R. (1989/1990). Predictors of responsible environmental behavior in three Wisconsin conservation organizations. *The Journal of Environmental Education, 21*(2), 35–40.

Smith, G. (2002). Place-based education: Learning to be where we are. *Phi Delta Kappan, 83*, 584–594.

Stapp, W. B., & Wals, A. E. J. (1994). An action research approach to environmental problem solving. In L. V. Bardwell, M. C. Monroe, & M. T. Tudor (Eds.), *Environmental problem solving: Theory, practice and possibilities in Environmental Education* (pp. 51–65). Troy, OH: North American Association for Environmental Education.

The Watercourse & Council for Environmental Education. (2007). *Project WET*. Bozeman, MT: The Watercourse & Council for Environmental Education.

Tolley, K. (1994, April). *Study nature, not books: The nature study curriculum 1891–1932*. Proceedings of the Annual Meeting of the American Educational Research Association, New Orleans, LA. (ERIC Document Reproduction Service No. ED37400)

UNESCO. (1978, September). The Tbilisi declaration. *Connect: UNESCO-UNEP Environmental Education Newsletter, 3*(1), 1–8.

UNESCO-UNEP. (1976, January). The Belgrade charter. *Connect: UNESCO-UNEP Environmental Education Newsletter, 1*(1), 1–2.

Wals, A. E. J., Beringer, A., & Stapp, W. B. (1990). Education in action: A community problem-solving program for schools. *The Journal of Environmental Education, 21*, 13–19.

Wigginton, E. (Ed.). (1971–2004). *Foxfire* (Vol. 1–12). New York: Anchor Books.

Winther, A. A., Volk, T. L., & Hungerford, H. R. (1994). Issue investigation and action training: An instructional model for EE. In L. V. Bardwell, M. C. Monroe, & M. T. Tudor (Eds.), *Environmental problem solving: Theory, practice and possibilities in Environmental Education*. Troy, OH: North American Association for Environmental Education.

Winther, A. A., Volk, T. L., & Shrock, S. A. (2002). Teacher decision making in the 1st year of implementing an issues-based EE program: A qualitative study. *Journal of Environmental Education, 33*(3), 27–33.

Wyzga, M. C. (1995). *Homes for wildlife*. Concord, NH: New Hampshire Fish and Game Department.

Yager, R. E., & Akcay, H. (2008). Comparison of student learning outcomes in middle school science classes with an STS approach and a typical textbook dominated approach. *Research in Middle Level Education Online, 31*(7), 1–16.

Yager, S. O., Lim, G., & Yager, R. E. (2006). The advantages of an STS approach over a typical textbook dominated approach in middle school science. *School Science & Mathematics, 106*(5), 248–260.

Environmental Education Within Early Childhood

Linda Plevyak and Amy Mayfield

Vignette (Part I)

Right before the winter break, Mrs. Clark's third-grade class was taking a nature walk near their elementary school that was situated close to a creek. The students were observing how the environment changes from season to season. This was their second walk of the year as they had observed the change from summer to fall in September. They had brought their journals they had started on the previous walk that contained pictures, poems, stories, data on animal tracks, water depth and speed, soil erosion, etc. The students broke into groups and began observing the creek for changes from the previous walk. Students added more data and information to their journals on water depth and speed, erosion of soil, and animal tracks. They concluded that compared to their September walk, there was less water in the creek and the flow had slowed down, more soil had fallen from the sides of the creek, and there were fewer animal tracks – especially bird tracks.

Conclusions were drawn from the data that focused on seasonal information such as less rainfall, animals hibernating or migrating to warmer places, etc. Each of the students had planted a tree during the fall trip and had tagged each one with the tree name and student's name on a protective laminated card. The students returned to their tree to take height measurements and to look at overall quality. They recorded data and their observations in their journals.

Water samples from the creek were taken back to the classroom. Using droppers and slide protectors, the students viewed the creek water with large-screen microscopes. They were very interested to see if there were any active organisms in the water. Terms such as organism, macroinvertebrates, mayfly larva, stonefly larva, and dragonfly nymph were introduced during the viewing of the water and students were asked to include the terms when talking about the water samples. As this water was mainly runoff from local homes, no live organisms were found.

L. Plevyak (✉) and A. Mayfield
University of Cincinnati, One Edwards Center, Room 2150Q, Cincinnati, OH 45221-0002, USA
e-mail: Linda.plevyak@uc.edu

A.M. Bodzin et al. (eds.), *The Inclusion of Environmental Education in Science Teacher Education*, DOI 10.1007/978-90-481-9222-9_4,

[Safety precautions were taken as students wore latex gloves and eye protection. They were also told not to drink the water.] It was interesting to note that the students felt the water was clean because it was clear and they did not see anything living in the water samples. This became a "teachable moment" to discuss chemical point and nonpoint source pollution and how life cannot be sustained in this type of environment. Knowing that the water was not going to have much life in it, Mrs. Clark had sample slides of the larva and insects for them to view. [We will take a look at Mrs. Clark's classroom later in the chapter.]

Goals and Objectives of This Chapter

The goal of this chapter is for the reader to gain a better understanding of how environmental education (EE) can play an important role in early childhood education in preschool through primary classrooms with children of ages 4–8. Based on research of adult environmental activists, Chawla (1998) found that environmental experiences early in life were precursors to their activism later in life. This implies that children can learn to make a difference about the environment and that these learning experiences will carry into adulthood.

The objectives of this chapter emphasize early childhood education and include the following:

- Defining environmental education
- Explaining how EE can be incorporated into the early childhood curriculum
- Describing how EE can contribute to other subject areas within early childhood
- Supporting students in acquiring knowledge about the environment
- Assisting students in developing an environmental ethic
- Adapting EE activities to promote the inclusion of students with special needs
- Developing authentic assessment techniques that encompass EE within early childhood

Definition of Environmental Education

Much of what we do within both EE and science relates to the processes of human and animal life, how nature works, and how technology impacts the world. Interweaving EE and science makes sense, as the knowledge, skills, and attitudes being used in both areas overlap. For example, students develop scientific knowledge by studying the interdependence of environmental systems. When investigating water quality they are developing skills in data collection and analysis, and communicating results.

The following definition of EE is included in this chapter in part because it contains a reference to making decisions about environmental issues based on "scientifically sound information" (US Environmental Protection Agency (EPA) 2008, ¶ 2). When helping children understand environmental issues, it is important that what is being discussed is supported with reliable scientific evidence. The US EPA Office of Environmental Education defines EE as:

> [i]ncreasing public awareness and knowledge about environmental issues and providing the skills necessary to make informed decisions and taking responsible actions. It is based on objective and scientifically sound information. It does not advocate a particular viewpoint or course of action. It teaches individuals how to weigh various sides of an issue through critical thinking and it enhances their own problem-solving and decision making skills.
>
> (US EPA 2008, ¶ 2)

Incorporation of EE into Early Childhood Curriculum

In a National Environmental Education and Training Foundation Report (2001), 95% of Americans supported the inclusion of EE in K-12 schools. Schools have been answering the call with over half of the nation's schools including some type of environmental teaching throughout the school year.

Promoting EE in early childhood involves the intrinsic motivations of the students. Children are naturally curious about their surroundings and teachers can use this curiosity to highlight the natural world (Chalufour and Worth 2003). Part of helping children to see the beauty and wonder of the world is that they begin to better understand themselves in the process (Davis and Elliott 2003).

Wilson (1999) created guidelines for developing and implementing an EE program within early childhood based on the understanding of how children learn:

- Begin with simple experiences.
- Keep children actively involved.
- Provide pleasant, memorable experiences.
- Emphasize experience versus teaching.
- Involve full use of the senses.
- Provide multimodal (learning through more than one avenue of information) learning experiences.
- Focus on relationships.
- Demonstrate a personal interest in and enjoyment of the natural world, and model caring for the natural environment.
- Maintain a warm, accepting, and nurturing atmosphere.
- Introduce multicultural experiences and perspectives.
- Focus on the beauty and wonder of nature.
- Go outside whenever possible.
- Infuse EE into all aspects of an early childhood program.

The above guidelines should be promoted as part of an overall emphasis on EE within the early childhood classroom. Using supplemental materials, including the *Project* curricula that support EE knowledge and skills, can also be successfully incorporated into early childhood classrooms. The activities in these curriculum materials integrate numerous subject areas so they can be used in either theme-based (all subjects are taught under an umbrella topic such as the "water cycle" or "living things" that offers an organizing framework for the implementation of an interrelated series of lessons) or more traditional classrooms where subjects are taught separately. *Project WET* (2010), *Project Learning Tree* (2006), *Project WILD* (2000), and *Windows on the Wild: Biodiversity Basics* (World Wildlife Fund 1999) are examples of EE supplemental curricula. In addition, the Council on Environmental Education recently developed a curriculum guide specifically for early childhood, *Growing Up WILD* (2009).

Contribution of EE to Other Subject Areas

Learning about the environment involves "knowledge and skills from all disciplines" (Grant and Littlejohn 2005, p. xi). Research has found that using the environment as an integrating context for learning is the most effective way to teach EE (Lieberman and Hoody 1998; Lonning et al. 1998).

Science

"Doing science" instead of "learning about science" has become the theme for many educators. Students are naturally fascinated with the outdoors and will ask questions that are often quite complex. Environmental investigations that focus on the students' surroundings can reinforce science process skills such as observing, measuring, predicting, and describing. The students can begin to problem solve by refining the question to be studied. For example, they might ask whether earthworms prefer light or dark areas. They might begin with generating a hypothesis that states what the students initially believe, such as "the earthworms like to be in dark places because they live in soil." The students can then set up an experiment by placing earthworms in a box with half of it covered by black construction paper and the other half open to the light. The earthworms can be observed and data taken on the earthworms' movements. For example, the students could create a checklist with two columns (light/dark) with 5-min intervals. They could check off which side the earthworms prefer over a 1-h period. Finally, the students can communicate their results through graphs, journal writings, poems, or presentations.

This investigation could be expanded to decide on the best environment for earthworms. This might include a study into soil types: sand, gravel, soil, or clay to discover earthworms' preferred habitat. The students can observe worms outside in their natural environment as well. They might further their investigation by asking what would happen to birds if earthworms were no longer around. The important point is that the investigative ideas come from the students with the educator acting as a facilitator throughout the process.

Mathematics

Measurement, traditionally a mathematics topic, is a natural tool used to study the environment. For example, when focusing on weather, students can measure temperature, precipitation, humidity, and wind velocity (Engleson and Yockers 1994). Estimations can be made when trying to find tree height, width of rivers, or depth of a lake. In the vignette, Mrs. Clark's class estimates the amount of soil erosion along the creek bank. When studying plant development, students can measure growth in relation to days, weeks, and months. Time can also be studied in relation to animal movement, for instance, how long it takes a turtle to eat a small handful of berries.

Geometry also has clear connections to the natural world. Objects in nature have geometric shapes that can be explored and classified. These objects can be two- and three-dimensional with properties such as round, square, large, small, curved, etc. Students can also find simple patterns of symmetry in the environment such as the wings of a butterfly or certain tree leaves.

Arithmetic, such as addition, subtraction, multiplication, and division of whole numbers, can be used when counting the number of flowers on a plant, estimating the number of trees in a wooded area, or amount of fruit being produced by a grove of apple trees. Gathering data by measuring, counting, or performing simple experiments will help students gain skills in statistical analysis. These exercises can be kept simple, for instance, students can count the number of different birds that come to a birdfeeder. These data can be used to construct a bar graph or pictogram that highlights bird type and number of birds eating a particular food.

Language Arts

Emphasizing communication through writing, speaking, listening, and reading is crucial to a successful language arts curriculum. The purposes of communication which includes informing, expressing, transmitting feeling, and persuading contribute nicely to EE (Engleson and Yockers 1994; NSTA 2003). Students can express their feelings through stories, songs, or poems, keep a journal for recording EE investigations, write letters to companies and elected officials, or put on a play for another group of students.

Observation of plants and animals can help in the development of visual discrimination that supports emergent readers. For instance, naming items on a hike for each letter in the alphabet, counting a particular species of plants, classifying, or sequencing can assist in the development of beginning readers. Reading nonfiction and fiction books that highlight the environment can introduce students to different ecosystems such as a mangrove swamp, rainforest, or coral reef. Students might then act out different scenarios from the ecosystems, such as rain providing water to dried-out plants or fish swimming amongst the reefs. Students who can write and are beginning to read for understanding could select an object from the environment and list descriptive words that highlight this object without showing anyone. Peers can then attempt to identify the object through the descriptive clues.

Social Studies

Relating EE to the students' daily lives allows them to develop a deeper, more meaningful understanding of geography, history, economics, sociology, and political systems. They see the interrelationships and connections made between individuals and communities when they have the chance to apply their knowledge to real-world settings. For example, students could help a local farmer plant crops such as corn, tomatoes, or watermelon. They could also weed around the plants, keep records on amount of rainfall, pick the fruit and vegetables, sell them in a market, and eat them. This process would help them to better understand economics through pricing and competition, geography through land use and weather, and natural resources through soil quality.

The ultimate goal of social studies is to help students become effective citizens in a democratic society. To do this, students need to be informed, committed, and involved in their local community. There are many different types of EE projects that can be developed that encompass the different parts of social science. For instance, students might petition the local government to create a green space in a vacant lot. The students will need to understand how government functions to know whom to contact. They will also need to know about environmental justice issues and understand economics to find funding to support their project. Once the project has been developed, they will need to focus on stewardship of the space as well as sustainability after they no longer can care for it themselves (possibly placing the care of the lot with a neighborhood nature organization).

Music

Using the sounds of nature such as crashing waves on the shore, the hammering of a woodpecker, the songs of birds, or the wind rustling tree leaves will help students to see music as something broader than the music created by humans with musical instruments. The students can use natural objects such as rocks, sticks, sand, or seedpods to create different musical sounds. Providing numerous materials (i.e., natural objects and materials for the containers: paper towel holders, plastic bowls with lids, etc.) and allowing the students to decide what to create will help to keep this activity student-directed. They can combine the sounds to create a concert. Different orchestral instruments mimic sounds of nature (flute = bird, harp = flowing water, maracas = water hitting rocks). The students can experiment to see which instruments mimic sounds in nature or they can create new instruments that relate to the ocean, birds singing, or backyard sounds.

Expressive dance or movement can be combined with nature sounds. Each student can move to express the way they feel about what is being played. This exercise can also be done outdoors where there are nature sounds. After studying an environmental topic such as the water cycle or plant growth, the teacher can choose a common tune and have the students rewrite the lyrics to relate what they just learned.

Art

Teachers can focus on texture, color, lines, shapes, spaces, balance, contrast, rhythm, movement, and repetition while viewing the environment (Engleson and Yockers 1994). Students can focus on individual senses such as touch or sight to explore their surroundings. For example, touching tree bark, stones, leaves, flowers, soil, or twigs can show students the diversity of the natural world. The students can go outdoors when there are leaves on the trees and discuss the different greens they see. Students can use different shades of green to capture the variations. Students can also observe the attraction of insects to the bright colors of flowers. They can watch as the insects move from flower to flower which helps the pollination process. Color is important to animals as well. The class can visit a nature center or a zoo so the students can observe how animals use color to attract a mate, hide from predators, regulate temperature, etc.

Watching a line of ants as they take bits of leaves back to their colony or following a butterflies' path from flower to flower can promote a sense of wonder of the outdoors (Wilson 2000). Verbalizing what the students see can also help to internalize an appreciation for nature. Some of the projects that students can complete include the creation of dioramas, pictures, collages, poems, and stories that highlight the aesthetics (beauty), interdependency, and change within the environment.

Physical Education

Developing physical skills that children can use throughout their lives is an important part of EE. Paddling a canoe, safely traversing a creek, taking a hike with needed supplies in a backpack, digging a hole to plant a tree, are all desired skills that can be a part of the physical education curriculum. These needed skills can also be accompanied by discussions on how to support a quality environment through putting trash in a receptacle, keeping a vegetable garden, or creating green space in urban areas.

Health

There are two ways to look at health education, through the physical health of human beings and through the health of the environment. Many people would say that human health is intertwined with the quality of our environment. The air quality in a wooded park can be compared to the air near a highway busy with cars. Foods with chemical additives can be compared to organic foods. Students can keep a record of the types of foods they consume over the course of a day. They can then study ingredients and food labels to get an idea of how healthy they are eating. The health of a community is tied to proper waste and sewage disposal that for primary grade students can be highlighted through excess packaging of toys and landfill problems.

Vignette (Part II)

On the students' third walk to the creek in early spring, their attention turned to the small buds on the trees. The trees were still bare, but they could see that in a week or two the leaves would be out. Pushing away the rotting leaves on the ground they also saw that there were plants beginning to break through the soil. The students worked in small groups to record how many plants were budding in a 2 ft^2 area. They concluded that the areas closer to the creek had more budding plants because there was a ready supply of water.

As the students were keeping track of the soil erosion by the creek, they again measured the width of the creek in the same area as the previous two trips. Not much erosion had taken place since the winter trip. The students thought that little water had flowed during the winter so little soil erosion took place. Water speed was picking up since their last trip and the students hypothesized that the next time they came to the creek it would be much deeper with spring rains coming soon. They found more animal tracks and took a plaster cast of them so they could find out what types of animals were around.

The students also went back to the trees they had planted to observe changes. They again measured tree height and the beginning buds. They wrote an additional journal entry about their observations. This was a time for the students to use their sense of sight, smell, and sound to further their understanding of nature.

Acquiring Knowledge About the Environment

Research has shown that humans learn mainly from their own direct experiences by using their senses, interacting with others, as well as manipulating and observing materials (Barratt-Hacking et al. 2007; Etkina and Mestre 2004; Littledyke 2008). At younger ages, students need more concrete, hands-on experiences and less abstract experiences. Abstract experiences, that emphasize the use of textbooks, visual or verbal symbols, may be used because content can be compressed into a shorter amount of time than when utilizing concrete experiences. Teachers though, should not make the mistake of selecting abstract experiences solely for the purpose of "covering" material since this does not allow the learner time to meaningfully relate to the knowledge being emphasized.

Students must first begin to construct knowledge by observing objects and events (Piaget 1983; Vygotsky 1978). To further the process of observation, students can use skills such as sequencing, measuring, predicting, or interpreting. They will begin to process and internalize knowledge by having numerous direct experiences that relate to a particular idea. These ideas are emphasized in the Science, Technology, and Society (STS) approach to teaching that promotes using higher-order thinking skills such as evaluation, synthesis, and analysis (Krathwohl et al. 1973), using student-generated questions, and accentuating real-world problems

(Yager 2007). For example, when focusing on the concept of landfills and waste removal, students can:

1. Observe the composition breakdown of different materials such as an apple, plastic, wood, and metal over time
2. View a video on how landfills manage waste and how certain materials are recycled
3. Keep track of the amount of garbage they create over several days and discuss what can and cannot be recycled
4. Reduce their garbage by not buying items with wasteful packaging, reuse items, or begin a recycling campaign at the school or in their neighborhood

The more direct and concrete the experiences, the more permanent and complete the learning.

Developing an Environmental Ethic

The development of knowledge of the environment cannot be separated from attitude development. A student's feelings and emotions are tied to their content understanding. Promoting positive environmental experiences in the early grades supports both cognitive (knowledge) and affective (attitudes) domains (Hammond 1997). Teachers need to take into consideration what values and ethics are being promoted when dealing with environmental issues. The emphasis should not be placed on passing judgment or promoting a particular point of view, but on providing a process through which learners can actively engage in and better understand the world around them.

At the primary level, students may go for a nature walk observing the natural area through the use of their senses (such as looking around them, feeling objects, or smelling a flower). They could record sounds, write a description of the area, create a leaf print, or collect samples of items for further observation in the classroom. If there are negative impacts such as trash, neglect, or vandalism, students might attempt to correct the situation or record information for further study. Once back in the classroom, students can discuss their experience by thinking about what a squirrel or a tree might think of the area. Would they want to live in this area? Would people want to come and visit this place? How could it be improved? What good experiences have we had in this area? Students could then write to someone who has never seen this site and describe their experiences. These types of activities help students to value nature and give them the opportunity to view the outdoors from different perspectives.

Adapting EE Activities for Inclusion of Students with Special Needs

Focusing on what is of interest to the students is the best way to incorporate nature into the classroom. Some children who may have attention problems in the classroom find the outdoors exciting and interesting. These children look under rocks for bugs or observe the rushing water in a stream. If the students are outside to observe birds but find the squirrels more interesting, do not feel compelled to force them back to bird-watching. Highlight the teachable moment by having them ask questions and closely observe the squirrels' behavior. Their excitement may carry over into the classroom where books and journaling can be emphasized.

Children who are visually impaired can be guided by others and can use other senses to relate to nature. They can lie in the grass and feel the warmth of the sun on their face, bury their hand in sand or soil, feel the texture of bark on different trees, or compare two flowers by feeling the size difference in petals or sepals. Children who are hearing-impaired can use their sense of touch and sight to observe nature. This can include throwing leaves up in the air and watching them glide to the ground; viewing a hummingbird feed on sugar water at a feeder outside the classroom window; or touching items in different bags such as soil, bark, leaves, and fruit, that relate to a particular tree.

Children with severe physical disabilities will need support with planting seeds or plants, using tools or utensils. This does not mean that the students should not participate, but may need guidance and/or adaptations of equipment. Most park trails can accommodate a wheelchair, but if possible the children should also be lifted out of the chair to have direct experiences with nature such as lying in the grass or on a pile of leaves. Items should be brought close to the wheelchair and they should be allowed the opportunity to hold animals or feel their fur.

Some children may have certain fears of nature including spiders, snakes, thunderstorms, darkness, and others. Do not underemphasize or belittle these concerns. Try to find out specifically what causes the fear so it can be addressed. The more the students know about what they fear, such as snakes, the more likely they are to value its part in nature. Some activities to do before the introduction of an animal include viewing pictures and hearing stories, creating a play, and pretending to be a particular animal. Observation of an animal can start indirectly by watching others' interactions. It is always best to invite trained naturalists into the classroom when introducing animals as they know the animals and can give detailed information about habitats and niches.

Vignette (Part III)

On the students' fourth and final walk to the creek right before the end of the school year, they found that all of the trees and plants had leaves and many were in full bloom. Mrs. Clark asked them to use their senses to observe the area. The students quietly walked around watching butterflies, feeling the bark of different trees and listening to the birds and wind rustling the leaves. In one part of the creek, the students could see that a large tree branch had fallen and the water had become pretty deep in that area. There was still water getting by the branch, but it had stopped some of the creek flow. There had been a fair amount of spring rain and the students could see that by the fallen branch a fair amount of soil erosion had taken place. They decided that the branch falling had quickened the erosion since the deeper water was now higher on the creek bank.

The animal tracks were still around the creek. Since they had investigated the tracks in the winter they knew that raccoons, white-tailed deer, squirrels, cardinals, robins, and other birds frequented the creek area.

The trees that the students had planted had grown and produced leaves. They took height measurements and fertilized each tree. Many students said they wanted to continue coming to the creek next year and said they were going to ask their fourth-grade teacher to include the walks when possible. They sat next to their trees and wrote their final journal entry about their nature walks to the creek.

Assessing Environmental Education

Assessment should be tied to measurable learning objectives that are part of a well-designed curricular plan (Tierney 2006). Using authentic assessment allows the teacher to focus on individual learning styles and rates of growth. It also provides the flexibility for multiple approaches to problem solving within investigations. Incorporating actual authentic assessment techniques might also provide for a learning experience. Some examples of authentic assessments include observation, rubrics, checklists, interviews, performances, portfolios, projects, oral or pictorial responses, self-reports, journals, and presentations (Enger and Yager 2001).

It is important to involve the students as much as possible in their own self-assessment. They can assess their progress through interviews or self-reports. The teacher should make sure that expectations are clearly communicated to the students prior to the assessment activity. This might include handing out rubrics that focus on learning outcomes or providing journal-writing tasks that will be a compilation of their investigations during an environmental project. Using a diversity of assessment types that highlight the complexity of learning will help with understanding how far students have progressed over time.

Throughout this chapter, Mrs. Clark's third-grade class has been taking trips to a creek to observe soil erosion, water depth, and plant growth. During their investigation they wrote in their journals about the different observations. Table 1 is an example of a scoring guideline that clarifies Mrs. Clark's expectations for these journal entries.

Summary

This chapter focused on how to incorporate environmental education into either a theme-based or traditional preschool through primary (ages 4–8) classrooms. Environmental education is a topic that easily fits with other subject areas and promotes both knowledge and skill development in relation to the outdoors. Allowing

Table 1 Scoring guideline for journal entries

	Level 1	Level 2	Level 3	Level 4
Journal Writing:	Unacceptable performance	Approaching acceptable performance	Acceptable performance	Exceeds acceptable performance
Content and Organization	Did not define the idea	Defined but not thoroughly developed the topic	Mostly a developed topic	Developed topic
	Supporting details are missing	Supporting details are minimal	Supporting details are relevant but limited	Supporting details are relevant and provide important information about the topic.
	No organization, may be a brief list of items	Little organization to topic idea	Organization is appropriate, but may falter in logic	Develops the topic in a logical, organized way
	Ideas are unconnected	Ideas are not connected to the specified purpose	Ideas are connected	Connects ideas to the specified purpose
Style and Fluency	Writing is unreadable or messy	Portions are unreadable or messy	Writing is mostly readable and neat	Writing is readable and neat
	Word choice or vocabulary is inappropriate	Selects words that are not accurate or appropriate	Mostly selects words that are accurate or appropriate	Selects words that are accurate and appropriate and may use a variety of sentence structures
	Errors severely impede communication	Errors may impede communication	A few errors, but doesn't impede communication	Nearly error free
Effort	No effort was made to write	Little effort was made to write	Effort was made to write	A great deal of effort was made to write

student interest to drive the environmental investigations will promote learning that is internalized and valued by the students. Specifically:

- EE should be defined in the context of the environment where the students live, as this will be where it is the most meaningful.
- EE is important to include in the early childhood curriculum because it emphasizes children's natural curiosity of the world around them and promotes the foundation for their future experiences with the outdoors.
- Integrating EE into all subject areas and relating it to the real world is the most effective way to help students develop knowledge and skills in relation to the environment.
- When students have direct, concrete experiences with the environment their learning is more permanent and complete.
- Knowledge and attitude develop simultaneously and promoting positive environmental experiences in the early grades supports both cognitive (knowledge) and affective (attitudes) domains.

- Adapting EE activities to promote the inclusion of students with special needs is crucial. Students should be allowed and encouraged to participate in outdoor experiences. They can use different senses such as touch, smell, sight, or hearing to experience nature.
- There are numerous examples of authentic assessment techniques that can be utilized with EE activities. Focusing on individual learning styles and rates of growth should be the main factors in deciding what assessment strategies should be used.

References

Barratt-Hacking, E., Scott, W., & Barratt, R. (2007). Children's research into their local environment: Stevenson's gap, and possibilities for the curriculum. *Environmental Education Research, 13*(2), 225–244.

Chalufour, I., & Worth, K. (2003). *Discovering nature with young children*. St. Paul, MN: Redleaf Press.

Chawla, L. (1998). Significant life experiences revisited: A review of research on sources of environmental sensitivity. *Environmental Education Research, 4*(4), 369–383.

Davis, J., & Elliott, S. (2003). *Early childhood environmental education: Making it mainstream*. Australia: Early Childhood Australia.

Enger, S. K., & Yager, R. E. (2001). *Assessing student understanding in science*. Thousand Oaks, CA: Corwin Press.

Engleson, D. C., & Yockers, D. H. (1994). *A guide to curriculum planning in environmental education*. Madison, WI: Wisconsin Department of Public Instruction.

Etkina, E., & Mestre, J. (2004). *Implications of learning research for teaching science to non-science majors*. Harrisburg, PA: National Science Foundation, SENCER. (Grant DUE-0088753)

Grant, T., & Littlejohn, G. (Eds.). (2005). *Teaching green: The elementary years*. Toronto, Canada: Green Teacher.

Hammond, W. F. (1997). Educating for action: A framework for thinking about the place of action in environmental education. *Green Teacher, 50*, 6–14.

Krathwohl, D. R., Bloom, B. S., & Masia, B. B. (1973). *Taxonomy of educational objectives, the classification of educational goals. Handbook II: Affective domain*. New York: David McKay.

Lieberman, G. A., & Hoody, L. L. (1998). *Closing the achievement gap: Using the environment as an integrating context for learning*. San Diego, CA: State Education and Environment Roundtable and The Pew Charitable Trusts.

Littledyke, M. (2008). Science education for environmental awareness: Approaches to integrating cognitive and affective domains. *Journal of Environmental Education Research, 14*(1), 1–17.

Lonning, R. A., DeFranco, T. C., & Weinland, T. P. (1998). Development of theme-based, interdisciplinary, integrated curriculum: A theoretical model. *School Science and Mathematics, 97*(6), 312–319.

National Environmental Education and Training Foundation. (2001). *Lessons from the environment*. Washington, DC: NEETF.

National Science Teachers Association. (2003). Retrieved March 16, 2008, from http://www.nsta.org/about/positions/environmental.aspx

Piaget, J. (1983). Piaget's theory. In P. Mussen (Ed.), *Handbook of child psychology* (4th ed.). New York: Wiley.

Project Learning Tree (PLT). (2006). *Project learning tree: PreK-8 environmental education activity guide* (6th ed.). Washington, DC: PLT/American Forest Foundation.

Project WET. (2010). *Project WET (water education for teachers): Curriculum and activity guide*. Bozeman, MT: The Watercourse and Western Regional Environmental Education Council.

Project Wild. (2000). *Project WILD: Curriculum and activity guide*. Houston, TX: Western Regional Environmental Education Council.

Project Wild. (2009). *Growing up WILD*. Houston, TX: Environmental Education Council.

Tierney, R. D. (2006). Changing practices: Influences on classroom assessment. *Assessment in Education, 13*(3), 239–264.

U.S. Environmental Protection Agency Office of Environmental Education. (2008). Retrieved September 2, 2008, from http://www.epa.gov/enviroed/basic.html

Vygotsky, L. S. (1978). *Mind in society*. Cambridge, MA: Harvard University Press.

Wilson, R. (1999). *Starting early: Environmental education during the early childhood years*. Columbus, OH: ERIC Clearinghouse for Science Mathematics and Environmental Education.

Wilson, R. (2000). The wonders of nature: Honoring children's ways of knowing. *Early Childhood News*, pp. 6–9, 16–19.

World Wildlife Fund. (1999). *Windows on the wild: Biodiversity basics*. Tustin, CA: Acorn Naturalists.

Yager, R. E. (2007). STS requires changes in teaching. *Bulletin of Science, Technology & Society, 27*(5), 386–390.

Environmental Education Service-Learning in Science Teacher Education

Teddie Phillipson-Mower and April Dean Adams

Introduction

The purpose of this chapter is to explore service-learning in environmental education (EE). We have borrowed from the literatures of EE, service-learning, and science education to weave together a story that will serve to inform readers of the historical, theoretical, and practical nature of using environmental service-learning to enhance the development of teachers and ultimately K-12 students. Our objectives are to provide readers with a rationale for including environmental service-learning in their courses, offer models of environmental service-learning, and give specific examples of service-learning projects and resources.

What Is Service-Learning?

Service-learning is a form of experiential learning that consists of planning, action, and reflection. Unlike community service or volunteerism, this type of service is bound to student learning through the authentic and meaningful application of knowledge. According to the National and Community Service Act of 1990, service-learning:

- Promotes learning through active participation in service that meets the needs of a community
- Is coordinated collaboratively between the school and community

T. Phillipson-Mower (✉)
292 CEHD, University of Louisville, Louisville, KY 40292, USA
e-mail: t0phil01@louisville.edu

A.D. Adams
Department of Natural Sciences, Northeastern State University, College of Science and Health Professions, 611 N. Grand, Tahlequah, OK 74464-2302, USA
e-mail: adams001@nsuok.edu

A.M. Bodzin et al. (eds.), *The Inclusion of Environmental Education in Science Teacher Education*, DOI 10.1007/978-90-481-9222-9_5,

- Helps promote civic responsibility and sense of caring for others
- Enhances the academic curriculum of the students
- Provides structured time for participants to reflect on the experience, and
- Provides opportunities to use newly acquired skills and knowledge in real-life situations

In 2004, there were an estimated 4.5 million U.S. K-12 students engaged in some form of service-learning. This pervasiveness can be explained, in part, by the wide-ranging benefits that administrators and teachers associate with this pedagogical technique. These benefits include academic achievement, personal and social skill development, citizenship, and school-community relationships (Kielsmeier et al. 2004). To obtain these results, it is necessary to ensure that those that will guide these experiences are comfortable and qualified to implement high-caliber programs. Participation in effective authentic learning through service during teacher preparation will enable future teachers to implement engaging programs that closely resemble the definition above (Erickson and Anderson 1997). EE shares many of the same goals as service-learning and offers an effective context, the environment, for teacher preparation and implementation. Both science content reinforcement and pedagogical exploration can be emphasized. The natural world is a web of complexity that offers ill-structured problems that rely on a variety of perspectives and disciplinary knowledge to solve. Environmental education is interdisciplinary, promotes critical thinking skills and respect for multiple perspectives, and offers students and teachers authentic and meaningful ways to make a difference in their own local area.

Theoretical Support

Anderson and Guest (Erickson and Anderson 1997) sorted the many theoretical approaches to service-learning into five categories: experiential learning, transformational (social reconstructionist) theory, critical reflection, multicultural education, and civic responsibility education. Of these, experiential learning is often credited as the foundation of the learning approach used in service-learning (England and Marcinkowski 2007; Schneller 2008). Dewey's theory of experience focuses on the nature of human experience that results in learning and is often used as a framework for service-learning. His emphasis on the role of deep reflection after action and before subsequent action can be seen in current models of service-learning.

Kolb's experiential learning cycle serves as a learning tool to organize learning activities and assessment (Erickson and Anderson 1997). Eyler and Giles (1999) describe using Kolb's model to facilitate reflection since it "suggests how the integration of feeling and action with abstract and systematic thought might be accomplished." The four stages begin with the concrete experience (CE) followed by reflective observation (RO) in which students describe their experiences. This prepares students for abstract conceptualization (AC) where meaning making and connection to other knowledge take place. This is followed by an action plan or hypothesis that is carried out in active experimentation (AE). As Eyler and Giles

(1999) point out, knowledge is constructed through a cycle of feeling, observing, thinking, and doing that leads to the service-learning reflection mantra of "What? So what? Now what?"

Research and Models

Although research on service-learning has grown substantially since 1995, our understanding is still quite limited. Explicit calls to advance both the quality and quantity of research can be viewed at every major service-learning web site (for example, National Youth Leadership Council at www.nylc.org; National Service-Learning Clearing House at www.servicelearning.org; and the National Service-Learning Partnership at www.service-learningpartnership.org). Much of what we "know" about service-learning is based on anecdotal observations, practitioner program sharing, and program evaluations that use self-report measures. However, even with limited quality research literature, certain generalizations can be made. S. H. Billig, in her role with the RMC Research Corporation, provides updated summaries of research conducted on the impact of service-learning on K-12 students at the Learn and Serve America's National Service-Learning Clearinghouse Website (www.servicelearing.org/instant_info/fact_sheets/k-12_facts/impacts/index.php). The studies included here suggest that service-learning has a positive impact on academic achievement, cognitive engagement, problem-solving, and motivation for learning. Positive impacts were also found in personal and social areas such as self-efficacy, respect for diversity, relationship skills, and responsible decision-making. Mixed results are noted in the area of citizenship that is rationalized by variation in program objectives and quality. It is important to point out that this work builds off an earlier paper where Billig (2000) describes the limitations of the research that is presented. Perhaps more important, work has begun in identifying effective service-learning characteristics. Billig and Weah (2008) sifted through service-learning literature to find a research base to build the K-12 Standards and Indicators for Quality Service-Learning Practice. While this can be considered a great accomplishment for the service-learning field, the emphasis is on practice in K-12 settings and not on teacher preparation.

Root and Furco (2001) suggested that with the increased use of service-learning in the preK-12 classrooms, there is a need for beginning teachers to have the skills for implementation. However, research indicates that there is a general lack of recognition and understanding of the concept of service-learning among teacher educators even though many of the strategies and methods taught in preservice teacher education are connected to this approach (i.e. constructivism, experiential and active learning, project-based learning). Research findings on K-12 learner outcomes are often applied to postsecondary learners as well. However, there have been a few studies specific to preservice teachers. This research suggests that quality engagement in service-learning enhances preservice teachers' self-efficacy, confidence, professional participation skills (reliability, follow-through, attentiveness to job quality), relationship skills (diversity, patience, assertiveness), teaching skills (management and planning), community building and knowledge, and course content knowledge (Hart and King 2007; Swick and Rowls 2000; Wade 1995). Wade et al. (1999) found that novice teachers were more likely to use service-learning in their teaching if they had these experiences during their preservice teacher education. Research needs to be conducted on how to maximize key variables involved in ensuring highly competent facilitation of this instructional methodology.

As with service-learning in teacher education, environmental service-learning (service-learning specifically using an environmental context) is also in an early stage of development. However, researchers have found that environmental service-learning "can increase awareness of environmental issues and community awareness; locus of control; environmental consciousness; personal and social development; student motivation in school; personal environmental actions; and enjoyment of nature" (Schneller 2008).

One area that offers great promise with environmental service-learning is eco-justice. This justice embodies the commitments of both human justice and ecological sustainability. Many teacher educators are familiar with social justice and there has been some work done in social justice and service-learning in teacher education (Eyler and Giles 1999). This work tends to emphasize teachers' needs to understand diverse backgrounds and undergo "perspective transformation" from attributing social problems from individual to more systemic causes. As science education has embraced the importance of teaching nature of science, the social and cultural embeddedness of science has become more salient. Environmental justice, because of its grounding in justice surrounding environmental science related events, is probably a more comfortable justice to include in instruction. However, while environmental justice and social justice carry similar western assumptions of economic equality, globalization, and resource use, eco-justice goes further by questioning these assumptions of hyperconsumerism. Eco-justice also advocates for the ability of future generations to enjoy the same or higher level of quality of life and honors social and cultural traditions (the cultural commons) as well as ecological interdependence of human and nonhuman entities (the environmental commons). Science educators have begun to look at what eco-justice education might look like within the context of science education (i.e. Mower et al. 2009). Eco-justice service would connect well to Cobern's (2000) worldview theory and Aikenhead's (2006) humanistic science education.

Service-Learning and National Science Education Standards for Professional Development

The *National Science Education Standards* (NRC 1996) established four professional development standards that guide initial certification of preservice science teachers and advanced preparation for inservice science teachers. Two of these standards may be addressed by the inclusion of service-learning in teacher education.

"Professional Development Standard A: Professional development for teachers of science requires learning essential science content through the perspectives and methods of inquiry," (NRC 1996, p. 59) may be addressed by a service-learning project that is used to launch student inquiries. For instance, preservice teachers who monitor a local stream's water quality for a state agency may become interested in their own questions concerning the stream. These questions can lead to student-designed inquiries concerning stream health that are a result of their service-learning experience. These service-learning experiences address the following dimensions of Professional Development Standard A:

- Involve teachers in actively investigating phenomena that can be studied scientifically, interpreting results, and making sense of findings consistent with currently accepted scientific understanding.
- Address issues, events, problems, or topics significant in science and of particular interest to participants.
- Build on the teacher's current science understanding, ability, and attitudes.
- Incorporate ongoing reflection on the processes and outcomes of understanding through inquiry. (p. 59)

"Professional Development Standard B: Professional development for teachers of science requires integrating knowledge of science, learning, pedagogy, and students; it also requires applying that knowledge to science teaching," (p. 62) is addressed by service-learning projects that provide an opportunity for preservice teachers to engage K-12 students in science-related projects. The following dimensions of Professional Development Standard B are addressed:

- Connect and integrate all pertinent aspects of science and science education.
- Occur in a variety of places where effective teaching can be illustrated and modeled, permitting teachers to struggle with real situations and expand their knowledge and skills in appropriate contexts.
- Address teachers' needs as learners and build upon their current knowledge of science content, teaching, and learning.
- Use inquiry, reflection, interpretation of research, and guided practice to build understanding and skill in science teaching. (p. 62)

Inquiry and Service-Learning

Adams et al. (2007) found that open inquiries, in which learners engage in questions concerning science, give priority to empirical evidence, formulate explanations that are supported by evidence, connect their explanations to current scientific knowledge, and present their findings to others (National Research Council 2000), can be facilitated by first engaging the learner in a rich learning experience. In this study, which documented the actions of teachers engaged in inquiry-based instruction, rich learning environments were used to facilitate inquiry in grades 1–9. This engagement strategy would seem to apply to learners at all grade levels, including preservice and inservice teachers. Service-learning with its real-world complexity and relevance may provide this rich experience. It may activate interest on the part of the learner and generate curiosity, which seems to be a fundamental component for inquiry experiences that are meaningful to the learner.

Learners may become interested in a local stream as a result of monitoring its water quality. If their test results are not in the normal range, they may question why this occurred. They may become concerned about water quality in other bodies of water that they are familiar with. As a result of their learning through service experience, they may feel more empowered and confident in their skills to further investigate these concerns. If a goal of the service-learning experience is to enhance citizenship, learners can present their data and conclusions from the inquiry to decision-makers and advocate for a change in policy.

Assessing Service-Learning

As with all effective assessments, the assessment must align with the goals of the project. The impact of service-learning can be assessed for multiple participants including students, faculty, community, and the institution (Gelmon et al. 2001). In addition, Giles and Eyler (1998) suggest that the impact of service-learning on society should also be considered. For the purpose of this chapter, we will focus on the impact of service-learning on students.

Gelmon et al. (2001) published a five-point Likert scale survey that assesses student impact. The scale range includes Strongly Disagree, Disagree, Neutral, Agree, and Strongly Agree. In addition to measuring student perceptions of the student learning experience (LE), the survey evaluates student perceptions of how the service-learning project affected their attitude toward community involvement (CI), their major and profession (MP), and their self-understanding (SU). Sample items include the following:

- The community participation aspect of this course helped me to see how the subject matter I learned can be used in everyday life. (LE subscale)
- The community work involved in this course helped me to become more aware of the needs of my community. (CI subscale)
- Performing work in the community helped me to define my personal strengths and weaknesses. (MP subscale)
- I can make a difference in my community. (SU subscale) (pp. 32–34)

This survey could act as a starting point for assessment of service-learning in teacher education. However, the goals of service-learning experiences for preservice and inservice teachers should guide the assessment items. Although the particular experience will affect the specific goals, we offer the following potential service-learning goals for preservice and inservice teachers:

1. Understand the processes and products of scientific knowledge.
2. Develop the skills needed for scientific inquiry and inquiry-based instruction.

3. Understand the relationships among science, society, and personal well-being.
4. Articulate nature of science connections.
5. Increase interest in science and science-related issues.
6. Encourage collaborative skills and the development of a community of learners.
7. Develop science-teaching skills.
8. Integrate knowledge of science, teaching, and students.
9. Encourage reflection on practice.

Conducting open inquiries on streams as previously described has the potential to address many of these goals if the project is coupled with discussions and reflections concerning the nature of science and scientific inquiry, inquiry-based instruction in classrooms, and science-related personal decision-making.

The goals of a service-learning experience may be assessed through a variety of means. Student surveys can be used to assess a large number of students in an efficient manner, but may be difficult to interpret. Student interviews and focus groups can provide more in-depth knowledge, but are difficult to administer to large numbers of students. A small sample size can affect results. Other means include document analysis, observations, and open-ended surveys with follow-up interviews. In general, it seems better to have a variety of methods, both quantitative and qualitative, that are aligned with the goals of the service-learning experience.

Service-Learning Models

There are many service-learning models. The most frequent categories found on university service-learning web pages are direct (students are directly engaged with community, face to face), indirect (infrastructure, support, work towards broad goals), research (students collaborate with community to identify research question and implementation), and advocacy (educating others about topics of public interest). Other service-learning models mentioned in service-learning circles are civic-based, problem-based, consultant-based, and community-based. EE offers issue investigation models that, in general, share the same purpose as the service-learning models; to move students to action through understanding and develop responsible citizens capable of making well-founded decisions. The Tbilisi conference declaration (1978), one of the founding documents of EE, calls for "utilizing the findings of science and technology [to] play a leading role in creating an awareness and a better understanding of environmental problems. It must foster positive patterns of conduct towards the environment and nations' use of their resources." The document outlines the objectives of EE (awareness, sensitivity, attitudes, skills, and participation), which help to operationalize the concept of an "environmentally responsible person." Hungerford and Volk (2005) point out that the traditional simplistic view of behavioral change (Knowledge → Awareness → Action) does not hold up to research. They introduced a much more complex model of environmental behavior with three categories of variables (entry-level, ownership, and empowerment) each with major and minor contributory variables. From this work, they offer two models, the "Issue Investigation and Action Model" and the "Extended Case Study Model."

The "Issue Investigation and Action Model" involves three phases (Ramsey et al. 2005; Winther 2005). Students are first introduced to issue identification by separating problems from issues and then analyzing issues based on the stakeholders and their positions, beliefs, and values. In this phase, they are also taught investigation skills (identifying variables, sampling techniques, etc.). Next, the students use their new knowledge to investigate a new issue and formulate research questions. This is considered the practice phase. The final phase, or apply phase, is citizenship action in which the students develop an action plan based on their research, which they may or may not actually carry out. This involves considering various types of actions (persuasion, consumerism, political action, eco-management, and sometimes legal) and the level of action (local, state, or national). Chapter 3 provides more details about this model. "The Extended Case Study Model" involves many of the same skills as the "Issue Investigation and Action Model," but is limited to an issue that is predetermined, usually by the teacher. Hungerford and Volk (2005) note that while this model is successful, it is not as powerful as the investigation and action model.

Earth Force Community-Based Problem-Solving (CAPS) involves six steps and contains many of the components found in the two EE models. Instead of using direct instruction as designed in the Issues Investigation and Action Model, a 5-E sequencing strategy can be seen in each of the six steps. The steps are:

- Community Environmental Inventory – students define and take an inventory of the community to identify problems and collect information.
- Issue Selection – students select an environmental problem for further study based on the results of the Community Environmental Inventory. They research, narrow, and refine the problem.
- Policy and Community Practice Research – students identify and analyze relevant public and private policy and community practices to find out who makes decisions and how.
- Options for Influencing Policy and Practice – students identify possible project options for affecting change in either policy or practice. Goals and strategies are determined after research.
- Planning and Taking Civil Action – students develop and implement a plan of action.
- Reflecting, Going Public, and Planning for the Future – students assess and reflect on their work and the process, report their findings, and determine next steps for action or next question to be addressed. Entire class celebrates what was learned.

All steps involve student choice, voice and investigation from the beginning and research skills are developed and utilized throughout. Celebration helps the students understand that the process and learning are as important as the actual result of their work on the project. It allows the students and teachers the option of thinking about the same problem in a different way the next time issue selection takes place.

Examples of Environmental Education Service-Learning in Teacher Education

Oklahoma Blue Thumb

Oklahoma Blue Thumb is a water quality monitoring project of the Water Quality Division of the Oklahoma Conservation Commission (http://www.ok.gov/okcc/Agency_Divisions/Water_Quality_Division/Blue_Thumb/). The mission of Blue Thumb is to improve Oklahoma water quality through education. The program recruits citizens to monitor local streams monthly. Volunteers perform chemical and physical monitoring of the site and participate in quarterly quality assurance sessions. In addition, Blue Thumb personnel with the help of stream monitors collect and count invertebrates twice a year and count fish every 5 years at every monitoring site. Training and materials are provided free of charge. The program has become popular with Oklahoma schools. Therefore, engaging preservice teachers enrolled in the second author's elementary education science methods course in monitoring a stream that runs through campus was a natural fit. The course integrates science content and how to teach science in first through eighth grade. The class meets 16 weeks, three times per week for 80 min each day. In seeking more open inquiry experiences for her preservice teachers, the second author found that Blue Thumb could meet this necessary pedagogical and content learning experience and provide an opportunity for her preservice teachers to incorporate this program in their future classrooms.

During the service-learning project, preservice teacher groups conduct the following tests on a monthly basis: dissolved oxygen, ortho-phosphate, nitrogen ammonia, pH, nitrate/nitrite, chloride, and E. Coli/Coliform count (during warm months). Each group conducts one of the tests each month, and the tests are rotated among groups so that the preservice teachers have experience with as many tests as possible during the semester. The test protocols are practiced, but the meaning of the tests are not discussed before the first data collection. After the first stream monitoring, the preservice teachers generally begin to ask, "What do the results mean? What is a normal reading? Why do we care about dissolved oxygen levels?" At that point, each group is asked to research three questions concerning the test that they conducted during the first month. They are encouraged to use Internet and print resources and are given prompts and questions to go beyond initial simplistic answers. The questions are:

- What can your test indicate about the health of a stream?
- What do your results indicate about the health of our stream?
- If the test were to indicate that a stream is not healthy, what might be the cause?

The groups then present their results to the class.

After an additional month's monitoring, the preservice teachers usually seem to be more confident in their skills and the instructor starts to hear preservice teachers wondering about the stream and about other streams near their homes. Most of the preservice teachers live in or come from rural areas of Oklahoma. The stream that they monitor flows through the campus. These factors contribute to preservice teacher curiosity. With these "wonderings," preservice teachers working in groups develop testable questions concerning stream health and sometimes Blue Thumb testing protocols. Some of the questions that preservice teachers have investigated include:

- Is the dissolved oxygen content of favorite fishing sites high enough to support bass?
- How does the *E. coli* count vary at several locations along the Town Branch creek?
- How does the amount of stream turbulence affect dissolved oxygen measurements?
- What can the Blue Thumb chemical tests tell us about Tar Creek (an Oklahoma Superfund site)?

After the preservice teachers develop their questions, they plan their inquiry-based investigation, collect relevant data, interpret the data, and make an oral presentation of their findings to their peers. Many travel to sites near their home to investigate their questions.

The instructor has noted that the preservice teachers appear to gain confidence in their ability to conduct chemical tests, learn about stream health and ecology, learn how schools can become Blue Thumb monitoring sites at no cost, learn about open inquiries, and become more aware of the importance of water quality. These are important gains for preservice grades 1–8 teachers. In addition, the community benefits from the data collected and accumulated on the water quality of the campus stream and quite possibly additional streams around the state through the inquiries the preservice teachers complete and the inquiries their students may conduct in the future. This service-learning experience could be considered an example of a research model. It also could be considered an example of the Extended Case Study Model in that the instructor chooses the project and leads students through skill development before asking them to apply their knowledge to a new but similar situation.

Louisville Environmental Youth Summit

In Louisville, Kentucky the Partnership for a Green City's Environmental Youth Summit is held in the fall and spring every academic year. The Partnership is a collaboration between the Louisville Metropolitan Government, the University of

Louisville, and Jefferson County Public Schools. The *Summits* are a gathering of students who are concerned about their environment and are committed to working toward change. Currently, each *Summit* is limited to 37 schools and 300 students and registration usually reaches full capacity prior to the deadline. This program offers many layers of service-learning that benefit the participants and the community.

In an effort to improve the framework of the *Youth Summit*, Kentucky Green and Healthy Schools (KGHS) and Earth Force initiatives were added to the program. Prior to this, students who came to the Youth Summit had a "great time" according to the program surveys but there were few students that followed up by taking the information from the *Summit* back to their schools to conduct service-learning action projects. KGHS (www.greenschools.ky.gov) is a new state program that engages students in moving to action by working to make their school run at peak efficiency. Students and their teachers choose from nine inventories (Energy, Health and Safety, Solid Waste, Green Spaces, Indoor Air Quality, Transportation, Hazardous Chemicals, Water, and one for just teachers – Instructional Leadership), and complete the inventory questions designed to help them learn more about how their school operates in that area. When they have completed the inventory, they design and implement an action project (service) based on what they found out.

The first year, teachers struggled with facilitating the action projects. An opportunity to connect to the Earth Force CAPS program, outlined above, added the essential steps to action that were needed. Six inservice and two preservice teachers were trained in the Earth Force program. Prior to the *Youth Summit*, these eight "teacher leaders" led an Earth Force training session at a teacher professional development institute designed to prepare other inservice teachers for their roles as facilitators for the inventories and action projects their students would select during the academic year. In addition, all teachers were introduced to the breakout sessions that their students would be participating in during the Youth Summit and how they would connect to the KGHS program. The breakout sessions were designed to develop the necessary skills to conduct environmental inventories and prepare for the service-learning events that they would engage in at their home schools. Data were collected, analyzed, and conclusions were drawn in each breakout session to model components of inquiries/action projects that the students would be expected to carry out. A discussion on instructional strategies to promote students' inquiry development took place. The preceding week, experts from the community and volunteer preservice teachers (from Science Methods for Elementary Teachers and Exploring Teaching courses at the university) engaged in the same training as they prepared for the *Youth Summit* and to take mentor roles in schools during the academic year. The breakout topics included carbon sequestration, energy, transportation, storm water, and resource management (dumpster diving) in the fall and an outdoor classroom session replaced carbon sequestration in the spring.

The day of the *Youth Summit* the students go through one of the breakout sessions, meet university researchers who are investigating climate change or sustainability issues, meet other students who have completed service-learning projects, and brainstorm about what they learned during the day and how this knowledge might transfer to projects at their home schools. The students indicate that this last part is what energizes them the most. In between the fall and spring *Youth Summits*, mentors (including some preservice teachers) meet with the classroom teachers at least once a month in support of the projects the students have chosen. During the spring *Summit*, students share not only what they did during the year but also their thinking behind it.

Individuals in all groups, including mentors, preservice and inservice teachers, and the students, have noted success in terms of number and the quality of the projects completed. Assessment of the participating schools' table displays and presentations indicated understanding of scientific investigation methods and data collection skills. Students present were able to articulate how they arrived at the question and addressed it as well as carry on conversations about the environmental topic(s) with invited guests (university faculty, staff and administrators, and community members). Overall, the schools that had an Earth Force trained teacher completed more inventories and action projects. Anecdotal information from teachers support notions that students are more engaged in science and connections to the community, have greater interest in environmental topics, and feel a sense of empowerment. Earth Force trained teachers tend to need less motivational support than other teachers and are able to move ahead on their own. Preservice teachers have noted their appreciation of having participated in this project and their intention to participate in this and other service-learning experiences through learning logs in the science methods course and an assigned creative project presentation in the exploring teaching. They have shared their surprise at how much students at the lower grade levels know and can do. In addition, they show an understanding of the difference between an environmental "problem" and "issue," interest in further application of their newly acquired knowledge and skills, and creative thinking with ideas for other service-learning experience for students. All participants look forward to hearing what impact the service-learning projects have made on the community. The university Vice President of Business has taken action on some of the recommendations that students in the summer program made as they were piloting the projects to be used in the Youth Summit.

Concluding Remarks

Implementing environmental service-learning into teacher preparation offers a vehicle for deeper science, citizenship, pedagogical, and professional development through reflection and active service. To meet the environmental decision-making challenges that global citizens are facing, teachers need to be prepared in using this increasingly popular instructional method to achieve the positive outcomes that have been suggested by its advocates. While there is a lack of quality research, we both have experienced increased student engagement, questioning, empowerment, "ah ha" moments, knowledge acquisition and application, environmental awareness and behavioral change, and participation and professional skill development in our service-learning programs.

Service-learning connections to curriculum are as varied as are the topics covered. We suggest starting with the course objectives to identify specific content and skills preservice teachers should be able to exhibit by the end of the course. From here, discuss these needed outcomes with community agencies and organizations to emphasize the necessary learning that is to result from the service. This will allow the focus of the project to remain on the service-learning instead of community service. Preservice teachers are resourceful in determining service-learning opportunities when expectations for learning are made clear. Including them in project planning allows for them to learn skills of planning and designing for outcomes that will reinforce quality lesson-planning techniques. Some other suggestions for environmental service-learning projects that would be appropriate for the science or science education methods courses are:

- Mapping trees for city planning and/or maintenance schedules.
- Designing, implementing, and assessing science programs for a science fair at a local school, a science museum, or a planetarium.
- Monitoring air quality and reporting the results to the local newspaper or city council.
- Developing or adding to a community climate change resource document that outlines and quantifies specific contributions of greenhouse gas emissions in the community, documents changes in biodiversity and ecology in the community, lists organizations and governmental resources for reducing or sequestering carbon and/or other greenhouse gases, and offers suggestions for reducing personal carbon footprints.
- Partnering with a local public school to educate the community about the benefits of going beyond recycling (reduce, reuse); preservice teachers could be paired with each classroom to facilitate a class service-learning project on this topic.
- Developing a booklet of environmental science writing pieces that demonstrate a variety of genres to be included in middle school writing and reading instruction.

We encourage others to begin to use this instructional method and share their experiences through practitioner and research networks to build knowledge of best practices in this very exciting area.

Some of this material is based on work supported by the National Science Foundation under grant number ESI-0455573 and a Northeastern State University Service-learning Grant. Any opinions, findings, and conclusions or recommendations expressed in these materials are those of the author(s) and do not necessarily reflect the views of the National Science Foundation.

References

Adams, A. D., Macklin, M. J., Cambiano, R., Oliver, J., Hurst, V., Willingham, S., & Underwood, M. (2007). *The development of the analysis of inquiry rubric based on observations of practicing teachers and its implications for science teacher preparation*. Proceedings of the National Association for Research in Science Teaching Annual Conference, New Orleans, LA.

Aikenhead, G. S. (2006). *Science education for everyday life: Evidence-based practice*. New York: Teachers College Press.

Billig, S. H. (2000). Research on k-12 school-based service-learning. *Phi Delta Kappan, 81*(9), 658–664.

Billig, S., & Weah, W. (2008). K-12 service-learning standards for quality practice. In *Growing to greatness 2008: The state of service-learning*. St. Paul, MN: National Youth Leadership Council. Online <PDF> 2008-07-10, from http://www.nylc.org/pages-newsevents-news-K_12_Service_Learning_Standards_for_Quality_Practice?oid=6091

Cobern, W. W. (2000). *Everyday thoughts about nature: A worldview investigation of important concepts students use to make sense of nature with specific attention of science*. Boston: Kluwer.

England, Y. A., & Marcinkowski, T. (2007). Environmental service-learning program in Florida high schools and colleges: Nature, status, and effects as determined by a statewide program census. *Journal of Environmental Education, 38*(4), 51–60.

Erickson, J. A., & Anderson, J. B. (1997). *Learning with the community: Concepts and models for service-learning in teacher education*. Washington, DC: American Association for Higher Education.

Eyler, J., & Giles, D. E. (1999). *Where's the learning in service-learning?* San Francisco: Jossey-Bass.

Gelmon, S. B., Holland, B. A., Discoll, A., Spring, A., & Kerrigan, S. (2001). *Assessing service-learning and civic engagement principles and techniques*. Providence, RI: Campus Compact.

Giles, E. G., & Eyler, J. (1998). A service-learning research agenda for the next five years. In R. A. Rhoads & J. Howard (Eds.), *New directions for teaching and learning*. San Francisco: Jossey-Bass.

Hart, M. H., & King, J. R. (2007). Service learning and literacy tutoring: Academic impact on preservice teachers. *Teaching and Teacher Education, 23*, 323–338.

Hungerford, H. R., & Volk, T. L. (2005). Changing learner behavior through environmental education. In H. R. Hungerford, W. J. Bluhm, T. L. Volk, & J. M. Ramsey (Eds.), *Essential readings in environmental education*. Champaign, IL: Stipes Publishing.

Kielsmeier, J. C., Scales, P. C., Roehlkepartain, E. C., & Neal, M. (2004). Community service and service-learning in public schools. *Reclaiming Children and Youth, 13*(3), 138–143.

Mower, T. P., Benze, L., Alsop, S., Mueller, M. P., & Tippins, D. (2009, April). *Eco-justice in and through science education: A community discussion*. Proceedings of the National Association for Research in Science Teaching, Garden City, CA.

National and Community Service Act of 1990. Public Law 106-170. 42 USC 12401:104 Stat. 3127 (1991). Online <PDF> 2008-07-10, from http://www.csc.ca.gov/aboutus/documents/ncsa1990.pdf

National Research Council. (1996). *National science education standards*. Washington, DC: National Academy Press.

National Research Council. (2000). *Inquiry and the National science education standards*. Washington, DC: National Academy Press.

Ramsey, J. M., Hungerford, H. R., & Volk, T. L. (2005). A technique for analyzing environmental issues. In H. R. Hungerford, W. J. Bluhm, T. L. Volk, & J. M. Ramsey (Eds.), *Essential readings in environmental education*. Champaign, IL: Stipes Publishing.

Root, S., & Furco, A. (2001). A review of research on service-learning in preservice teacher education. In J. B. Anderson, K. J. Swick, & J. Yff (Eds.), *Service-learning in teacher education:*

Enhancing the growth of new teachers, their students, and communities. Washington, DC: American Association of Colleges for Teacher Education.

Schneller, A. J. (2008). Environmental service-learning outcomes of innovative pedagogy in Baja California Sur, Mexico. *Environmental Education Research, 14*(3), 291–307.

Swick, K. J., & Rowls, M. (2000). The "voices" of preservice teachers on the meaning and value of their service-learning. *Education, 120*(3), 461–468.

Wade, R. C. (1995). Developing active citizens: Community service-learning in social studies teacher education. *The Social Studies, 85*, 122–128.

Wade, R. C., Anderson, J. B., Yarbrough, D. B., Pickeral, T., Erickson, J. B., & Kromer, T. (1999). Novice teachers' experiences of community service-learning. *Teaching and Teacher Education, 15*(6), 667–684.

Winther, A. A. (2005). Investigating and evaluating environmental issues and actions: An instructional model for environmental education. In H. R. Hungerford, W. J. Bluhm, T. L. Volk, & J. M. Ramsey (Eds.), *Essential readings in environmental education*. Champaign, IL: Stipes Publishing.

Beyond *Terra firma*: Bringing Ocean and Aquatic Sciences to Environmental and Science Teacher Education

Diana L. Payne and Timothy D. Zimmerman

In this chapter, we highlight the importance of teaching and learning a particular aspect of environmental and science education in K-12 classrooms – ocean and aquatic ecosystems. We contend that ocean and aquatic systems concepts are critically important to understanding our planet as a system and in promoting an environmentally literate society. Finally, we reason that one cannot be science or environmentally literate without being literate in ocean and aquatic concepts, as these are the "conceptual glue" (Hoffman and Barstow 2007, p. 7) that bind together much of the Earth science systems content.

Our objective is to emphasize the critical role teaching and learning about ocean and aquatic environments plays in promoting a scientifically and environmentally literate society. Our theoretical framework draws from research in the learning sciences and focuses on deep conceptual understanding, learning in addition to teaching, and the critical role of prior knowledge. In addition, our framework draws broadly from the environmental education (EE) literature, including notions of environmental sustainability (Tilbury 1995), environmentally sound decision-making (Hungerford and Volk 1990), and embeddedness within social and cultural practices (Roth and Barton 2004; Roth and Lee 2002). Much of the work in both science education and EE incorporates a hands-on, scientific inquiry-based approach to teaching. This method often includes "real-world experience" and is viewed as a productive pedagogical approach to teaching and learning (Gough 2002; Hmelo-Silver, Duncan and Chinn 2007; White and Fredricksen 1998). We adopt such an approach for learning and teaching about ocean and aquatic systems in an environmental context.

D.L. Payne (✉)
Neag School of Education Department of Educational Psychology, Connecticut Sea Grant, University of Connecticut at Avery Point, 1080 Shennecossett Road, Groton, CT 06340, USA
e-mail: diana.payne@uconn.edu

T.D. Zimmerman
University of Rutgers, 10 Seminary Place, New Brunsuick, NJ 08901-1183
e-mail: timothy. zimmerman@gse.rutgers.edu

A.M. Bodzin et al. (eds.), *The Inclusion of Environmental Education in Science Teacher Education*, DOI 10.1007/978-90-481-9222-9_6,

Earth Systems Science Education: A Conceptual Understanding

Earth systems science as a discipline, and ocean and aquatic issues in particular, are poorly represented in national and state frameworks and standards (Hoffman and Barstow 2007). These standards and frameworks often drive the curriculum, instruction, and assessment at the local, state, and national levels. However, recent research indicates that a large-scale coherent theme (e.g., ocean and aquatic science) can be used as a model to assist in student understanding of complex systems if integrated in to the curriculum and instruction (Fortner et al. 2005; Lambert 2006). This theme is also reflected in the North American Association for Environmental Education (NAAEE) guidelines for preparing environmental educators (2004), which encourage a systems-based approach with a specific focus on human interdependence with the environment and the importance of place-based education. Table 1 outlines this core vision of EE with examples of relevant ocean and aquatic education topics.

Table 1 Instructional vision of environmental education and connections to ocean and aquatic education

Idea	Instructional vision of environmental education	Connections to ocean and aquatic education
Systems	A system has parts that can be understood separately, but the whole cannot be understood completely without recognizing the relationships among its parts. Systems are nested within other systems.	The ocean as a system drives other systems, including the Earth's climate, cycles, and intense weather (e.g., hurricanes, floods, drought).
Interdependence	Human well-being is inextricably bound with environmental quality. We and the systems we create – our societies, politics, economics, cultural activities, technologies – affect the systems and cycles of the rest of nature. We are challenged to recognize the ramifications of our interdependence.	Major cities are often settled beside waterways or along the coast and can affect water quality, human health, and natural habitats. Water is a precious global resource present in a limited quantity.
Importance of where one lives	Environmental education begins close to home, encouraging learners to explore and understand their immediate surroundings. The sensitivity, knowledge, and skills gained provide a base for moving out into larger systems, broader issues, and a lifetime of learning about causes, connections, and consequences.	Exploration of local aquatic ecosystems and watersheds can build the foundation for an understanding and appreciation of the global ocean, including the ocean's influence on us and our influence on the ocean.

Idea and instructional vision from NAAEE (2004)

Particular issues continue to arise regarding misconceptions of large-scale environmental issues. Topics such as ozone depletion, global climate change, biodiversity reduction, and water quality can arguably be traced to a lack of understanding of content in Earth systems and specifically in ocean and aquatic sciences concepts (Kim and Fortner 2006; Summers et al. 2000; Summers et al. 2001).

Ocean Education and Awareness

The general lack of attention to ocean concepts led two national panels to highlight the need for ocean-focused curricula as a central component for raising ocean awareness and understanding. The U.S. Commission on Ocean Policy (2004) noted:

> Strengthening the nation's awareness of the importance of the oceans requires a heightened focus on the marine environment, through both formal and informal education efforts. School curricula, starting in kindergarten, should expose students to ocean issues, preparing the next generation of ocean scientists, managers, educators, and leaders through diverse educational opportunities. (p. 122)

The Pew Oceans Commission (2003) concluded that "[r]estoring and sustaining the oceans require broad public support" (p. 92). However, in line with notions of environmental literacy, the Commission called for support on the part of the "public" that involves increased knowledge, concern, and action. Indeed, the Pew Ocean Commission recognized the connection between ocean knowledge and ocean protection stating:

> We must build a national constituency for the oceans that includes all Americans...[and] links people to the marine environment. Through enhanced marine education and awareness, we can inspire the next generation of scientists, fishermen, farmers, business, and political leaders—indeed all citizens—with a greater understanding and appreciation for the oceans. (p. 91)

The Need for Ocean and Aquatic Science Education

For the purposes of this chapter, we define scientific and environmental literacy as terms that include notions of science as a practice conducted within rich social and cultural settings and involve a complex interplay of content knowledge components (e.g., ecology, human interactions), affective components (e.g., attitudes, values), and behaviors (e.g., personal action, investment, and responsibility) (Roth and Lee 2002; Roth 1992). This definition is clearly domain general and can be easily applied to concepts in ocean and aquatic sciences. Ocean and aquatic science education explicitly involves physical, life, and Earth sciences as well as technology, society, culture, and the history and nature of science (Lambert and Sundburg 2006;

Payne 2006). For example, the 90% rapid decline in large predatory fish populations (Myers and Worm 2003) could have a cascading effect and lead to serious changes in global fish populations. Thus, to be science or environmentally literate on this or any ocean or aquatic topic, one must have some level of content knowledge, understand how attitudes and values impinge upon the topic, and be empowered to take action around the topic (Strang et al. 2007). Unfortunately, researchers have repeatedly noted the lack of understanding of ocean and aquatic systems by both teachers and students (Brody and Koch 1989–1990; Fortner and Meyer 2000; Fortner and Corney 2002; Lambert 2006; Tran, Payne and Whitley 2010).

In K-12 schools, ocean and aquatic concepts are infrequently taught and rarely appear in K-12 curriculum materials, textbooks, assessments, or standards (Hoffman and Barstow 2007; McManus et al. 2000). Additionally, educational research has paid little attention to teaching and learning ocean and aquatic science concepts in contrast to other areas of science such as chemistry, physics, and biology. The small body of literature regarding the teaching and learning of ocean and aquatic science topics reveals a wide array of nascent and nonnormative ideas. Brody and Koch (1989–1990) report that more than 86% of the elementary, middle, and high school students they studied in Maine do not know concepts essential to understanding ocean science and ocean resources. The students in this study also held nonnormative ideas that would significantly impact their ability to make informed decisions about ocean resources. For example, few students knew the role of nutrients in ocean ecosystems and at least 50% of students believed that ocean resources are limitless. Brody (1996) reported a comparable lack of conceptual knowledge and adherence to misconceptions amongst similarly aged students in the state of Oregon, while Ballantyne (2004) found that students in South Africa had similar difficulties understanding ocean concepts such as sources of salinity, wave propagation, and human impacts. Based on teaching experience with undergraduate students, Nelson, Aron, and Francek (1992) report student misconceptions in physical oceanography including the location of continents, and thus the size and location of ocean basins, concepts related to sea level, and freezing and melting of ocean water. Working with high school students, Lambert (2005, 2006) noted that many students, even after participating in an integrated science curriculum focused on ocean science, held nascent or nonnormative ideas about seafloor spreading, density, the water cycle, light waves, and matter.

The result of the omission of ocean and aquatic content in science education (and science education research) has been predictable. While Americans are expected to comprehend and respond to increasingly complex issues like global climate change, environmental pressures on coastal and ocean resources, and biotechnology potential within the ocean, we often do so with no more than a sixth-grade understanding of how the natural world works and often with a nonscientifically accurate understanding of the ocean. Leaving ocean and aquatic science out of science education has resulted in a general public that is not well informed on many ocean and coastal policy issues (Steel et al. 2005). Indeed, public opinion surveys have documented a lack of ocean knowledge associated with robust ocean misconceptions. A report for the Ocean Project (Belden Russonello and Stewart and American Viewpoint 1999) revealed that 46% of Americans surveyed did not know enough about the ocean to offer an opinion regarding the health of the ocean. A new and more extensive survey was conducted for the Ocean Project in 2008, indicating little change since 1999 (Boyle and Mott 2009). For example, 35% of the respondents were unable to identify any ocean related issue affecting the United States, and the overall perception regarding climate change was that it is almost entirely a terrestrial issue (Boyle and Mott 2009). A SeaWeb survey found that although many people believe that the ocean is in need of protection, most incorrectly think that pollution is the greatest problem (Mellman Group 1999). Finally, Steel et al. (2005) report that of 1,233 citizens surveyed, the average score on a short quiz about general knowledge of the ocean was a mere 2.23 out of a possible 5 points. This lack of knowledge persists despite the fact that over half of the U.S. population lives within 50 miles of the coast; the ocean provides commerce, trade, mineral resources, and much of the oxygen we breathe; hosts great biological diversity; and harbors vast medical potential (U.S. Commission on Ocean Policy 2004).

Currently, there is little educational research regarding teaching and learning of ocean and aquatic science (Tran, Payne and Whitley 2010). Few studies have considered the relationship between understanding of science and tendencies for espousing an ocean protection edict. In one study, Zimmerman (2005) found that students with a deeper, more integrated understanding of science concepts developed better arguments about the need for ocean conservation. It is very likely, therefore, that a correlation may exist between ocean science knowledge and improved reasoning about ocean protection as found with general environmental behavior (Frick et al. 2004). Another study describes research on ocean environmental knowledge, attitudes, and behavior in relation to dolphins (Barney et al. 2005). The researchers found that respondents with a more expert knowledge of dolphins espoused a more environmentally responsible attitude and were less likely to engage in potentially harmful behavior to the dolphins. These few studies indicate that stewardship of ocean and aquatic resources could benefit from environmental and science education that includes ocean and aquatic sciences concepts.

Incorporating Ocean and Aquatic Science into Environmental and Science Teacher Education: A Process of Teaching and Learning

Although teachers indicate positive attitudes toward teaching environmental concepts (Ko and Lee 2003; McCaw 1979–1980), they do not often feel comfortable with the complexity of what may be considered contentious issues. Teacher preparation in environmental education must include specific pedagogical training as well as issue-specific content knowledge (Kim and Fortner 2006). Complex and possibly controversial issues, including global climate change and threats to human health (e.g., water pollution, contaminants in food and drinking water), require an understanding of the role of ocean and aquatic systems as part of the broader environment. On a global scale, the international economy, agriculture, weather, biodiversity, and where people can live are all affected by changes in ocean and atmospheric processes (Hoffman and Barstow 2007).

Preparation in Environmental Science Concepts

Two studies on teacher preparation and environmental education point to a low incoming level of environmental literacy and a need for resources devoted to both teacher preparation and classroom instruction. Goldman, Yavetz, and Pe'er (2006) investigated the environmental literacy of students entering a teacher training program in Israel. With surveys of students' knowledge of, commitment to, and behaviors around environmental issues, the researchers sought to assess the environmental literacy of a broadly representative sample of students. Using behavior as a proxy for environmental literacy, the researchers found that incoming students had low environmental literacy. It was also noted, not surprisingly, that student enrollment in an environmentally affiliated field of study (as opposed to a *non*environmentally affiliated field of study) were significantly more likely to engage in environmental behavior. Finally, the study concluded that students in teacher training programs who grew up in nonurban environments were more likely to engage in environmental behavior, pointing to a connection between direct experiences with nature as formative for environmental behavior (Goldman et al. 2006).

Heimlich, Braus, Olivolo, McKeown-Ice, and Barringer-Smith (2004) report the results of a national survey of teacher preparation programs in the United States and the inclusion of EE in those programs. The researchers ascertained that awareness of environmental education resources was low, and there was a strong need for educational resources for both teacher preparation and teacher preparation curriculum. Surveys of leaders of teacher preparation programs also revealed that a major barrier to the incorporation of EE into preservice programs was a lack of expertise of the faculty (Heimlich et al. 2004). Similar findings are prevalent regarding the lack of knowledge in terms of ocean and aquatic science content and resources (Pew Oceans Commission 2003; US Commission on Ocean Policy 2004; Walker et al. 2000).

Preparation in Ocean and Aquatic Science Concepts

Several studies are specifically relevant to the teaching and learning of ocean and aquatic science concepts. For example, Goldman et al. (2006) noted the lack of environmental literacy amongst incoming preservice students, the connection between students' prior science coursework (environmentally affiliated or nonenvironmentally affiliated fields) and past exposure to natural environments. All three factors have significant implications for preparing teachers to teach ocean and aquatic sciences. First, studies of the learning of ocean and aquatic science concepts point to a poor general understanding of these topics through adulthood. Students entering teacher preparation programs with a low level of environmental literacy (and thus probable extremely low level of ocean and aquatic science literacy), are most likely not equipped to teach ocean and aquatic science concepts. Second, with the consistent reduction in students majoring in Earth (including ocean and aquatic sciences), biological, and agricultural sciences (National Science Board 2004), it is plausible that fewer students in teacher preparation programs will have completed such environmentally affiliated majors. Third, although 50% of the U.S. population lives within the coastal counties, 50% have had little to no exposure to the ocean other than second-hand images (U.S. Commission on Ocean Policy 2004). Without personal experiences or adequate resources, it is likely that future teachers will be unable to personally engage, and therefore engage their students, in the teaching and learning of ocean and aquatic science, and related environmental stewardship and behavior.

Ocean and Aquatic Science Education Initiatives and Resources

Clearly, there is a need to provide quality resources to educators at all levels, including those who strive to increase the specific content expertise of preservice and inservice teachers (Heimlich et al. 2004). Within the ocean and aquatic science education community, several such resources do exist from entities including the National Oceanic and Atmospheric Administration (NOAA), Sea Grant, the National Marine Educators Association (NMEA), the Centers for Ocean Sciences Education Excellence (COSEE), the National Ocean Sciences Bowl (NOSB), Project WET (1995), and Project WILD Aquatic (2004). A collaborative effort of many of these entities produced the Ocean Literacy Essential Principles and Fundamental Concepts (OLEPFC) (Schoedinger, Tran and Whitley 2010). An overview of several key resources is provided below.

Ocean Literacy Essential Principles and Fundamental Concepts and Scope and Sequence Conceptual Flow Diagrams

The development of the OLEPFC (Schoedinger, Tran and Whitley 2010) began with a small group of teachers, informal educators, and scientists in November 2004. The resulting document, first published in 2005, was the result of a consensus within the ocean science education community on what constitutes an ocean literate person. Following intense discussion and review, the team developed seven essential principles and associated fundamental concepts which are aligned with the *National Science Education Standards* (NRC 1996). This effort has inspired a variety of projects including designing ocean-related curricular materials, providing professional development opportunities for educators to assist in incorporating ocean sciences concepts into their classrooms, partnering with informal education and environmental education centers to improve ocean education outreach, and connecting advanced ocean observation technology tools to classrooms for educational purposes. The essential principles of ocean literacy and their connection to EE are outlined in Table 2. Efforts are currently underway to expand the OLEPFC to include more aquatic and watershed content.

Table 2 The essential principles of ocean literacy and connection to EE instructional vision

EE instructional vision	Essential principles
Systems	1. The Earth has one big ocean with many features.
Systems	2. The ocean and life in the ocean shape the features of the Earth.
Systems; interdependence	3. The ocean is a major influence on weather and climate.
Systems	4. The ocean makes Earth habitable.
Systems; interdependence	5. The ocean supports a great diversity of life and ecosystems.
Interdependence; place	6. The ocean and humans are inextricably interconnected.
Systems; interdependence	7. The ocean is largely unexplored.

Upon completion of the OLEPFC, the team began development of a scope and sequence to assist educators in the implementation of ocean and aquatic sciences in their curriculum. The complete OLEPFC as well as scope and sequence conceptual flow diagrams for grade level bands K-3, 3–5, 6–8, and 9–12 were published in 2010 (Strang and Tran, 2010) and are available at www.coexploration.org/oceanliteracy.

National Programs

Centers for Ocean Science Education Excellence (COSEE)

Partially in response to concerns raised by the U.S. Ocean Commission and the Pew Oceans Commission, and in some measure due to fewer students going into ocean science fields, the National Science Foundation's (NSF) Geosciences Directorate, with additional funding and support from NOAA and Sea Grant, undertook a bold mission to improve ocean sciences education nationwide. Through the COSEE program (www.cosee.net), several centers are now established around the United States. COSEE serves a catalytic role by engaging ocean scientists directly with teachers, nonformal educators, aquariums, environmental education providers, and curriculum designers.

Several COSEE offer opportunities for teachers to improve ocean and aquatic science education. Two specific examples are the COSEE-Networked Ocean World (COSEE-NOW, formerly COSEE-Mid Atlantic or COSEE-MA) professional development workshops and COSEE-California's (COSEE-CA) Communicating Ocean Sciences (COS) Instructor's Workshops. These programs provide K-16 educators with tools to increase the ocean sciences content knowledge of teachers, their students, and a broad audience of learners.

COSEE-NOW conducted professional development workshops focused on training educators in the use of real-time data derived from ocean observing systems (OOS). These week-long sessions were conducted as hands-on experiences. While working with OOS scientists and educators, teachers used real-time and near-real-time data sets. A variety of resources were utilized to highlight ocean science concepts. For example, a density tank (a large tank filled with salt water on one side and fresh water on the other with a removable divider) offered participants a hands-on opportunity to visualize density-driven boundary layers that form in the ocean. The concepts were then connected to real-time and near-real-time data collected by ocean-going "gliders" – autonomous computerized robots that fly through the ocean collecting data on ocean density. The activities imparted ideas, curricula, and tools for teaching ocean science concepts while expanding participants' understanding of these concepts (Parsons 2008). Many of the COSEE-NOW real-time data lessons are delivered via the Coastal Ocean Observation Laboratory (COOL) Classroom, an online, free ocean observing curriculum delivery platform designed for middle school students at http://www.coolclassroom.org/.

COSEE-CA offers COS Instructor's Workshops for university-level instructors seeking to improve their students' abilities to "communicate" ocean science topics. University instructors take part in COS workshops as preparation to offer a semester long course at their home institution, where undergraduate science students work with K-5 teachers and their students. The undergraduate students taking the course design ocean science lessons for use in K-5 classrooms. The lessons are used by the undergraduate students in their mentor teachers' classroom, providing new ocean science tools for the teachers and new ocean sciences learning experiences for the K-5 students. In the process of working with the K-5 classrooms, several generations (undergraduate students, K-5 teachers, and K-5 students) improve their knowledge and understanding of ocean science concepts. In 2006, the COS model was expanded to nonformal audiences through the Communicating Ocean Sciences to Informal Audiences (COSIA) program. Instead of placing undergraduate students in K-5 classrooms, higher education instructors are paired with educators at informal science learning institutions (e.g., aquariums, science centers, nature centers). Together, the COS and COSIA programs have led to a better understanding of the teaching and learning of ocean sciences, particularly for undergraduate students majoring in science. In fact, many of these students have, upon graduation, decided to pursue teaching as a career instead of science, thus providing a new recruiting opportunity for ocean-literate teachers.

National Marine Educators Association

The National Marine Educators Association (NMEA) brings together people interested in the teaching, study and enjoyment of both fresh and salt water, and provides a focus for marine and aquatic studies all over the world. In addition to providing print resources in the form of *Current: The Journal of Marine Education* and *NMEA News*, NMEA supports its regional chapters, co-sponsors the BRIDGE website, and provides scholarships to attend its annual conference (www.marine-ed.org). Recently, the NMEA Ocean Literacy committee worked with the publisher of a high school textbook *Life on an Ocean Planet* (Alexander et al. 2010) to align content with the OLEPFC.

National Ocean Sciences Bowl

The National Ocean Sciences Bowl (NOSB) is a high school competition where students test their knowledge of marine sciences including biology, chemistry, physics, geology, as well as marine policy, geography, and economics. Teams consist of four to five students and a coach (teacher). The NOSB mission is to enrich science teaching and leaning across the US through a high-profile national competition that increases high school students' knowledge of the oceans and enhances public understanding and stewardship of the oceans (oceanleadership.org/education/national-ocean-sciences-bowl/).

National Oceanic and Atmospheric Association

The National Oceanic and Atmospheric Association (NOAA)'s vision is of an informed society that uses a comprehensive understanding of the role of the oceans, coasts, and atmosphere in the global ecosystem to make the best social and economic decisions (www.noaa.gov/about-noaa.html). A plethora of education resources and funding opportunities exist through several NOAA programs as outlined in Table 3.

Table 3 Marine and aquatic resources available through NOAA

NOAA program	Highlights of ocean and aquatic education resources	Web link
Coral Reef Conservation	Coral education resources CD	www.coralreef.noaa.gov/outreach
Estuarine Research Reserves	Estuaries 101, a high school level curriculum	www.estauries.gov
National Marine Sanctuaries	Curriculum, lesson plans, events, and activities	www.sanctuaries.noaa.gov/education
National Ocean Service	Online lessons for grades 3–5 and 6–12 in different topical areas	www.oceanservice.noaa.gov/education/welcome
Office of Education	B-WET and Environmental Literacy grants	www.oesd.noaa.gov
Office of Ocean Exploration	Lesson plans; curriculum; expedition education modules	www.oceanexplorer.noaa.gov/edu
Teacher at Sea	Places teachers on a NOAA vessel for a first-hand experience of science and life at sea	www.teacheratsea.noaa.gov
Sea Grant	Links to ocean and aquatic education resources in 32 coastal and Great Lakes state programs	www.seagranted.net

Curricular Resources

Curricular resources for ocean and aquatic topics are available in print and web format. Project WET: Water Education for Teachers (1995) and Project WILD Aquatic (2000) are available through local entities after a full day professional development session led by a trained facilitator. Each guide contains a variety of activities for students in grades K-12. Project WILD Aquatic emphasizes aquatic wildlife and ecosystems, while Project WET focuses on the importance of water as a local and global resource.

The Marine Activities, Resources and Education (MARE) curriculum immerses people of all ages in ocean content. Focused on teacher professional development and curricular resources, MARE also provides opportunities to include entire schools and families in ocean-related activities (www.lawrencehallofscience.org/mare).

The BRIDGE, an ocean education web site funded by NOAA, NMEA and maintained by Virginia Sea Grant, provides teacher-approved resources (www.marine-ed.org/bridge) in a variety of topic areas. Links to lesson plans, research and data, professional development opportunities, and information for students (e.g., careers, summer programs) are highlighted.

Integrating Ocean and Aquatic Sciences in Environmental and Science Teacher Education

In this chapter, we contend that ocean and aquatic science concepts must be included in environmental and science education efforts in order to develop an environmentally literate society. This need is reflected in the (1) recognized role of ocean sciences concepts as the "glue" that helps learners understand and contextualize earth systems science concepts, (2) concomitant dearth of research on the learning and teaching of ocean sciences concepts, and (3) documentation of consistent and persistent alternate conceptions of ocean science related concepts. As an emerging subdomain of both science education and EE, ocean and aquatic sciences can be taught using exciting, hands-on, inquiry-based approaches that not only provide opportunities for improved content knowledge, but also have affective and actionable components. In his Keynote Address to the Fourth World Environmental Education Congress, Scott (2007) challenged the environmental education community to consider "...what insights does the environmental education research community provide that will help us... ?" We believe one component that will help us is to ensure inclusion of the largest geological surface feature on our planet: the ocean.

Acknowledgments The authors thank Rosanne Fortner, Julie Lambert, Meghan Marrero, and Sharon Walker for their guidance and support.

References

Alexander, L., Desonie, D., Kelchner, C., Lambert, J., Leaney, L., Menzel, T., et al. (2010). *Life on an ocean planet* (2nd ed.). Rancho Margarta, CA: Current Publishing.

Ballantyne, R. (2004). Young student's conceptions of the marine environment and their role in the development of aquaria exhibits. *GeoJournal, 60*, 159–163.

Barney, E. C., Mintzes, J. J., & Yen, C.-F. (2005). Assessing knowledge, attitudes, and behavior toward charismatic megafauna: The case of dolphins. *Journal of Environmental Education, 36*(2), 41–55.

Belden Russonello & Stewart and American Viewpoint. (1999). *Communicating about oceans: Results of a national survey*. Washington, DC: The Ocean Project.

Boyle, P., & Mott, B. (2009). *America, the Ocean, and Climate Change: New Research Insights for Conservation, Awareness, and Action – Results of a National Survey*. Providence, RI: The Ocean Project.

Brody, M. (1996). An assessment of 4th-, 8th-, and 11th-grade students' environmental science knowledge related to Oregon's marine resources. *Journal of Environmental Education, 27*(3), 21–27.

Brody, M., & Koch, H. (1989–1990). An assessment of 4th-, 8th-, and 11th-grade students' knowledge related to marine science and natural resource issues. *The Journal of Environmental Education, 21*(2), 16–25.

Council for Environmental Education. (2000). *Project WILD aquatic* (4th ed.). Hasten, TX: Author.

Fortner, R. W., & Corney, J. R. (2002). Great Lakes educational needs assessment: Teachers' priorities for topics, materials, and training. *Journal of Great Lakes Research, 28*(1), 3–14.

Fortner, R. W., Corney, J. R., & Mayer, V. J. (2005). Growth in student achievement as an outcome of inservice environmental education using standards-based infusion materials. In B. Simmons (Ed.), *Preparing effective environmental educators* (pp. 73–89 NAAEE Monograph 2). Washington, DC: NAAEE

Fortner, R. W., & Meyer, R. L. (2000). Discrepancies among teachers' priorities for and knowledge of freshwater topics. *The Journal of Environmental Education, 31*(4), 51–53.

Frick, J., Kaiser, F. G., & Wilson, M. (2004). Environmental knowledge and conservation behavior: Exploring prevalence and structure in a representative sample. *Personality and Individual Differences, 37*, 1597–1613.

Goldman, D., Yavetz, B., & Pe'er, S. (2006). Environmental literacy in teacher training in Israel: Environmental behavior of new students. *Journal of Environmental Education, 38*(1), 3–22.

Gough, A. (2002). Mutualism: A different agenda for environmental and science education. *International Journal of Science Education, 24*(11), 1201–1215.

Heimlich, J. E., Braus, J., Olivolo, B., McKeown-Ice, R., & Barringer-Smith, L. (2004). Environmental education and preservice teacher preparation: A national study. *Journal of Environmental Education, 35*(2), 17–21.

Hmelo-Silver, C., Duncan, R.G., & Chinn, C.A. (2007). Scaffolding and achievement in problem-based and inquiry learning: A response to Kirschner, Sweller, and Clark (2006). *Educational Psychologist* 42, 99–107.

Hoffman, M., & Barstow, D. (2007). *Revolutionizing Earth System Science Education for the 21st Century, Report and Recommendations from a 50-State Analysis of Earth Science Education Standards*. Cambridge, MA: TERC.

Hungerford, H., & Volk, T. (1990). Changing learner behavior through environmental education. *Journal of Environmental Education, 21*(3), 8–21.

Kim, C., & Fortner, R. (2006). Issue-specific barriers to addressing environmental issues in the classroom: An exploratory study. *The Journal of Environmental Education, 37*(3), 15–22.

Ko, A. C., & Lee, J. C. (2003). Teachers' perceptions of teaching environmental issues within the science curriculum: A Hong Kong perspective. *Journal of Science Education and Technology, 12*, 187–204.

Lambert, J. (2005). Students' conceptual understandings of science after participating in a high school marine science course. *Journal of Geoscience Education, 53*(5), 531–539.

Lambert, J. (2006). High school marine science and scientific literacy: The promise of an integrated science course. *International Journal of Science Education, 28*(6), 633–654.
Lambert, J., & Sundburg, S. (2006). Using an integrated ocean science approach to chart a new course in high school science curriculum. *The Science Teacher, 73*(6), 40–43.
McCaw, S. C. (1979–1980). Teacher attitude toward environmental education. *The Journal of Environmental Education, 11*(2), 18–23.
McManus, D. A., Walker, S., Cuker, B., Goodnight, P., Humphris, S., Keener-Chavis, P., et al. (2000). *Center for Ocean Sciences Education Excellence (COSEE): The report of a workshop sponsored by the National Science Foundation*. Long Beach, MS: Gulf Park.
Mellman Group. (1999). *Top-line analysis of public attitudes toward National Marine Sanctuaries and ocean protected areas*. Silver Spring, MD: Sea Web.
Myers, R. A., & Worm, B. (2003). Rapid worldwide depletion of predatory fish communities. *Nature, 423*, 280–283.
National Research Council. (1996). *National science education standards*. Washington, DC: National Academy Press.
National Science Board. (2004). *Science and engineering indicators 2004*. Arlington, VA: National Science Foundation.
Nelson, B. D., Aron, R. H., & Francek, M. A. (1992). Clarification of selected misconceptions in physical geography. *Journal of Geography, 91*(2), 76–80.
North American Association for Environmental Education (NAAEE). (2004). *Guidelines for the preparation of Environmental Educators*. Rock Spring, GA: NAAEE.
Parsons, C. (2008). *COSEE Mid-Atlantic: Taking the pulse of the ocean teacher professional development 2003-2007 - Final evaluation report*. Retrieved March 16, 2009, from http://coseenow.net/resources
Payne, D. (2006). Under pressure: A study of issues in oceanography. *The Science Teacher, 73*(6), 30–35.
Pew Oceans Commission. (2003). *America's living oceans: Charting a course for sea change*. Arlington, VA: Author.
Project WET: K-12 Curriculum and Activity Guide. (1995). Bozeman, MT: The Watercourse and Western Regional Environmental Education Council.
Project WILD Aquatic: K-12 Curriculum and Activity Guide. (2000). Houston, TX: Council for Environmental Education.
Roth, C. E. (1992). *Environmental literacy: Its roots, evolution and directions in the 1990s*. Columbus, OH: ERIC Clearinghouse for Science, Mathematics and Environmental Education.
Roth, W.-M., & Barton, A. C. (2004). *Rethinking scientific literacy*. New York: Routledge.
Roth, W.-M., & Lee, S. (2002). Scientific literacy as collective praxis. *Public Understanding of Science, 11*, 33–56.
Schoedinger, S., Tran, L.U., & Whitley, L. (2010). From the principles to the scope and sequence: A brief history of the ocean literacy campaign. *NMEA Special Report #3: The Ocean Literacy Campaign*, 3–7.
Scott, W. (2007). *Environmental education research: Thirty years on from Tbilisi*. Keynote Address to the 4th World Environmental Education Congress.
Steel, B. S., Smith, C., Opsommer, L., Curiel, S., & Warner-Steel, R. (2005). Public ocean literacy in the United States. *Ocean and Coastal Management, 48*, 97–114.
Strang, C., DeCharon, A., & Schoedinger, S. (2007). Can you be science literate without being ocean literate? *Current: The Journal of Marine Education, 23*(1), 7–9.
Strang, C., & Tran, L.U. (Eds.). (2010). *NMEA Special Report #3: The Ocean Literacy Campaign*.
Summers, M., Kruger, C., & Childs, A. (2000). Primary school teachers' understanding of environmental issues: An interview study. *Environmental Education Research, 6*, 293–312.
Summers, M., Kruger, C., Childs, A., & Mant, J. (2001). Understanding the science of environmental issues: Development of a subject knowledge guide for primary teacher education. *International Journal of Science Education, 23*(1), 33–53.
Tilbury, D. (1995). Environmental education for sustainability: Defining the new focus of environmental education in the 1990s. *Environmental Education Research, 1*, 195–212.

Tran, L.U., Payne, D.L., & Whitley, L. (2010). Research on learning and teaching ocean and aquatic sciences. *NMEA Special Report #3: The Ocean Literacy Campaign*, 22–26.

U.S. Commission on Ocean Policy. (2004). *An ocean blueprint for the 21st century: Final report of the U.S. Commission on Ocean Policy.* Washington, DC: Author.

Walker, S. H., Coble, P., & Larkin, F. L. (2000). Ocean sciences education for the 21st century. *Oceanography, 13*(2), 32–39.

White, B. Y., & Fredricksen, J. R. (1998). Inquiry, modeling, and metacognition: Making science accessible to all students. *Cognition and Instruction, 16*(1), 3–118.

Zimmerman, T. D. (2005). *Promoting knowledge integration of scientific principles and environmental stewardship: Assessing an issue-based approach to teaching evolution and marine conservation.* Unpublished doctoral dissertation, University of California-Berkeley, Berkeley, CA.

Part II
Environmental Education Pedagogy

Promoting the Use of Outdoor Learning Spaces by K-12 Inservice Science Teachers Through an Outdoor Professional Development Experience

Mark A. Bloom, Molly Holden, April T. Sawey, and Molly H. Weinburgh

Unlike our European and Australian counterparts, environmental education does not yet constitute a formal component of the public school curriculum in the United States (American Association for the Advancement of Science 1993; Connell 1999; Hicks & Bord 2001; National Research Council 1996). However, national interest in the environment has increased over the past few decades for a myriad of reasons particular to the variety of groups who express the interest. Despite the complexity surrounding environmental issues, they are all closely interrelated and, left unchecked, have the potential for disastrous economic, health, and biological consequences. Given these concerns, environmental topics are gradually being incorporated into national and state standards, mostly in the sciences and social studies (North American Association for Environmental Education 2004). Likewise, opportunities for environmental instruction in informal settings (natural history and science museums), outdoor spaces (school grounds, parks, other native land), and through environmental project-based community learning have increased significantly, as has research associated with these alternative teaching models (Corcoran 1999; Louv 2003; Sobel 2004).

Regardless of the means, the relevance of environmental issues to students' lives is thought to most typically occur when learning experiences are situated in a meaningful context, and when human-environment connections are well understood (Connell 1999; Littledyke 2008), instruction addresses cognitive, affective, and behavioral domains (Littledyke 2008; Loughland et al. 2002; Loughland et al. 2003; Martin & Brouwer 1991), and environmental problems are presented as worthy of serious consideration and action but not as insurmountable. Sadly, few U.S. students today are exposed to natural spaces except for short periods of time.

M.A. Bloom (✉), M. Holden, A.T. Sawey, and M.H. Weinburgh
Andrews Institute of Mathematics & Science Education, Texas Christian University, TCU, Box 297920, Fort Worth, TX 76129, USA
e-mails: M.Bloom@tcu.edu; M.Holden@tcu.edu; A.T.SAWEY@tcu.edu; M.Weinburgh@tcu.edu

A.M. Bodzin et al. (eds.), *The Inclusion of Environmental Education in Science Teacher Education*, DOI 10.1007/978-90-481-9222-9_7,

School curricula, particularly based on textbook learning, generally do not support the use of outdoor spaces for instruction. Recess, now thought to be a partial antidote to such childhood epidemics as childhood obesity and attention deficit/ hyperactivity disorder, has been shortened or cut altogether as a result of *No Child Left Behind*. A study by the Center for Public Education (2008) reported that while most children, regardless of location, continue to get recess on a regular basis, children who attend high-minority, high-poverty, or urban schools are far more likely than other children to get no recess at all. School administrators (and teachers) commonly have concerns about such issues as liability, safety, and actualization of academic benefits in many outdoor/off-campus activities, which are commonly perceived as time-consuming and not beneficial to students in terms of achievement gains for the effort expended.

The theoretical basis for the professional development (PD) program described in this chapter is that solutions to these issues cannot be conceived and acted upon without concern for the natural world and re-evaluation of our role in it. One way that this concern develops is from a deep understanding of the earth, its systems, and their interrelationships, an understanding that is significantly enhanced by direct experiences in the environment(s) one is attempting to comprehend. This chapter describes our observations and experiences in developing and conducting a PD program to educate teachers in the use of outdoor learning spaces (OLS) for environmental science instruction. An underlying assumption of the PD design is that, when properly conceptualized, planned, and aligned with appropriate standards, outdoor instruction can be beneficial to students and result in long-term gains that go beyond academic achievement. The first step in that direction is to help teachers identify and find ways to overcome the challenges they face in using outdoor learning spaces. The overarching goal of this PD experience was to "re-introduce" teachers to outdoor environments to: (1) supplement their content knowledge about environmental issues and their related scientific concepts; (2) improve their pedagogical skills by experiencing teaching strategies for transferring scientific knowledge in unique ways; and (3) align outdoor lessons with national and state teaching standards, with the short-term outcome of improving teacher efficacy concerning the use of outdoor spaces. What we report here are some of the challenges and successes we and our teacher-participants encountered as we worked to achieve these goals.

Background and Description of PD

This chapter discusses the work we conducted with K-12 science teachers during a 2-week, introductory component of an outdoor PD experience conducted from late July through early August 2008. It should be noted that only preliminary results of the experience are reported; we continued conducting 1-day workshops with these teachers once a month for the remainder of the 2008–2009 academic year. Most of the study participants were teachers from a local district "pyramid," divided into two distinct cohorts for much of the workshop. The first cohort consisted of 18 teachers from five local elementary schools. The second cohort consisted of 18 science teachers from the two middle schools and the high school to which the other seven schools feed. By working with the teachers within this pyramid of schools, we hoped to foster a consistent environmental education (EE) emphasis to be used throughout the K-12 progression. The schools within the pyramid are generally characterized as urban, economically disadvantaged, with a high number of English language learners and a low passing rate on the state standardized assessment for knowledge and skills in science.

The high school selected for this study was recently chosen by the district as one of 15 schools to participate in the Public Educators Accelerating Kids (PEAK) pilot program that encourages educator collaboration, cooperation, and professional growth over competition. PEAK is based on rewards and incentives and is consistent with the philosophy that teaching is a team sport. The program stresses that leadership capacity through development of teacher knowledge and skills will ultimately have the greatest impact on student achievement. One of the long-term objectives for the program at this high school is to create an "environmental curriculum" to improve student retention and prepare many graduates for viable postsecondary academic/professional careers with an environmental focus.

Our goals for the PD summer component with the teachers were to (1) provide integrated instruction on environmental issues, (2) model the use of OLS to help learners contextualize large-scale environmental issues in relation to their immediate surroundings, (3) provide guidance in aligning outdoor education experiences with state and national standards to fulfill curriculum requirements as mandated by many school districts, and (4) build a foundation for the academic year follow-up component of the PD.

Preassessment Results: Incentives and Challenges in Using OLS

To best achieve the initial goal of identifying challenges teachers face in using OLS for instruction, we conducted multiple preassessments to determine the factors that encourage or discourage them to teach outdoors. We hoped that by identifying both the challenges and the factors encouraging OLS use, we could tailor our PD in such a way that the teachers could realize ways to overcome the challenges and recognize the benefits of outdoor teaching already enjoyed by those who utilize this teaching strategy.

The first preassessment was administered to the teacher participants via an online survey before the first day of the PD. The survey included a section devoted to the teachers' frequency and purpose of OLS use. On reviewing the answers provided, a distinct difference between the two teacher cohorts became apparent. The majority of the elementary teachers reported using OLS to teach, and did so on a regular basis, whereas only a few secondary teachers reported using OLS at all (and then only infrequently) and half reported never using them.

The elementary teachers listed several incentives for using OLS to teach. These ranged from pleasure for the students ("children like to learn outdoors," "allowing students to explore," "getting fresh air") to practical academic advantages ("making science less intimidating," "reinforcing classroom content," "connecting science to real-world experiences"). All of these seemed to align with our goal of connecting students with natural spaces to encourage development of an appreciation for their environment. The elementary teachers identified only a few challenges: difficulty with classroom management, inhospitable weather conditions, and lack of time (though one teacher qualified her response to the latter with "but probably not").

In contrast, the secondary teachers listed fewer incentives and significantly more challenges. Only two of these teachers indicated that they took their students outside at least once a month; the motivating factors listed for these events included: "students like it," "experiential learning," and the teacher's own "passion for teaching outdoors." The remaining 16 secondary teachers either took their students outdoors "seldom" or "never." All but one listed challenges, which included those presented by the elementary teachers as well as potential liability, expense, need for chaperones, administrative disapproval, and complicated logistics.

The second preassessment, implemented during the first day of the PD, was used to further explore teacher OLS use and perceived challenges. The teachers reviewed a set of statements posted around the classroom; these statements covered a wide range of issues involving use of OLS for teaching. Some were positive ("I like the outdoors," "I regularly use outdoor spaces to teach," "Students enjoy using outdoor spaces to learn") and others were negative ("There is not enough time to use the outdoors to teach," "My administration does not support using outdoor spaces to teach," "liability is a concern"). Each teacher was provided five stickers and instructed to "vote" by placing the stickers on the statements that most strongly corresponded with their thoughts and beliefs about using OLS. The goal of this activity was to help us determine what they considered most important about OLS use, particularly "new" incentives and challenges not mentioned in the online presurvey responses. The new challenges were identified as being important by both cohorts of teachers. Consistent with the survey, the elementary teachers identified time, weather, and classroom management as challenging issues, but also "voted" for safety, lack of a specific curriculum, and inadequate pedagogical content knowledge (PCK). The secondary teachers also identified time, classroom management, logistical problems, and liability, but voted for "lack of a specific curriculum" as a factor discouraging OLS use.

Immediately after the voting activity, a third preassessment measure was implemented. The teachers responded to the following prompt in small focus group discussions: "*You just voted on statements about using OLS including your schoolyard, parks, an Outdoor Learning Center, other outdoor areas. Talk about why you do, or do not use OLS to teach, including the perceived/actual benefits and limitations of their use.*" These conversations were audio-recorded for the purpose of further "unpacking" what the teachers considered their most significant challenges. As a result of analyzing the teacher responses in conjunction with the preassessment measures, three major themes became apparent:

- Logistics – i.e. permission to leave campus, transportation, chaperones, potential liability

Table 1 Identified challenges of OLS use

	Elementary cohort			Secondary cohort		
	Preassessment			Preassessment		
Challenge/Obstacle	1	2	3	1	2	3
Time	✓	✓		✓	✓	
Weather	✓	✓		✓		
Management	✓	✓		✓	✓	✓
Liability		✓		✓		
Expense				✓		
Logistics			✓	✓	✓	✓
Administrative disapproval			✓	✓	✓	✓
Inadequate PCK		✓			✓	✓
No curriculum		✓			✓	✓
Geography (proximity, "primitiveness")			✓			✓

1: Online survey; 2: "Voting" on OLS statements; 3: Postvoting focus group discussion

- "Geography" – i.e. lack of suitable space close to school, outdoor spaces are too "primitive"
- Lack of administrative support – generally described as a narrow-minded focus on standardized testing, coupled with a lack of understanding about how outdoor activities can be successfully aligned with curricular frameworks and state science standards

Prior to our second meeting with these teachers, we reviewed the preassessment data to tailor the PD to participants' stated needs. We recognized that although both cohorts identified many of the same challenges to using OLS, their definition and perception of these challenges differed significantly, as shown in Table 1. Those reported by the elementary teachers were perceived more as *inconveniences* to be dealt with rather than as absolute barriers to OLS use. Those reported by the secondary teachers appeared, in fact, to be perceived as *obstacles* that *prevented* their use of OLS for instruction. Further evidence of the distinction between the cohorts' perceptions was found in the manner in which the secondary teachers repeated their challenges with each successive preassessment, whereas the elementary teachers identified new challenges with each preassessment, essentially "leaving behind" those mentioned previously.

Professional Development Interventions

We perceive that the participants in this PD represent a community of practice defined as a group of people who "...share a concern, a set of problems, or a passion about a topic, and who deepen their knowledge and expertise in this area by interacting on an ongoing basis" (Wenger et al. 2002). Subsequently, our PD was designed around the concept of providing several opportunities for members of the two cohorts to interact and thereby promote sharing of experiences, concerns, and ideas about OLS use. Specifically, we were hopeful that the elementary teachers could offer suggestions to the secondary teachers, since the former were, in large part, already using the outdoors for teaching. This section describes our efforts to foster a community of practice during the 2-week summer component of the PD experience.

Workshop activities included classroom presentations of environmental content followed by a gradual introduction of the teachers to increasingly "native" outdoor spaces, from a few minutes on the campus grounds, to a few hours at a local botanic garden, to two overnights at the school district's Outdoor Learning Center (OLC), a 228-acre native limestone prairie and post oak woodland on the edge of a lake. At the OLC, the teachers were introduced to various types of field sampling and navigation equipment, and encouraged to identify a particular area of study to pursue for the duration of the outdoor experience. Teacher discussion and reflective dialogue were used daily to address PCK being taught, as well as how the teachers' newfound skills (i.e. field equipment use) and knowledge (i.e. diversity and taxonomy of insects at the OLC) might fit the needs of their curriculum. Throughout the workshop, journaling was highly encouraged for documenting observations, making interpretations, and reflecting on experiences. The theoretical basis for this PD workshop is primarily based on Vygotsky's (1978) ideas about teaching and learning, whereby the "more knowledgeable other" helps learners to scaffold experiences, and learners answer their own questions through newfound skills. The teachers were expected to practice inquiry by formulating hypotheses, using new process skills, and making evidence-supported claims.

Many of the interventions conducted during the PD experience were specifically designed to address teacher challenges in using OLS. We introduced them to three continua from which they could view and interpret what constitutes an "outdoor education experience": geographical, temporal, and instructional content. Geographical pertains to the characteristics of a space suitable for outdoor instruction. The space may vary in size, proximity, and complexity from a small, manicured garden located just outside the classroom window to a highly primitive, natural space located far from campus. Temporal refers to the duration of any visit to an OLS, regardless of purpose, that may vary from a short trip to the school garden for a specific observation, to an entire class period in a local park, to a full school day at a more remote locality. Lastly, instructional refers to the manner in which an OLS is used as well as the "depth" of instruction. These may vary from the extremes of conducting a "traditional" classroom lecture under the shade of an oak tree to a well-planned, interdisciplinary lesson that looks at the concrete connections of plants, water, and soil in an outdoor space. Some of the teachers were biased toward one or more of these extremes. For example, a few teachers possessed the perspective that "outdoor education" necessitated that lessons be taught in their entirety in primitive outdoor settings located far from the school campus. Others recognized the benefit of OLS for instruction, whether introducing a topic in the physical environment in which it is located, teaching an entire lesson or unit there, or using the space as an alternative classroom. It was anticipated that bringing these continua to the teachers' attention would result in re-evaluation of their preconceived notions of "outdoor education" and the variety of ways in which it can be implemented.

Geographical Continuum

The PD component that most likely had the greatest impact on the teachers was helping them to recognize that outdoor learning does not necessarily involve primitive, native spaces far from their schools. We began the PD by assigning the first journal entry while sitting on the manicured lawns of our university campus (where the first few days of the PD took place). The next day we moved to a more "natural" (but still highly manipulated) space: the local botanic garden, and instructed them to consider and journal about their perceptions of "the garden as nature." Day four of the PD took us to the school district's OLC. By introducing the teachers to increasingly "primitive" outdoor spaces, we helped modify their perspective on what potentially constitutes an OLS, depending on the two other variables on the continuum (available time and instructional goals).

Throughout the PD, we reminded them of available alternatives to the district's OLC that require significantly less time and logistical effort to accomplish these goals. We posit that teacher recognition of something as seemingly insignificant as a patch of grass outside the classroom or an abandoned tennis court can help them address the challenges of *lack of appropriate space*, *time*, *classroom management* (particularly the need for chaperones), *liability* (typically a nonissue if they remain on school property), and *logistics* (such as securing permission and transportation), among others. If teachers who do not use OLS begin by utilizing proximal spaces for outdoor teaching, they will come to realize that many of the perceived challenges are more easily avoided and in time, they can venture to further, more natural spaces as they come to be perceived as more suitable.

Instructional Content Continuum

A similar distinction was made regarding how much of any particular lesson could potentially be taught in OLS. While some lessons, such as identifying plant growth forms to better understand the botanical taxonomy, could easily be transferred to an outdoor setting (and most appropriately should be, in our opinion), others, like cell respiration could be much more challenging. What we attempted to emphasize to the teachers was that, depending on the complexity and abstract nature of the content, differing levels of content depth should be considered when planning what topics to teach outdoors, and what that outdoor activity should be. All that may be necessary for certain content would be a quick trip to the garden to observe and/or document a particular phenomenon that has just been discussed in the classroom. For other content, a full outdoor immersion would be more beneficial. Because many of the teachers (at all grade levels) identified "lack of a specific curriculum" as a significant challenge to using OLS, we incorporated into our PD an assignment that they "transfer" their favorite science activity or lab to the outdoors. The purpose of this exercise was for the teachers to thoughtfully consider exactly what components of their own curricula most reasonably lend themselves to being taught outdoors, and when varying degrees of OLS use might be most fitting.

Temporal Continuum

To further help the teachers realize that their perceived challenges were not insurmountable, we emphasized that we were not advocating that *every* lesson be taught outdoors. Rather, we wanted them to perceive OLS as one of the several tools that can be utilized to teach science content (or nearly any other discipline, for that matter) more effectively. Once many of the originally reticent teachers became aware of the suitability and availability of many outdoor spaces and that certain science content naturally lends itself to outdoor instruction, their intimidation at the thought of using OLS, at least for short periods, began to decrease. This distinction allowed them to recognize the flexibility that OLS can add to content delivery, that just being outdoors has its own benefits, as does developing and teaching an entire lesson around a "natural" phenomenon.

Postassessment: Observed Effects of Professional Development

As mentioned previously, we observed marked distinctions between the two teacher cohorts throughout the PD experience. The elementary teachers generally conveyed a positive outlook on OLS use and reported utilizing outdoor spaces in their teaching prior to the PD, whereas the secondary teachers were generally less enthusiastic about the prospect of teaching outdoors and few reported ever using outdoor spaces for teaching prior to the PD. We attributed this observation to the realization that the challenges they reported reflect significantly different perceptions and beliefs about the use of OLS. The elementary teachers viewed their challenges more as inconveniences to be tackled, whereas the secondary teachers viewed them as obstacles preventing them from using outdoor spaces to teach.

At the conclusion of the second week of the PD, we administered postassessments, comprising written reflection and small group discussion (three to five teachers per group) on the teachers' PD experiences; the latter were audio-recorded. The groups were asked to discuss the challenges that the PD had helped them overcome and in what areas they felt they needed additional help or guidance. Because both cohorts reported all the challenges identified during the preassessment, all were addressed in the postassessment as well. These are shown in Table 2.

Table 2 Focus groups that identified solutions to challenges in OLS use

Challenge/Obstacle	Elementary focus groups	Secondary focus groups
Time	All	None
Weather	All	All
Management	All	Some
Liability, safety	All	Some
Expense	All	Some
Logistics – chaperones, transportation, permission	All	None
Administrative disapproval/State Science Standards	All	None
Lacking PCK	All	None
No curriculum; lack of/difficulty with TEKS alignment	All	None
Location (proximity, too "primitive")	All	None
Other teachers do not support my efforts[a]	All	All
Individual student challenges[a]	All	Not discussed

All = all focus groups within the cohort identified solutions; Some = some, but not all, focus groups within the cohort identified solutions; None = none of the focus groups identified solutions

[a]New challenge identified during postassessment focus group discussion

For the most part, the elementary teacher groups were able to perceive challenges as opportunities for improvement by the end of the PD. They also mentioned, and arrived at, possible ways to overcome issues that had not been previously reported: students with learning disabilities, physical disabilities, or English language deficiencies, and possible lack of support from fellow teachers.

We did not observe similar positive discussion within the secondary groups. Overall, they no longer perceived weather as a significant issue. They also expressed concern about the potential lack of support from their colleagues who participate with them in grade-level or cross-discipline teams. However, they acknowledged that significant effort is required for these collaborative teams to function effectively, regardless of subject or location. Some groups *were* able to identify possible solutions for classroom management, liability, and expense, but despite their PD experience, time, logistics (specifically transportation and lack of possible OLS near campus), administrative support, and lack of PCK and/or a specific "outdoor curriculum" continued to be expressed as troublesome or challenging for some of these teachers. These issues were further addressed during the academic year component of this PD through the instructors' continued modeling of the outdoors as an instructional resource, and continued collaboration within the learning community, whether informal (i.e., through e-mail communication) or formal (i.e., one of the teachers' "assignments" was to develop an outdoor lesson and teach it to us during a Saturday meeting).

Discussion

Reflection on our experiences and observations has informed us of differences in how teachers of varying grade levels can use OLS for teaching science. It further helped us understand what they perceive outdoor learning to be. These differences are rooted primarily in their past experience with learning and teaching outdoors, the content they teach, and how they respond to the pressure of high-stakes testing. The teachers who participated in this PD demonstrated varying levels of success in overcoming perceived and actual challenges.

Prior Learning and Teaching Experience Outdoors

Many of the elementary teachers were already well-versed in using OLS as part of their teaching practice before the PD experience; reporting frequent use of OLS throughout the school year in varying “intensities”; from a quick trip outside to see a bird’s nest to a comprehensive lesson to study habitat and observe/collect organisms. Because of this “on-the-job” experience, many already possessed the means to overcome challenges they faced when incorporating OLS in their instruction.

The majority of the secondary teachers lacked similar experiences. As a matter of fact, our intervention efforts may have increased their sense of futility by allowing them to expand their list of perceived challenges. It has been argued (Barrett 2007) that traditional Western education (of which most, if not all, of these teachers are a product) is antithetical to EE in general and *outdoor education* in particular. Such experiences are generally limited to recreational activities during the elementary years, and limited instruction as a component of an advanced elective in the natural sciences in secondary years. It is within this culture that the teachers in our PD now practice their craft and are being asked to modify it to include elements that are relatively foreign to them. Since many reported that they lacked the experience of being a student in outdoor environments themselves, it came to us as a little surprise that they have not included outdoor instruction in their own teaching.

Content Differences

Another explanation for the observed discrepancy between the cohorts’ use of OLS and their perception of challenges in using them pertains to the differences in what they teach. Elementary science content is relatively straightforward, concrete, and descriptive (i.e., observation of characteristics, and changes in these over time), can often be presented in terms of binary opposites (i.e. living/nonliving, rough/smooth, shiny/dull), and readily lends itself to direct, physical observation in natural environments. For example, a lesson on plants can be easily adapted to include observations in a school garden, landscaped areas on campus, or a nearby park. On the other hand, secondary science content, particularly that of a theoretical or abstract nature, is not as readily adapted (i.e., properties of elements, cell structure and function, photosynthesis and cellular respiration). In fact, further scrutiny into our data revealed that the commonly referenced challenge of "lack of curriculum" had different meanings between the two teacher cohorts. For the most part, the elementary teachers hoped to receive from us general "guidelines" to help them implement more interactive and educationally meaningful ways to teach outdoors, to improve on what many of them already practiced. On the other hand, the secondary teachers were seeking a set of "instructions" on how to even approach teaching their subject matter outdoors. Some of these teachers could envision teaching outdoors only by taking advantage of the space's physicality (i.e., role-playing how chemical bonds between salt ions, played by some students, can be separated by water molecules, played by other students) rather than its naturalistic components (i.e., study of a topic at the "microlevel" through observation of its macromanifestations in nature). While this type of outdoor teaching does not fulfill the primary objectives of our PD, it does provide teachers with an idea of the range of possible opportunities for taking their students outdoors and is, in the authors' opinion, better than being taught exclusively in an indoor classroom. At the very least, it is one step in the direction we hope to see our teachers move in their use of outdoor spaces for instruction.

State Science Standards

One of the most notable observations during our experience is the fear of administrative accountability that secondary teachers face when asked to employ a new teaching strategy such as teaching in an OLS. Consistently, throughout our assessments, we found that the secondary teachers cited administrative disapproval as a major challenge to using OLS to teach. We found this strange considering that the high school included in our study was emphasizing EE (in all subject areas) and that the administration strongly encouraged our participating teachers to attend our PD for the very purpose of enhancing their understanding of environmental issues and outdoor teaching. With more detailed analysis of the data, we realized that each time the secondary teachers referenced "administrative disapproval," they would connect this disapproval to a lack of alignment with State Science Standards on which their students are tested. Most of the teachers who are participants in the PD, regardless of grade level, referenced the Curriculum Frameworks (which outline the standards to be taught on any given day) as mutually exclusive to using outdoor spaces for their teaching. Even if they became skilled at translating their content to outdoor teaching, they still feared they would "fall behind" on the scope and sequence mandated by their administration and their students would not perform well on standardized tests, which would subsequently reflect poorly on them. Through our PD, these teacher-participants witnessed intensive, inquiry-based, EE taught in outdoor learning spaces and realized that such instruction requires a willingness to devote significant advance planning to successfully deliver meaningful content to their students in these settings. During the PD, they also witnessed, and realized, that successful outdoor educators must be flexible in their actions based on environmental variables, both abiotic (i.e., moving our activities indoors during the warmest part of the day to minimize health risks) and biotic (i.e., no tracks at the scent trap – how do we provide possible explanations with no data?). Teachers felt that they lacked this flexibility due to the rigid constraints of the Curriculum Frameworks. This lack of pedagogical flexibility, coupled with an inordinate focus on traditional instruction as the only way to successfully prepare students for standardized tests, results in teachers' perception of outdoor instruction as too challenging for all but the most die-hard outdoor enthusiasts.

Real Obstacles Overcome (A Brighter Note)

Our data may give the impression that the success of our PD efforts with the secondary teachers were minimal when compared to the elementary teachers. Nevertheless, we would be remiss not to mention what we consider real successes with this group. First, their stated challenges in using OLS – no matter how legitimate – have been brought to the fore, where they could be addressed directly through the remainder of the academic year. Second, most of the high-school teachers had not worked together since being hired to work at this PEAK school, but developed a genuine sense of community during the field immersion component of the PD experience. The teacher "teams" that coalesced had already begun discussing the incorporation of certain outdoor experiences into their science curriculum. We look forward to working with them further and watching their ideas become reality. In addition, we can infer improvements in efficacy, at least anecdotally. Several of the teachers, including a few in the secondary cohort, shared with us their initial apprehension about being "in the outdoors" or "fearful of insects and spiders," but later reported that their experiences were "exciting" and "reminded them of being a kid." One of these teachers, who admitted to being entomophobic, was so thrilled at the prospect of scorpion-hunting that she strapped a black light to her hat and hunted well into the night, despite exhibiting flu symptoms.

How These Observations Can Inform Other PD Providers

We have learned through this experience that development of PD aimed at encouraging teachers' use of outdoor learning spaces for instruction must first recognize qualitative differences between participants at the elementary and secondary levels. These are based primarily on the content and on the expectations placed on them by their administration due to *No Child Left Behind* and State/National Standards. We cannot approach outdoor teaching merely as a function of a teacher's knowledge, skills, and motivation, but rather in consideration of the potentially dramatic influence that these factors may have on a teacher's feeling of self-efficacy in utilizing outdoor strategies as one component of their teaching practice.

We further realize that many teachers in the U.S. feel compelled to explicitly follow the guidelines set forth in the National Standards due to the high-stakes nature of state-wide assessments and repercussions if their students fail to reach achievement goals. As a result, teachers are not compelled to explore beyond rigid curricular boundaries set forth in district-developed scopes and sequences. Without placing importance on student interactions with natural and outdoor environments, the State/National Standards will continue to restrict teacher ability to incorporate this element into their teaching practice. An omission on the part of these standards, then, is a mandate of exclusion.

We believe that meaningful experiences in natural environments is essential for development of a deep understanding of nature which, in turn, is crucial for its appreciation and the motivation necessary to affect positive change toward environmental issues. With this in mind, we promote the type of PD described herein as one way of increasing the use of outdoor learning spaces by K-12 science educators and suggest that perhaps the national and state standards should consider incorporation of these experiences into the K-12 curricula.

References

American Association for the Advancement of Science. (1993). *Benchmarks for science literacy*. New York: Oxford Press.

Barrett, M. J. (2007). Education for the environment: Action competence, becoming, and story. *Environmental Education Research, 12*(3), 503.

Center for Public Education. (2008). Time out: Is recess in danger? Retrieved September 30, 2008, from http://www.centerforpubliceducation.org/site/c.kjJXJ5MPIwE/b.1427855/k.FAA3/Welcome_to_the_Center_for_Public_Education.htm

Connell, S. (1999). If it doesn't directly affect you, you don't think about it: A qualitative study of young people's environmental attitudes in two Australian cities. *Environmental Education Research, 5*(1), 95.

Corcoran, P. B. (1999). Formative influences in the lives of environmental educators in the United States. *Environmental Education Research, 5*(2), 207.

Hicks, D., & Bord, A. (2001). Learning about global issues: Why most educators only make things worse. *Environmental Education Research, 5*(4), 353.

Littledyke, M. (2008). Science education for environmental awareness: Approaches to integrating cognitive and affective domains. *Environmental Education Research, 14*(1), 1.

Loughland, T., Reid, A., & Petocz, P. (2002). Young people's conceptions of environment: A phenomenographic analysis. *Environmental Education Research, 9*(1), 3.

Loughland, T., Reid, A., Walker, K., & Petocz, P. (2003). Factors influencing young people's conception of environment. *Environmental Education Research, 9*(1), 3.

Louv, R. (2003). *Last child in the woods: Saving our children from nature-deficit disorder*. Chapel Hill, NC: Algonquin Books. 334 p.

Martin, B. E., & Brouwer, W. (1991). The sharing of personal science and the narrative element in science education. *Science & Education, 75*(6), 707–722.

National Research Council. (1996). *National science education standards*. Washington, DC: National Academy Press.

North American Association for Environmental Education. (2004). *Environmental education materials: Guidelines for excellence*. North American Association for Environmental Education. http://www.naaee.org/publications/guidelines-for-excellence

Sobel, D. (2004). *Place-based education: Connecting classrooms & communities*. Great Barrington, MA: The Orion Society.

Vygotsky, L. S. (1978). *Mind in society*. Cambridge, MA: Harvard University Press.

Wenger, E., McDermott, R., & Snyder, W. M. (2002). *A guide to managing knowledge: Cultivating communities of practice*. Boston: Harvard Business School.

Integrating Environmental Education Field Trip Pedagogy into Science Teacher Preparation

Bryan M. Rebar and Larry G. Enochs

"Go my Sons, buy stout shoes, climb the mountains, search the valleys, the deserts, the sea shores, and the deep recesses of the earth. ... for in this way and no other will you arrive at a knowledge of nature and the properties of things"

P. Severinus (1571)

Few teacher education programs prepare preservice teachers to lead effective and meaningful field trips (Griffin 2007). Yet, there is substantial research on the preparation and delivery of field trips that may be used to enhance environmental learning and awareness in ways not replicable in the secondary school classroom. Rather than acting as an additional expectation, incorporating field trip pedagogy into preservice programs provides a means to accomplish most, if not all, of the existing goals common among exemplary programs while simultaneously enhancing the preparation of teachers. For example, field trips may be used to address science content standards and, when infused into a preservice program, provide an opportunity for preservice teachers to lead themed and inquiry-based lessons. Carefully framed, such a field trip focus in preservice programs might include all the characteristics of excellent science teacher preparation programs as outlined by the National Science Teachers Association (2004). This idea to integrate environmental education (EE) into science methods courses has been suggested previously by Heimlich et al. (2004) as a means to overcome the barriers posed by the many requirements that must be addressed in preservice teacher preparation. Specifically, we suggest introducing field trip pedagogy as a means to support EE.

B.M. Rebar(✉)
Oregon State University, 2046 Cordley Hall, Corvallis, OR 97331, USA
e-mail: rebarb@onid.orst.edu

L.G. Enochs
Science and Mathematics Education Department, Oregon State University, 237 Weniger Hall, Corvallis, OR 97331, USA
e-mail: enochsl@onid.orst.edu

A.M. Bodzin et al. (eds.), *The Inclusion of Environmental Education in Science Teacher Education*, DOI 10.1007/978-90-481-9222-9_8,

While recognizing that EE is sometimes treated as a methodology and at other times treated as something to be taught (Swan 1975), we assert that preparing teachers with the strategies to facilitate and optimize their students' personal experiences is essential to ensuring that quality education about, in, and for the environment takes place in formal education. Field trips, we believe, provide the ideal shared student experience for teaching and learning with our environments as advocated by McInnis (1975b). Continuing with her argument for this sort of contextual approach, McInnis (1975a) declares "rather than being one more egg for the overcrowded curriculum basket, environmental education provides a more adequate basket for the existing curriculum" (p. 51). Field trips naturally form a large part of the basket, while teaching strategies are eggs already included in preservice methods courses. By using field trips as a centerpiece for science methods, teaching skills may be enhanced and developed more completely.

In the following chapter, we describe the key teacher strategies for facilitating field trips and suggest how they may be integrated into science teacher preservice programs at the middle- and high-school levels. Including field trip pedagogy in such programs is a significant step in addressing the goals of EE as put forth in the landmark Tblisi Declaration (UNESCO/UNEP 1978). In this chapter, discussion begins with the characteristics that define a field trip. Next, discussion centers on the research evidence, learning theory, and rationale supporting the inclusion of field trips in science education. Subsequently, research-based field trip strategies are introduced. Finally, discussion concludes with ideas regarding how these recommended strategies for leading EE field trips might be infused into a science methods course.

What Is a Field Trip?

To teachers, the term "field trip" often connotes a major undertaking involving extra time and effort. Generally, all indications, such as museum attendance records, suggest that the number of field trips teachers lead declines with age level such that, by the time the students are in high school it is very likely that they will not experience educational field trips. Our extensive experience working both as teachers and with teachers suggests that there may be several factors contributing to this decline: (1) additional challenges in logistics posed by secondary schools' multicourse multi-teacher typical school day; (2) pressures placed on teachers to cover the required curriculum (which, in all but the rarest of cases, does not explicitly or inherently support field trips); (3) students' increased involvement in other conflicting after-school activities such as sports, clubs, and jobs; and (4) the (inappropriate) assumption that field trips are educationally most effective for students at an earlier developmental stage.

We take a cosmopolitan view on field trips that should help assuage teachers who fear or would rather avoid grand endeavors with their students. Put simply, we define field trips as any educational activity that teachers guide or direct in a setting outside the classroom. Given this view, there is reason to believe that every teacher has the minimal resources needed to lead field trips. A field trip might be as close as a short walk to the schoolyard. As such, none of the factors mentioned above pose realistic barriers to planning and realizing field trips. This is an important point to bear in mind as we discuss why field trips are needed, how they may simultaneously enhance EE and science education efforts, and, moreover, why and how they might be supported as a method within preservice science programs.

Why Are Field Trips Necessary?

There are many well-justified reasons for including field trips in the curriculum, and they derive from research of student learning on field trips, learning theories, and the underlying principles behind formal education. Field trips have recently been recommended as a way to teach science and conduct inquiry (National Research Council 1996, 2001). However, the notion that out-of-school sites can enhance education is not new. In 1917, Twiss asserted in a book on science teaching that "in spite of all the difficulties, therefore, it ought, in any school, to be possible to have in every subject some field observation in which a considerable portion of the class can participate" (p. 145). Concurring with Twiss's view, the preeminent educational philosopher John Dewey argued that all genuine education comes through experience (Dewey 1938). Today, field trips of all types are common practice, at least at the primary level. However, in practice, teachers often fail to maximize learning opportunities afforded by exhibits, models, live specimens, natural settings, experts, and other resources not accessible in their classrooms. This failure may be attributed in part to the general absence of proper teacher professional development for such events.

The potential contributions of field trips to students' achievements are well documented. Eshach (2007) summarizes the research literature by concluding that "children enjoy going on scientific field trips. They are aware that they are expected to learn from the trip, and that it should not only be a 'fun day', but rather a day where they enjoyably learn science" (p. 177). In a metastudy reviewing the research on outdoor learning, Rickinson et al. (2004) conclude that "substantial evidence exists to indicate that well-taught and effectively followed up [outdoor lessons] offer learners opportunities to develop their knowledge and skills in ways that add value to their everyday experiences in the classroom" (p. 24). Rickinson et al. further add that "there is substantial research evidence to suggest that outdoor adventure programmes can impact positively on young people's: (1) attitudes, beliefs, and self-perception – examples of outcomes include independence, confidence, self-esteem, locus of control, self-efficacy, personal effectiveness, and coping strategies; (2) interpersonal and social skills – such as social affectiveness, communication skills, group cohesion and teamwork" (p. 32). The rationale for utilizing field trips is supported by Braund and Reiss (2006a) who maintain that when science is introduced in an out-of-school real-world context, it is more "authentic" and may be recognized by students as having more relevance.

From a theoretical perspective on learning, the sociocultural school of thought most closely associated with Vygotsky (1986) draws attention to the importance of social interactions and, specifically, to the significance of peers or teachers enabling students to grasp new and more complex ideas by means of facilitated experiences. Field trips, which, by definition change the setting from a formal to an informal context, are particularly well suited to such interactions because they better allow for social behaviors characteristic of everyday learning experiences outside of school time. As a result, field trips not only provide for valuable social-learning opportunities, they do so in a way that helps students connect their school learning to their everyday life learning. The North American Association for Environmental Education (NAAEE) EE guidelines explicitly support this outcome of connecting learning with the real world as one of the stated essential underpinnings (Simmons et al. 2004).

In accordance, research studies consistently reveal that students show a positive attitude toward all types of field trips (e.g., Falk and Balling 1982; Flexer and Borun 1984; Falk and Dierking 1997; Pace and Tesi 2004). Students' positive attitude toward field trips contrasts with their increasingly negative attitude toward school science as a factor of age (Braund and Reiss 2006a) and therefore leads to the suggestion that students might be engaged in school science when it is purposefully and intricately linked with out-of-school science activities (Braund and Reiss 2006b) such as field trips. Including field trips in the science curriculum may improve students' attitudes toward science because doing so compels teachers to vary their teaching strategies. A variety of strategies has been shown to increase the efficacy of teaching and therefore has been used to argue for the inclusion of more informal science learning experiences, especially field trips (Hofstein and Rosenfeld 1996). Finally, research studies have demonstrated how to build successful field trip models that actively involve students in environmental learning (e.g., see Enochs and Kean 1999; Orion 1993). In these models, field experiences are intentionally linked to school science, again addressing the issue of poor student engagement in science. It is worth noting that in his model, Orion (1993) discusses how field trips provide hands-on experiences that, drawing on a Piagetian view, facilitate the transition from concrete to more abstract levels of cognition. In summary, the rationale for using environmental field trips to support school science is well founded.

What Strategies Do Science Teachers Need to Learn?

Familiar Strategies Applied to Field Trips

Although most research and theory points to the overriding message that field trips have enormous potential to enhance school science, and specifically EE, many studies reveal numerous missed learning opportunities on field trips. For example, when organizing field trips, teachers do not often plan how to monitor the effectiveness of their students' experiences or how to build on these experiences

(Amos and Reiss 2006). Griffin and Symington (1997) have shown that teachers often fail to identify or clearly communicate their instructional goals for the trip to their students and, moreover, that teachers often fail to recognize the extent of their influence on the teaching strategies and the content of field trips. In other words, many teachers do not recognize field trip settings as appropriate environments for planning and facilitating organized lessons centered on specific learning objectives.

In many ways, a number of research supported strategies for enhancing field trip learning parallel strategies identified as the best practices in the classroom. For example, research suggests that teachers should first determine the learning objectives, and then develop appropriate activities for the trip (Rennie and McClafferty 1995). For teacher educators, this means demonstrating how the curriculum can be used to guide field trips rather than showing how the curriculum fit (that is, the trip's relevance to class topics) may be used to justify the experience; in practice, the latter is the norm (Anderson et al. 2006). Often the destination drives the activity and teachers tend to view trips as general enrichment (e.g., see Gottfried 1980; Griffin and Symington 1997). Ideally, planned field trip activities, similar to classroom activities, should align with learning objectives, connect to the curriculum (Finson and Enochs 1987; Guisasola et al. 2005; Wolins et al. 1992), and support science standards (Cox-Petersen et al. 2003) and EE guidelines. Moreover, the same methods of inquiry used to teach classroom science may be applied to out-of-the-classroom settings (National Research Council 2001). Studies of field trips consistently reveal a pattern in which teachers frequently do not approach, frame, and facilitate field trips as an integral part of the curriculum (for example, Anderson et al. 2006). Within the context of preservice programs, simply introducing the use of field trips as learning events that can be utilized to support classroom teaching would begin to address these missed opportunities. When compared with the preparation required for normal classroom lessons, field trips require considerably more time, effort, and expense. Therefore, it is imperative that teachers are provided training based on the use of research-supported teaching strategies for field trips.

Strategies for Out-of-the-Classroom Challenges

Additional, perhaps less familiar, recommended strategies have been identified to help teachers prepare for the common challenges (unique when compared with classroom challenges) posed by the informal settings where field trips take place. These strategies and the associated specialized knowledge for facilitating learning contrast with formal teaching strategies (Cox-Petersen and Pfaffinger 1998; Griffin 1994). Studies consistently support the conclusion that these strategies and this knowledge are not common or instinctual among the majority of teachers who lead field trips. Thus, the unique challenges for teachers that field trips present may be used to organize the research-recommended strategies that should be included in a preservice program. These challenges are (1) students' overstimulation caused by new surroundings on field trips (and the chaos that often results); (2) limited time available to take advantage of unique opportunities, (3) difficulty in creating a suitable learning tool such as a worksheet; (4) unknown nature of new settings that leads to surprises; and (5) preparation and management of additional adult chaperones (see Table 1). The following passages briefly describe these challenges and some of the research-recommended strategies for handling them.

The challenge of students' overstimulation is really a factor of novelty. In other words, when a field trip setting contains too many new stimuli, students are unable to focus sufficiently to engage in meaningful learning. This problem may be addressed in several ways. Foremost, it is essential for teachers to recognize the developmental level (Taylor et al. 1997) and assess the experience of their students to plan an appropriately stimulating trip. The goal should be to introduce moderate amounts of novelty such that students are neither disinterested due to the familiarity of the setting and/or activities nor overstimulated due to the novelty of the experience, but rather optimally engaged and focused (Falk 1983; Falk and Balling 1982). Teachers may reduce novelty by orienting their students to the trip ahead of time by concentrating on three domains: cognitive (students' relevant knowledge level), geographic (students' familiarity with the setting), and psychological (students' "mental readiness for a field trip") (Orion 1993). Another strategy to reduce novelty (thereby enhancing students' ability to learn) is to repeat visits to the same site. Such a strategy often poses logistical and financial challenges, but is certainly possible for schoolyard field trips. A repeat visit strategy is presented by the National Research Council (2001) as a model way to use inquiry methods in which students are guided to conduct an investigation of water quality at a nearby pond over the course of several months. First students become familiar with the site. Then, in the classroom, students work on developing an investigative question and tools needed to conduct their project. Students return regularly to the site to gather data that they eventually compile into a final report. Field trips and classwork complement each other such that students continue to be stimulated, but not overwhelmed or bored, both in the field and in the classroom.

A wealth of opportunities and limited time to explore them leads some teachers to attempt to squeeze every possible experience into a trip by exposing students to as many places, exhibits, people, and/or presentations as possible in a tightly structured schedule. However, research suggests that students will retain more when field trips focus on resources, activities, and content that is closely tied to the curriculum (Finson and Enochs 1987; Guisasola et al. 2005; Wolins et al. 1992), fewer new items are introduced (Barnard et al. 1980), and students are allowed time to explore in small independent groups (Cox-Petersen and Pfaffinger 1998). The goal on field trips should be to take advantage of the unique resources not available in the classroom. Therefore, activities and tasks that can be completed in the classroom should not be imposed on students while in the field. One way to extend a field experience is to use web resources, particularly when the field trip site supports its own web site (Cox-Petersen and Melber 2001). Notably, this same recommendation to use a destination's web site with students is also a suitable strategy to reduce novelty when used as an advance (pretrip) organizer.

Table 1 Common field trip challenges that require the use of strategies as recommended in the research

Challenge	Recommended strategies
Chaos/Overstimulation	• Use previsit lessons specifically related to the site's topics. • Plan trips that introduce moderate novelty; use pretrip orientation to reduce the novelty of new settings. • Prepare for novelty: cognitively, geographically, psychologically.
Limited time	• Link the content to the curriculum to improve students' retention. • Use the site's web site to plan logistics and extend lessons in the classroom. • Incorporate science standards in lesson planning. • Limit the stimuli, such as number of exhibits visited, to improve learning. • Allow time for small group exploration.
Teaching tools (such as tasks, worksheets, or prompts)	• Consider students' input, interests, and abilities in planning your trip. • Give students choice in exploring. • Allow for some less structured time. • If you use worksheets, emphasize concepts rather than a broad survey of the content, and preference questions that prompt students to interact with resources and allow some degree of choice in response. • Encourage social interactions, even while using worksheets.
Surprises	• Determine the trip's purpose first, then plan the setting. • If you choose a museum destination, consider how it supports your agenda. • Visit the field trip site ahead of time and coordinate with staff on safety, logistics, expectations, and learning.
Chaperones	• Recognize and support multiple roles of chaperones and encourage chaperones to use new approaches to facilitate learning. • Encourage chaperones to promote conversations among students (because most of students' talk in learning settings is learning talk), and ask questions that require students to explore the available resources. • Encourage chaperones to interact with students in a family-like way in small groups. • Consider providing chaperones with a list of questions and a bag of props they can use to draw students' attention and inquiry. • Model interest in exhibits. • Prepare chaperones with an understanding of students' current ideas, thinking, values, and learning needs.

In keeping with the notion that field trips should highlight unique resources, it is vital that teachers choose or develop appropriate strategies for engaging and focusing students. Too often on field trips, teachers impose formal classroom structures that do not suit or optimally take advantage of the setting (Griffin and Symington 1997). While acknowledging that there are a variety of teacher motivations for leading a field trip – several categories of motivations have been identified (see Kisiel 2005) – there are, nonetheless, certain guidelines that should inform the teacher's selection of teaching strategies in all cases:

1. Research clearly points to the benefits of valuing students' interests and choices with respect to where the trip takes place or to which exhibit or organism an individual focuses on in order to answer a question or complete a project (Gilbert and Priest 1997; Kisiel 2003; Mullins 1998; Orion and Hofstein 1991).
2. Social interactions should be encouraged especially during assigned tasks or activities (Hofstein and Rosenfeld 1996; Watson et al. 2002).
3. Prompts or questions, be they verbal, written, or otherwise, should target responses that promote conceptual learning and require interaction with unique resources (Kisiel 2003, 2007).

Providing choices allows students to draw from their intrinsic motivation to make discoveries and learn. Promoting social behaviors allows for many varied forms of learning, including peer-to-peer learning talk, sharing activities, cooperative manipulations, observing others engage in learning activities, peer-to-peer teaching, and creative play (Watson et al. 2002). Assignments and tasks that require students to read considerable text, for example, rather than focusing on observations and interactions with their surroundings on a field trip fail to offer a truly unique experience that cannot be replicated in the classroom using books or other texts. Therefore, teachers' use of prompts that require student interactions with each field trip site's unique resources are key to capitalizing on the learning potential presented by out-of-school settings.

With respect to limiting and mitigating unwanted surprises on field trips, the research points to several effective strategies. The destination should reflect the preidentified purpose or learning goals for the trip rather than choosing goals to suit a predetermined setting (Rennie and McClafferty 1995). This is a subtle but consequential point. Many schools continue traditional field trips long after the original guiding purpose and relevant curriculum ceases to be in place. Such cases are not ideal for optimizing learning opportunities and can lead to unintended situations (often because teachers do not take full ownership of the trip). Once a clear purpose and learning goals have been identified, the next step is to consider the agenda of potential destinations (Kisiel 2005) should the site have one. The site's agenda should support the teacher's agenda; if it does not, students may be introduced to unrelated content at best or, at worst, students may be subjected to inappropriate propaganda. A site's agenda may be considered on several levels. It is worth learning about the site's mission as well as their methods and programming. Organizations that are likely to have their own agendas, such as museums, interpretive programs, and managed natural areas, tend to have easily accessible online mission statements, thereby allowing teachers to screen them for significant conflicts. But even when there is agreement, on-site educators may favor lectures rather than interactive activities and, moreover, may highlight specific topics or concepts that do not support the teacher's goals. Thus, the importance of communication and collaboration between teachers and other educators who may be involved is indispensable (Tal et al. 2005). Ideally, teachers should visit field trip sites ahead of time to avoid these undesired surprises. Previsits allow teachers to plan and coordinate for safety, logistics, expectations, and learning activities (Anderson et al. 2006; Cox-Petersen and Melber 2001; Martin and Seevers 2003). In practice, much of this planning may take place using all other tools available: web sites, brochures, email, and phone conversations with the site's staff, etc.

The informal context of field trips often requires the assistance of additional adults. Managing and preparing these chaperones is a new task that few teachers perform in their classrooms. Consequently, teachers may not recognize the extent to which they should prepare chaperones in order to best ensure learning while on the trip. Without such guidance, chaperones may struggle with their role. Therefore, it is as equally important that the chaperones understand the goals for the trip as the students. Moreover, the teacher needs to communicate an understanding of students' current ideas, thinking, values, and learning needs with chaperones (Schauble et al. 2002). Teachers can make use of chaperones' individual skills by encouraging them to facilitate learning in their own ways (Sedzielarz 2003). By organizing students into small groups with chaperone leaders, teachers can promote family-type interactions among students and adults, thereby encouraging informal learning conversations. Such an approach has been suggested as ideal for promoting learning in informal settings (Griffin and Symington 1997; Parsons and Muhs 1994). Chaperones should be prepared to ask questions that require students to explore their surroundings (Watson et al. 2002); this may be accomplished by providing chaperones with a list of questions or a bag of prompts (Cox-Petersen et al. 2003). Finally, teachers must encourage chaperones to model the sort of interactions they expect students assume (Griffin and Symington 1997).

Assessment

Research-based recommendations for assessment of field trips are limited. However, given that learning in field trip settings differs from usual classroom learning, it follows that appropriate assessment would, similarly, differ. As discussed earlier, research on the use of worksheets indicates that text-based, fact-focused assignments do not suit informal settings (Kisiel 2003, 2007). Because interactions are the ideal goal of field trips, Parsons (1999) suggests evaluating the learning process as well as the product; this might be accomplished by using conversations as evidence of learning. Depending on the purpose of the field trip, many varied outcomes might be considered for assessment, and many of these may be gathered after the trip has concluded and students have returned to the classroom.

Clearly, assessments should not interfere with students' opportunities to interact with their surroundings while on a trip; if anything, in-trip assessments should be designed to enhance these interactions. One such method that can achieve this goal is the requirement that students create a photo journal documenting their observations (perhaps including specific expectations). If carefully planned, such an assignment has the potential to focus students' attention on details of interest and relevance to the trip's purpose and, moreover, might be employed to facilitate connections once students return to the classroom. An additional application of principles discussed earlier would be the use of group assessments rather than individual evaluations. Since field trip learning should be social, it follows that assessments of this learning would, most appropriately, involve peers working together. Therefore, group projects, presentations, or reports, for example, seem to align well with this type of learning experience. Open-ended, alternative measures of learning such as free-writing, drawings, and paired interviews in response to simple prompts have been demonstrated as effective means to capture students' conceptual growth resulting from outdoor field trips (Rebar 2008). In summary, as with all assessment, measures should reinforce and reflect the nature of the learning expected.

Integrating Field Trips into Preservice Programs

Why Science Methods?

Colleges and universities need to make field trip planning and methods an integral part of their preservice programs. Introducing field trip pedagogy in science methods courses should result in both more and better field trips led by science teachers. This claim is supported in part by research revealing that teachers cited a lack of preservice preparation for planning, enacting, and evaluating student learning in the field as major reasons for not taking field trips (Mason 1980). Additional benefits of including field trip training in science methods courses may be drawn from the results of a study in which 715 institutions were surveyed regarding preservice EE. In her study, McKeown-Ice (2000) found that when EE was included in teacher preparation, it was primarily included in science education, although institutions generally rated their delivery of EE instruction methods as only adequate (32%) to poor (33%). In other words, institutions that have integrated EE to some degree have found that its best fit is in science education and they recognize the need to improve these programs. Integrating EE field trip pedagogy in science methods would accomplish this. McKeown-Ice also found that most students specializing in EE (and not necessarily preparing for classroom certification) received their teaching methods in science methods courses. It is quite likely that many of these specialized EE students will pursue careers in which they are involved in coordinating and facilitating field trips in roles other than that of the classroom teacher. McKeown-Ice concluded that even those teacher education programs including EE were not adequately preparing preservice teachers to effectively teach about the environment (including the use of field trips). Clearly, preservice programs in science education are well positioned to fill this niche of better serving environmental educators while simultaneously better preparing all preservice science teachers for EE with the inclusion of field trip pedagogy as an integral part of their curriculum because, as discussed earlier, field trips are ideal for introducing EE and field trip leading skills are essential to leading learning-focused trips.

Representative Science Methods Course Objectives

The overall goal of a science methods course should be to increase student competence and confidence in teaching science. The objectives of the course should be designed to develop students' knowledge of science teaching and learning. Such objectives of a representative science methods course provide numerous opportunities to introduce research-based field trip pedagogy (see Table 2). When considering ways to include more field trip connections in science methods courses, we encourage teacher educators to begin by examining their science methods course objectives and aligning field trip practices as appropriate. Table 2 illustrates this

Table 2 Example science methods course objectives and corresponding field trip connections

Representative science methods course objectives	Field trip connections
Discuss the research relevant to science teaching and its significance to preservice science teachers.	Provide field trip related articles as discussion topics. Reflect on theories, approaches, strategies that support field trip pedagogy.
Plan science curriculum including the development of lesson plans that include instructional objectives/outcomes.	Develop lesson and unit plans that include pre-, in-, and posttrip lessons. The site should be selected based on the identified objectives and outcomes.
Plan science curriculum including the development of appropriate assessments and assessment procedures for diverse learners.	Develop pretrip activities that introduce concepts to be addressed and orient students to the trip site. Develop authentic assessments for pre-, in-, and posttrip lessons that promote students' social and environmental interactions.
Construct and utilize plans for lessons involving science inquiry skills and/or generating science content through use of inquiry skills.	Design lessons that encourage students to ask testable questions, make observations and collect data in the field, draw conclusions, and present results.
Develop more favorable attitudes toward the teaching of science in addition to interests in science, which may carry over to future personal leisure time activities.	Engage preservice teachers by modeling interactive strategies that support field trips. Encourage preservice teachers to develop creative unit plans that include field trips.

using representative objectives typical of a science methods course. Subsequent steps of infusing field trip pedagogy should flow from each course's objectives. Naturally, the ways in which field trip pedagogy may best be introduced within a preservice program will depend on each program's existing structure. Regardless of the format, we hold that many opportunities exist for the seamless infusion of field trip pedagogy within these programs when examined closely. Including field trips in the preservice curriculum, we believe, will enrich the existing course preparations.

How Can Field Trip Preparation Be Integrated into a Science Methods Course?

The purpose of science methods courses is to prepare teachers to create learning environments for their students. Thus, preparing teachers to lead field trips centers on this same goal that has always guided science preservice programs. Preparing teachers for field trips requires many broad skills and strategies that are already integral to science methods courses. For example, from a broader perspective, approaches such as inquiry, project-based learning, and varying strategies in order to appeal to multiple learning styles are common inclusions in methods courses. Field trips and the identified strategies for optimizing learning discussed above are well suited to support each of these approaches. What better way to promote these approaches than to model their use within a science methods course by means of sample lessons from a unit including one or more field trips? One of the challenges is that organizing and leading out-of-the-classroom activities also requires additional skills, both managerial and pedagogical. How can these be infused into a methods course that already is limited by time and the topics that must be covered? Table 3 provides a representative model for how a science methods course might be restructured with the integration of field trips using a schoolyard garden trip as an example within a unit on botany and soil science.

Many science methods courses include a culminating assignment in which students are expected to design a unit plan including sample lessons. Thus, a logical extension of the outlined activities in Table 3 would be a unit plan assignment that requires preservice teachers to develop their own unit including one or more field trips. Sample lesson plans might be required for pre-, in-, and posttrip portions of the unit. Preservice teachers might be further expected to present a teaching demonstration that is taken from one of their component lesson plans. Within their teaching demonstrations, regardless of whether they choose a pre-, in-, or posttrip lesson, they must draw on some of the field trip strategies introduced in the course. Using such a course structure, science methods instructors may accomplish each of their course objectives while also preparing preservice teachers to introduce EE by means of field trips.

Table 3 Representative outline for a science methods course modeling a unit plan that includes field trip pedagogy

Sample unit plan: Soil science and field trip to the schoolyard garden	
Pretrip	1. *Model* a pretrip lesson to a botanic garden or vegetable garden (for example, an investigation of soil and how it affects plant growth). Include pretrip preparation by designating groups that work together to do a class activity and develop their own testable question. 2. *Model pretrip logistics preparation* by orienting students to the field trip destination by using the site's web site. Share a letter of invitation to parents including expectations (the time and in-trip responsibilities including teaching/group facilitation duties). 3. *Brainstorm* with preservice teachers other lessons they could do as a pretrip lesson as part of the unit. 4. *Assign* preservice teachers to develop one of the lessons that were brainstormed as a pretrip lesson.
In-trip	1. *Role-play* a field trip to the campus garden. Give "students" data tables with prompts to record observations. "Chaperones" help lead small "student" groups using cue cards with additional questions and a bag with tools for conducting the inquiry. "Students" investigate their question by planting seeds under appropriate conditions to address their group's question (if the site will be revisited), or collecting soils from various locations for further investigation in the classroom/laboratory. 2. *Reflect* on the field trip as a class (stepping out of assumed roles). Consider what would you do differently as a teacher? What did you notice about being a chaperone or student?
Posttrip	1. *Model* a posttrip lesson plan in which students analyze their data and prepare poster presentations as a form of authentic assessment. 2. *Reflect* as a class on pre-, in-, and posttrip lessons and the overall unit. 3. *Assignment*: (1) Preservice teachers design a posttrip lesson plan that fits with the overall unit. (2) Preservice teachers write a reflection on the overall experience and how they might use this type (or another) field trip.

Summary

Field trips are commonly used educational methods for teaching about the environment, yet teachers receive minimal, if any, preservice preparation for planning and leading trips. Not surprisingly, important missed learning opportunities on field trips are well documented in the research as a consequence. This line of research has also led to a wealth of recommended practices that would enhance field trips. In the preceding discussion, we have summarized these key recommendations and made the case that their inclusion in preservice courses, particularly science methods courses, would naturally integrate into the existing framework of such courses. Field trips provide an opportunity to apply many of the methods introduced in teacher training programs. However, in practice, teachers often struggle to take full advantage of field trips because although recommended field trip strategies share similarities with classroom strategies and they both support broader approaches such as inquiry, the informal context of outside-the-classroom settings requires different strategies and knowledge. By preparing aspiring teachers with these specialized field trip strategies and knowledge in their initial teacher training coursework, we believe that the quality of both university and secondary instruction will see improvements.

References

Amos, R., & Reiss, M. (2006). What contribution can residential field courses make to the education of 11-14 year-olds? *School Science Review, 87*(321), 1–8.

Anderson, D., Kisiel, J., & Storksdieck, M. (2006). School field trip visits: Understanding the teacher's world through the lens of three international studies. *Curator, 49*(3), 365–389.

Barnard, W. A., Loomis, R. J., & Cross, H. A. (1980). Assessment of visual recall and recognition learning in a museum environment. *Bulletin of the Psychonomic Society, 16*(4), 311–313.

Braund, M., & Reiss, M. (2006a). Towards a more authentic science curriculum: The contribution of out-of-school learning. *International Journal of Science Education, 28*(12), 1373–1388.

Braund, M., & Reiss, M. (2006b). Validity and worth in the science curriculum: Learning school science outside the laboratory. *The Curriculum Journal, 17*(3), 213–228.

Cox-Petersen, A. M., Marsh, D. D., Kisiel, J., & Melber, L. M. (2003). Investigation of guided school tours, student learning, and science reform: Recommendations at a museum of natural history. *Journal of Research in Science Teaching, 40*(2), 200–218.

Cox-Petersen, A. M., & Melber, L. M. (2001). Using technology to prepare and extend field trips. *Clearing House, 75*(1), 18–20.

Cox-Petersen, A. M., & Pfaffinger, J. A. (1998). Teacher preparation and teacher-student interactions at a discovery center of natural history. *Journal of Elementary Science Education, 10*(2), 20–35.

Dewey, J. (1938). *Experience and education*. New York: Kappa Delta Pi.

Enochs, L., & Kean, W. (1999). *Field geology for elementary teachers: A teacher enhancement study*. Paper presented at the Association for Education of Teachers of Science Annual Meeting, Austin, TX.

Eshach, H. (2007). Bridging in-school and out-of-school learning: Formal, non-formal, and informal learning. *Journal of Science Education and Technology, 16*(2), 171–190.

Falk, J. H. (1983). Field trips: A look at environmental effects on learning. *Journal of Biological Education, 17*(2), 137–142.

Falk, J. H., & Balling, J. D. (1982). The field trip milieu: Learning and behavior as a function of contextual events. *Journal of Educational Research, 76*(1), 22–28.

Falk, J. H., & Dierking, L. D. (1997). School field trips: Assessing their long-term impact. *Curator, 40*(3), 211–217.

Finson, K. D., & Enochs, L. G. (1987). Students' attitudes toward science-technology-society resulting form a visit to a science-technology museum. *Journal of Research in Science Teaching, 42*(7), 593–609.

Flexer, B. K., & Borun, M. (1984). The impact of a class visit to a participatory science museum exhibit and a classroom science lesson. *Journal of Research in Science Teaching, 21*(9), 863–873.

Gilbert, J., & Priest, M. (1997). Models and discourse: A primary school science class visit to a museum. *Science Education, 81*(6), 749–762.

Gottfried, J. (1980). Do children learn on school field trips? *Curator, 23*(3), 165–174.

Griffin, J. (1994). Learning to learn in informal settings. *Research in Science Education, 24*, 121–128.

Griffin, J. (2007). Students, teachers, and museums: Toward an intertwined learning circle. In J. H. Falk, L. D. Dierking, & S. Foutz (Eds.), *In principle, in practice* (pp. 31–42). Lanham, MD: Altamira.

Griffin, J., & Symington, D. (1997). Moving from task-oriented to learning-oriented strategies on school excursions to museums. *Science Education, 81*(6), 763–779.

Guisasola, J., Morentin, M., & Zuza, K. (2005). School visits to science museums and learning sciences: A complex relationship. *Physics Education, 40*(6), 544–549.

Heimlich, J. E., McKeown-Ice, R., Braus, J., Barringer-Smith, L., & Olivolo, B. (2004). Environmental education and preservice teacher preparation: A national study. *Journal of Environmental Education, 35*(2), 17–21.

Hofstein, A., & Rosenfeld, S. (1996). Bridging the gap between formal and informal science learning. *Studies in Science Education, 28*, 87–112.

Kisiel, J. (2003). Teachers, museums and worksheets: A closer look at a learning experience. *Journal of Science Teacher Education, 14*(1), 3–21.

Kisiel, J. (2005). Understanding elementary teacher motivations for science fieldtrips. *Science Education, 89*(6), 936–955.

Kisiel, J. (2007). Examining teacher choices for science museum worksheets. *Journal of Science Teacher Education, 18*, 29–43.

Martin, S. S., & Seevers, R. L. (2003). A field trip planning guide for early childhood classes. *Preventing School Failure, 47*(3), 177–180.

Mason, J. L. (1980). Field work in earth science classes. *School Science and Mathematics, 80*(4), 317–322.

McInnis, N. (1975a). Opportunities for change. In N. McInnis & D. Albrecht (Eds.), *What makes education environmental?* (pp. 49–51). Louisville, KY: Data Courier.

McInnis, N. (1975b). What makes environment educational? In N. McInnis & D. Albrecht (Eds.), *What makes education environmental?* (pp. 21–29). Louisville, KY: Data Courier.

McKeown, R. (2000). Environmental education in the United States: A survey of preservice teacher education programs. *Journal of Environmental Education, 32*(1), 4–11.

Mullins, J. A. (1998). *How field trips in natural areas associated with museums, arboreta, and aquaria impact the educational experiences of teachers and students.* Unpublished dissertation, University of Southern Mississippi, Hattiesburg, MS.

National Research Council. (1996). *National science education standards.* Washington, DC: National Academy Press.

National Research Council. (2001). *Inquiry and the National Science Education Standards: A guide for teaching and learning.* Washington, DC: National Academy Press.

National Science Teachers Association. (2004). *NSTA position statement: Science teacher preparation.* Retrieved April 14, 2008, from http://www.nsta.org/about/positions/preparation.aspx

Orion, N. (1993). A model for the development and implementation of field trips as an integral part of the science curriculum. *School Science & Mathematics, 93*(6), 325–331.

Orion, N., & Hofstein, A. (1991). The measurement of students' attitudes towards scientific field trips. *Science Education, 75*(5), 513–523.

Pace, S., & Tesi, R. (2004). Adult's perception of field trips taken within grades K-12: Eight case studies in the New York metropolitan area. *Education, 125*(1), 30–40.

Parsons, C. (1999). *Do self-guided school groups learn anything?* Paper presented at the American Zoo and Aquarium Association Annual Conference, Minneapolis, MN.

Parsons, C., & Muhs, K. (1994). Field trips and parent chaperones: A study of self-guided school groups at the Monterey Bay Aquarium. *Visitor Studies: Theory, Research, and Practice, 7*(1), 57–61.

Rebar, B. M. (2008). *Changes in children's conceptions of nature following a residential environmental education experience.* Paper presented at the National Association for Research in Science Teaching International Conference, Baltimore, MD.

Rennie, L., & McClafferty, T. (1995). Using visits to interactive science and technology centers, museums, aquaria, and zoos to promote learning in science. *Journal of Science Teacher Education, 6*(4), 175–185.

Rickinson, M., Dillon, J., Teamey, K., Morris, M., Choi, M. Y., Sanders, D., et al. (2004). *A review of research on outdoor learning.* Shrewsbury, UK: National Foundation for Educational Research and King's College.

Schauble, L., Gleason, M., Lehrer, R., Bartlett, K., Petrosino, A., Allen, A., et al. (2002). Supporting science learning in museums. In G. Leinhardt, K. Crowley, & K. Knutson (Eds.), *Learning conversations in museums*. Mahwah, NJ: Erlbaum.

Sedzielarz, M. (2003). Watching the chaperones: An ethnographic study of adult-child interactions in school field trips. *Journal of Museum Education, 28*(2), 20–24.

Simmons, B., Archie, M., Mann, L., Vymetal-Taylor, M., Berkowitz, A., Bedell, T., et al. (2004). *Excellence in Environmental Education: Guidelines for learning (Pre K-12)*. Washington, DC: North American Association for Environmental Education.

Swan, M. (1975). Forerunners of environmental education. In N. McInnis & D. Albrecht (Eds.), *What makes education environmental?* (pp. 4–20). Louisville, KY: Data Courier.

Tal, R., Bamberger, Y., & Morag, O. (2005). Guided school visits to natural history museum in Israel: Teachers' roles. *Science Education, 89*(6), 920–935.

Taylor, S. I., Morris, V. G., & Cordeau-Young, C. (1997). Field trips in early childhood settings: Expanding the walls of the classroom. *Early Childhood Education Journal, 25*(2), 141–146.

Twiss, G. R. (1917). *A textbook in the principles of science teaching*. New York: MacMillan.

UNESCO/UNEP. (1978). *Declaration of the Tbilisi intergovernmental conference on environmental education*. Paper presented at the UNESCO/UNEP Intergovernmental Conference on Environmental Education, Tbilisi, Georgia.

Vygotsky, L. (1986). *Thought and language*. Cambridge, MA: The MIT Press.

Watson, K., Aubusson, P., Steel, F., & Griffin, J. (2002). A cultural of learning in an informal setting. *Journal of Australian Research in Early Childhood Education, 9*(1), 125–137.

Wolins, I. S., Jensen, N., & Ulzheimer, R. (1992). Children's memories for museum field trips: A qualitative study. *Journal of Museum Education, 17*(2), 17–28.

"Eeew! There's Dew on My Toes": Common Characteristics of Preservice Elementary Teacher Learning in Environmental Education and Instructional Strategies for Science Teacher Educators

J. William Hug

Introduction

> Thirty pairs of eyes watched me closely. Pausing for effect after several introductory announcements to the elementary school science teaching methods class, I announced, "Let's go outside today!" Smiles broke out, voices rose in something like a cheer and their eyes communicated that my pedagogical decision met with their approval on this sunny spring day. We moved out of the university classroom, down the hall and outside. The excited voices continued as they followed me a few minutes later onto a manicured, weedless lawn glistening in the morning sun. I sensed them slowing down and turned to see many eyes filled with concern. Then I heard one flip flopped paused in mid-stride, exclaim, "Eeew! There's dew on my toes." Her body language suggested that this might be as far as she was going to go.
>
> I knew instantly that sound waves had already arrived to the rest of the class. They were assimilating the comment and choosing their complex social group behavior reaction. Would they all refuse to get their toes wet? Confronting me a few feet into the grass loomed a significant reflection-in-action (Schön1983) moment and a pedagogical decision. Do I coax them onto the dewy grass or not? Would forcing them onto the grass reinforce negative outdoor experiences in their minds and reduce the likelihood that they would take their future elementary students outdoors? Should I adapt the planned activities for the dry parking lot nearby? Should I explain how important it is for them to work to overcome their physical discomforts, perceived fears, and biophobias? What should I do? How should I choose?

This chapter explores the pedagogical circumstances described above: How do I as a science teacher educator, a person committed to environmental education (EE), understand the common characteristics of elementary preservice teachers and the pedagogical decisions that foster their increased EE teaching competence? In exploring this question, this chapter seeks to provide science teacher educators with descriptions of preservice teacher characteristics and discussions of science

J.W. Hug (✉)
Director, Center for Excellence in Elementary Science & Math Education,
California University of Pennsylvania, California, PA 15419, USA
e-mail: hug@calu.edu

A.M. Bodzin et al. (eds.), *The Inclusion of Environmental Education in Science Teacher Education*, DOI 10.1007/978-90-481-9222-9_9,

methods course instructional strategy suggestions for addressing these common characteristics.

The task before science teacher educators is daunting. In my experience, preservice elementary teachers exhibit five major characteristics that constitute the core of the challenge. First, many preservice elementary teachers exhibit low levels of basic science/environmental content knowledge. Second, an increasingly high percentage of students in my elementary science teaching methods classes exhibit science/bio/ecophobic attitudes and behaviors. Third, substantial numbers of preservice teachers complain about and avoid minor physical discomfort while engaged in EE activities. Fourth, many preservice teachers exhibit low confidence in their ability to successfully engage in mechanical or technological physical manipulations. Fifth, high percentages of preservice elementary teachers desire and often demand what I consider highly structured, prescriptive, detailed procedures for their teaching assignments. I interpret this characteristic to be a symptom from years of indoctrination within an educational system that often rewards passively following directions, accurately memorizing factual details and thinking within discrete disciplinary boundaries. These five characteristics: inadequate content knowledge, ecophobia, avoidance of minor physical discomfort, low confidence with physically manipulating mechanical devices, and a need for highly structured learning environments provide substantial challenges for teacher educators.

In spite of these challenges, every new semester begins with the optimistic hope that redesigned assignments, new class activities, and an increased commitment to providing preservice teachers the highest quality EE experiences possible will result in 30-year teaching careers spent bringing high-quality EE to thousands of school students.

The many pedagogical decisions encountered in an elementary school science teaching methods course are made within a curricular, academic, and professional context. As novice science teacher educators begin to teach elementary science methods courses, my observation has been that they progress along a similar developmental continuum with stages that could be labeled as initial preparation, induction, experienced, and master science teacher educator. I suggest that science teacher educators can engage in pedagogical reflection-in-action moments (Schön 1983, 1987, 1991) with greater depth, sophistication, and success if they have access to discussions of preservice elementary teacher common characteristics in addition to EE-specific instructional strategies. In this chapter, I seek to describe, reflect, and analyze my experiences in teaching EE with approximately 1,500 elementary preservice teachers in more than 50 sections of elementary science teaching methods courses at multiple universities with diverse sizes and geographic locations over 16 years.

My purposes in this chapter are to: (1) synthesize and discuss selected characteristics of preservice elementary teacher learning in EE in preservice science education methods courses and (2) synthesize and discuss science teacher education instructional strategies that are congruent with the learning characteristics of preservice elementary teachers, while developing increased EE pedagogical competence. The intent of the discussion that follows is to illuminate my lived experiences (van Manen 1990) as a science teacher educator teaching an elementary science methods course through the use of pedagogical practice vignettes and instructional strategies suggestions.

Theoretical Framework

The theoretical framework through which I engage in teaching preservice teachers is based on many influences. First among these, radical constructivism (von Glasersfeld 1989, 1995, 2008) informs my educational approach. This epistemology contributes the idea "that 'knowledge' is the conceptual means to make sense of experience, rather than a 'representation' of something that is supposed to lie beyond it" and "suggest[s] a theory of knowing that draws attention to the knower's responsibility for what the knower constructs" (von Glasersfeld 2008). This emphasis on learner responsibility for actively constructing knowledge that makes sense of their experience is well suited to the teacher educator. As a teacher educator, I strive to understand my learners' sense making. I then design course experiences that emphasize the learner's responsibility for creating understanding of EE pedagogical competence.

Similar to other teaching disciplines, the science teacher educator's pedagogical choices are informed through an understanding of several domains of teacher knowledge such as general content, general pedagogy, pedagogical content knowledge, and knowledge of learners (Shulman 1987). As a teacher educator, I develop pedagogical content knowledge about the nature of my learners (preservice elementary teachers), content (EE), and the instructional strategies that have been effective in teaching EE (Abell et al. 2009). Pedagogical decisions, as Schön (1983, 1987, 1991) described, are filtered through a reflection-in-action and reflection-on-action process that becomes more sophisticated in the way the teacher frames data from the classroom situation as the depth of their understanding in these teacher knowledge domains increases. While many elementary science teacher educators come from a background of doctoral preparation in science teacher education, a surprising number of science teacher educators come from science content areas. This chapter seeks to address knowledge of learner and pedagogical content knowledge deficiencies through emphasis on the nature of the preservice elementary teacher.

This chapter extends the thinking of Driver (1990; Driver et al. 1985, 1996, 1994) who described learners' prior conceptions and suggested that successful instruction depends on first understanding the preconceptions students bring with them into the learning experience. While it is generally accepted that school students have alternative conceptions in science, extending this idea to both preservice elementary teachers and science teacher educators is less commonly discussed. I assert that many science teacher educators operate from a set of conceptions (alternate or naïve or sophisticated) about the characteristics of their preservice teacher learners. This chapter seeks to help teacher educators identify characteristics and naïve conceptions of preservice elementary teachers, and contemplate instructional strategies that might be useful within their elementary science methods courses.

A high priority for environmental educators is to increase citizens' capacity to engage in social and individual decision-making about their behavior choices, especially as it relates to the relationship between human and natural systems. EE's definition, focus, and objectives have been discussed over many years (Disinger 1983; Hungerford and Volk 1990; Stapp 1969; UNESCO 1977). For the purposes of this chapter, I use an extremely broad conceptualization of EE inclusive of the wide range of conceptualizations. At the core of the many EE definitions is education of citizens about the environment. Research has shown that the state of citizens' understanding on key environmental content and issues is extremely low and has not improved much over the last 30 years (Coyle 2005; NEETF and Roper 2002). This suggests that we need different strategies for educating the next generation of citizens and their teachers.

A key focus area for EE is the development of preservice and practicing public school teachers' ability to teach EE with their students. While studies indicate that 96% of parents think that EE should be taught in schools (Coyle 2005), in practice there are not enough teachers implementing EE effectively in their classrooms. An effectively educated teacher who engages in high-quality EE has the potential to reach many children over many years.

Yet, substantial challenges exist in reaching public school teachers effectively (Ernst 2007). One of these challenges has been how to best go about incorporating EE into teacher education programs. Mckeown-Ice (2000) found that approximately half of teacher education programs surveyed (n = 446) "exposed" preservice teachers to some sort of EE, but it was not institutionalized and quality varied widely. She concluded that, "Preservice teacher education programs are not systematically preparing future teachers to effectively teach about the environment" (p. 10). One approach to require EE in preservice teacher education, where politically possible, has been to mandate EE through state teacher education policies (e.g., Wisconsin). Another approach has been to pass state academic standards that require every school district to teach environment and ecology in classrooms (e.g., Pennsylvania). The North American Association for Environmental Education (NAAEE) developed EE standards for use in teacher education accrediting bodies such as the National Council for Accreditation of Teacher Education (NCATE 2007). Powers (2002) described the challenges that dedicated teacher educators faced in incorporating EE into their preservice teacher methods courses. She found that the barriers to EE implementation are substantial. Not unlike public school teachers, the teacher educators who successfully implemented EE in their courses tended to be dedicated champions of EE and found solutions to their implementation issues.

Despite all these efforts to increase EE in teacher education, the teacher educator faces a basic dilemma: how to include all the worthwhile "needs" of future teachers within the limited credit hours in a typical university education curriculum. How do you encourage more people to become dedicated EE champions in their universities, schools, and classrooms? Do you infuse EE across all courses or require an EE specific course? For many teacher education programs, the elementary science teaching methods course is a common location for EE (Mckeown-Ice 2000).

The following sections describe selected preservice teacher characteristics accompanied by instructional strategies that address these characteristics. Each section begins with a fictionalized vignette based on actual classroom experiences. Each vignette is followed by a discussion of the characteristics and strategies for EE instruction.

Environment/Ecology Content Knowledge

Circulating among the groups, listening to their conversations provided key verbal assessment clues as to how these preservice elementary teachers progressed in their task. I stopped at one table to watch two students open a container of small objects and dump them on their notebooks. They then worked together to sort the objects into two circles they had drawn in the notebooks, one circle labeled "seeds" and one circle labeled "not seeds." We were in the midst of learning about science process skills and the intended instructional outcome was to demonstrate how a first-grade science activity could be used to explicitly deepen observation and classification skills. I listened closely to their conversation.

Student 1: Ummm. (picks up a kidney bean) Is it a seed or not seed?
Student 2: I'm not sure.
Student 1: I guess I'm not sure either. I should know this. Like, isn't this something from elementary school?
Student 1: Yes, we should know this. Oh, my gosh. Is it a seed or not a seed?
Student 2: I'm thinking it's not a seed because it doesn't look like the other seeds (she points to the sunflower and flower seeds). It's bigger than the rest of them.
Student 1: OK. (They put the kidney bean seed in the *not seed* circle and facial clues suggest they are not very sure of their decision.)
Student 2: Wait. Remember Jack and the Bean Stalk? In the book, they plant a bean and it grows into a plant. A bean must be a seed.
Student 1: Ya. But it's bigger than the others (seeds).
Student 2: I think it's a seed. (She moves it to the seed circle. They notice me listening and their faces suggest they want my confirmation that their decision was the right one).

The general science and more specifically the environment, ecology, and natural history content knowledge of many elementary preservice teachers is not much different than the common alternative conceptions held by their elementary students (Bleicher 2006; National Research Council 2007; Krall et al. 2009; Trundle et al. 2006). The preservice teachers illustrated above eventually worked out that a bean was a seed from their memory of a children's literature book. However, their understanding is not much different than the elementary school children they will teach. It is notable that these preservice teachers used a children's literature book for reference rather than their own experiences planting seeds. How is it possible that after 12 years of public schooling and the completion of prerequisite science content courses in teacher education programs that many preservice teachers' content knowledge does not meet basic proficiency levels?

Many preservice undergraduate elementary teachers dread and often delay scheduling their science teaching methods course because they think of themselves as "science dumb" or unable to understand science. In some cases, students describe poor experiences learning science in the past. As one student explained to me, "science just isn't my thing." This lack of confidence is often fueled by their initial conceptualization of a teacher as someone who knows the answers and tells them to students. They are fearful about the possible situation where a future student asks them a question and they will not know the answer. In a perfect conceptual trap, these future teachers think that they need to know science answers, they perceive themselves to not know the answers, and therefore they reason that they will not be a good teacher of science.

There are several approaches to addressing low content understanding of preservice elementary teachers. Bleicher's (2006) approach employs strategies to help preservice teachers develop content understanding through experiencing for themselves "hands-on, minds-on" in-depth inquiry activities. In my course instruction, I employ strategies to help preservice elementary teachers reconceptualize: (1) their image of teacher as knower and transmitter of science facts; and (2) their view of science as exclusively a body of knowledge to be memorized. EE provides important tools to address these preservice elementary teacher characteristics.

Initially, preservice teachers have a strong conception of teacher as knower of facts. Most likely developed over years of participation in schooling, this conception is deeply imbedded. My instructional approach, rather than to provide the facts they lack (and reinforce their image), is to address their conception of teacher. Incorporating EE activities early in the semester models the teacher's role as a facilitator and not as a knower of all science facts. The preservice teachers participate in outdoor activities, ask questions, and hear me respond with, "That's a great question. Let's look it up in this field guide together so you know how to answer your next question." I use other statements that model teacher as facilitator such as, "You know I have never explored that. Let's set up an experiment to see what happens." In a course activity, I use dichotomous keys with a primary focus on the process of observing tree characteristics (i.e. leaves or needles, opposite or alternate, simple or compound). This engages learners with obtaining content knowledge while engaged in the process of observing the tree. As I model the role of teacher as facilitator and engage learners in EE activities, preservice teachers can begin to broaden their conception of teacher and start to develop confidence in their ability to deepen their content knowledge.

Another way to use EE to help preservice teachers deepen their environmental content knowledge is to focus on inquiry. Emphasizing inquiry and science process skills provides preservice teachers with easier access to content knowledge. For many preservice elementary teachers, it does not cause anxiety to teach about the senses, record scientific observations of animals, or observe their community for signs of spring. Such activities engage preservice teachers with important facets of inquiry and can provide an entry point for additional learning. A focus on inquiry demonstrates to preservice teachers that they have what it takes to be effective science learners and teachers. Such experiences also lay the foundation for engaging preservice teachers with environmental issue investigations and environmental action projects.

One challenge to using inquiry is that some preservice teachers have a deeply ingrained conception of teacher as knower and disseminator of facts/knowledge that is quite resistant to change. Furthermore, their self-perception of their content knowledge is also low, so many often fail to comprehend that using science process skills may reduce their dependence on needing to know all the facts. To address these beliefs, I use systematic observation nature journals where prospective teachers spend time alone in an outdoor setting to record their observations (Leslie and Roth 2003). I also use a children's literature book critique assignment as a way to reduce anxiety with learning science content. In this assignment, the preservice teachers are given an academic standard related to EE. Next, they select a children's literature book that addresses the standard. Then, they review the book according to a list of quality criteria that includes the nature of science and the accuracy of science content. Since many preservice teachers enjoy children's literature books, this activity provides motivation and interest for learning science content.

In summary, EE provides one mechanism through which preservice teachers can begin to reconceptualize their image of teacher and the nature of science, which reduces the barriers to further developing their content knowledge. In addition, using inquiry-focused activities provides preservice teachers with an entry point for learning environmental science, ecology, and natural history content.

Ecophobia

"When you go owling you don't need words or warm or anything but hope. That's what Pa says. The kind of hope that flies on silent wings under a shining owl moon" (Yolen 1987, p. 32).

I pause and hold the children's literature book, Owl Moon, by Jane Yolen (1987) still for effect. Slowly I see some heads start to move from their straightforward, attentive positions, which is my signal to move on with directions for our next activity. I ask if anyone has seen an owl in the wild. No one has. After announcing my intentions, I gently produce a great horned owl study skin. Two preservice teachers in the front row physically recoil at the sight of the study skin and their chairs scrape on the floor as they push back from the front of the room. I ask, "Can you describe why you moved backwards?" One responds, "Eeew. Dead things freak me out." The other student asks, "Do I have to stay here?"

"Yes, when you are a teacher you have to learn to be brave." I say, adapting an earlier line in the book.

I ask for volunteers to describe their observations of external owl features that are especially adapted for where they live. Most don't know much about owls but we manage to notice and discuss wing feathers, feathered legs, talons, beak, and eyes. I then say, "Owls eat primarily rodents. They swallow them whole; digest most of the animal and then cough up the fur and bone in a little ball. I have one right here. An owl pellet."

"Yuck."
"Eeew."
"Gross."
I see one student shiver involuntarily.
My preservice elementary teachers are predictable. Now comes the real struggle. I pass out the foil wrapped owl pellets, describe how they have been sterilized, and ask that they begin to pull them apart, look for bones, see if they can assemble a full skeleton and collect data on the species in the pellets across the class. Moving group to group I notice that many preservice teachers refuse to touch the owl pellet. Even after I share soothing words and gently model how to proceed, some sit there fearful and defiant.

Whether it is owl pellets, macroinvertebrates, fungi, spiders, snakes, bees, hairless tails, mud, dirt, dead things, or decomposing anything, there are a high percentage of elementary preservice teachers who exhibit a range of phobias about all things gross, gooey, sticky, or creeping. This could be considered ecophobia, the sometimes-primal emotional experience of fear toward the natural world (Sobel 1999). Ecophobia can be considered the opposite of Wilson's (1984) biophilia. Ecophobia can physically manifest itself in ways such as an involuntary shudder at the sight of a snake; a reluctance to use a hand to pick up algae along a river bank or to touch the fur and bones in an owl pellet; panic-stricken swats, screams and sprints from insects; or simply avoiding "dirty" things.

Ecophobia can also mean a more general attitude of fear (mosquito bites, bear attacks, getting lost) toward the outdoors and nature. For example, during one of my outside EE class sessions, several gray squirrels went about their business 20 yards away while I provided directions for our next EE activity. I hardly noticed the squirrels. Habituated to receiving food from passersby, the squirrels all began simultaneously to hop toward our group looking for a handout. Half the class broke our circle and ran letting out a muted scream at the approaching squirrels. "The squirrels are attacking us," one person exclaimed. I took one quick step toward the squirrels and they all scampered up the nearest tree. I then listened to stories about friends or cousins who had bad encounters with squirrels. We discussed the reality of squirrel attacks, natural history, and the impact of feeding squirrels human junk food. In hindsight, a great strategy at that time would have been to engage in conducting behavioral observations of squirrels to enhance my students' comfort level with squirrels.

It is hard to assess where these strong ecophobic emotions come from, how they were created, and how to address them. Each preservice teacher is different. If these fears are socially learned, they can be unlearned. One can imagine a sibling chasing another around with a worm in his or her hand, thus creating a negative earthworm experience. In addition, people are often exposed to movie scenes where serial killers, aliens, and weird neighbors hide outdoors ready to strike the unsuspecting. Even nature shows emphasize the danger of being outdoors to hold viewers' attention.

There are several strategies that could help deal with ecophobia. One instructional strategy engages learners in a desensitization process. Similar to an allergist who administers small doses of an irritant until immunity is built up, preservice teachers with an aversion to "yucky" things can learn to reduce their perceived fears through appropriately phased experiences in a safe supportive environment. (I acknowledge that there are people with genuine clinical phobias, but I assert that they are rare in my classes.) Another aspect of this same strategy focuses on building extensive positive experiences that reduces the fear by reducing the amount of the unknown. For instance, taking many outdoor walks with a mentor provides the experience to feel comfortable outdoors. Along these lines, adventure education makes use of activities such as rock climbing to help people examine their perceived fears and then develop coping strategies to continue to function in the face of fear (Bacon 1983; Gass 1993; Schoel et al. 1988). Applied to ecophobia, a concentrated effort with preservice teachers to examine their perceived fears and develop coping strategies through positive social group processes could contribute to a reduction of ecophobic behavior. This would necessitate training teacher educators in adventure-based counseling techniques and transferring it to the science methods course context. It would also require voluntary personal commitment from preservice teachers and time frames longer than usually available in a university classroom context. Yet, beginning the process and setting the example are important. Within the university classroom, it is possible to start the process of helping future teachers address these ecophobias in a safe supportive environment through experiences such as modeling appropriate behaviors, structured "safe" activities, and gentle but firm encouragement.

I consistently model appropriate EE behaviors for the preservice teachers in my classes. Modeling is at the heart of teaching and learning, although I have come to believe that its usefulness is limited in my attempts to deal with ecophobia with my preservice teachers since I am perceived as being quite different from them. My appropriate modeling can be dismissed due to how I am perceived. "I could never do that," they say, or "I could never come up with something like that for my lesson." In other words, they attribute to me special characteristics that enable me to do things that the preservice teachers do not perceive themselves to be able to do. Instead of seeing the strength within themselves to become extraordinary environmental educators, they operate within safe self-perceived boundaries. The teacher educator's art is to find the right combination of strategies for each student to unlock their latent strengths and realize their potential as environmental educators.

One verbal strategy involves providing comfort and support for students. The working assumption is that preservice teacher's fear is socially learned and is a perceived fear rather than grounded in real experience. Reassuring words, a positive classroom climate, and acceptance of student attitudes can go a long way toward helping preservice teachers attempt something new within a safe supportive environment. While for some preservice teachers this strategy provides the atmosphere to explore new biophillic behaviors, a safe accepting environment may not provide enough motivation to tip others into cognitive dissonance, face the perceived fear, and attempt new behaviors.

A second verbal strategy points out to the preservice teachers that science and EE is for *all* children and as a teacher they do not have a choice but to learn how to desensitize their fears of nature. For example, I present to my students a scenario such as: "What would you do if a child came up to you on the playground and put an earthworm in your hand? Is it a viable option to scream and run away from your children? You have to teach yourself how to be brave for those students in your class who need you to help them learn about nature. It's a part of your job as a teacher just like changing a baby's diaper is for a mother." This strategy puts the behavior in the context of a requirement of the teacher's job. Accompanied with stories of other brave preservice teachers who have gone on to do amazing things with their fears sometimes can help provide enough motivation for current preservice teachers to try a new behavior in spite of their fear. Furthermore, another statement can be used: "These are the state standards and you are required to teach them to your students." This statement explains a key aspect of the job. The preservice teachers understand it intellectually, but in my experience this language does not really change behavior for most preservice teachers.

Verbal strategies should be supported by physical strategies to reduce ecophobia. For example, handling animals can be used to slowly desensitize fears through positive animal experiences. The selection of an animal and the progression to other animals is critical. Over the semester, preservice teachers can begin by handling a soft bunny. Holding a hamster is also perceived as a safe experience for many preservice teachers. Progressing to experiences handling birds, reptiles, insects, spiders, or macroinvertebrates are more challenging, but can be supported with a positive classroom atmosphere and gentle encouragement (Campbell, L. M., September 1992, personal communication).

Many people can learn to confront their perceived fears and triumph over them with enough time in a supportive environment. Most people with repeated positive experiences and a careful encouraging mentor can learn to confront their perceived fears. Snakes illustrate a particularly good example for this. Preparing people to touch a snake with strategies such as modeling slow movements and "gentle fingers" help children and adults to begin to confront their fears. Slightly reducing the snake's body temperature by putting it in a cool environment for a short time decreases its body movement and provides a less threatening animal for novices to handle or touch. A single instance of touching a snake will not remove the fear built up over a lifetime. However, incorporating many positive experiences to reduce ecophobia is crucial for preservice teachers to overcome their fears.

It is often easier to address preservice teacher ecophobia in a one-on-one situation rather than in a large social group. One-on-one contact increases the concentration on the desired behaviors, reduces the concern about the potential for embarrassing behavior in front of peers, and provides targeted verbal and physical support based on the preservice teacher's unique needs. Mentoring a student in a one-on-one context reduces their ability to not confront their ecophobia by relying on someone else in the group to touch an owl pellet or pick up an earthworm. Individual attention also allows the teacher educator to provide essential information to the preservice teacher that dispels any myths or fears they may have about the animal or object. The challenge in a typical preservice methods course with over 30 students is to find the time to address individual needs. One promising strategy that addresses both the role model issue and engages positive social peer pressure is to use peer role models to demonstrate appropriate environmental attitudes and behaviors.

The ecophobia characteristic in preservice teachers provides a substantial barrier for some students to participate fully in EE. Teacher educators need to consciously understand and explicitly plan for ecophobia in order to reduce preservice teachers' fears and encourage their participation in EE.

Physical Discomfort Avoidance

As described in the opening vignette, preservice elementary teachers often avoid minor physical discomforts that prevent them from fully engaging in environmental learning. Dew on their toes, rain on their hair, too much exercise for their muscles, insects on their arm, dirt on their clothes, and sun in their eyes are just some of the physical discomforts encountered by preservice teachers in EE. These behaviors either openly voiced as a verbal complaint or demonstrated through quiet noncompliance should be addressed.

One choice open to teacher educators is to move ahead and ignore complaints about physical discomfort. "Ah, come on. It's OK. The dew isn't going to hurt your toes," could be one response. Another choice is to increase the pressure to engage in the activity to the point where preservice teachers feel compelled to participate. The hope being that preservice teachers discover "it isn't as bad as I thought" resulting in a positive experience. This is a sensitive task for the teacher educator. As discussed earlier, providing enough encouragement to extend their perceived physical limitations is necessary without misusing course instructor power over preservice teachers in a way that results in coercion or a negative experience.

Another instructional strategy consists of avoiding physical discomforts by modifying EE activities. This approach assumes that providing learning tasks that take into consideration the comfort level of preservice teachers will develop the capacity for later risk-taking. For instance, playing a simulation game in the parking lot rather than in the wet grass. Another possibility is to reduce the anticipated physical discomfort through extensive prior preparation. Announcing in preceding class periods about the upcoming field trip, outdoor event, or hands-on activity in some cases can reduce issues with physical discomforts; however, in other cases it may serve to raise anxieties. For example, I have found it helpful to emphasize wearing appropriate outdoor clothing in preparation for a field trip to reduce potential complaints about physical discomfort due to weather conditions. Such strategies can be helpful to address issues with regard to preservice teachers' physical discomfort during EE activities.

Mechanical Disinclination

The sky was blue and clear as I looked at the 30 elementary preservice teachers assembled before me. We held class outdoors to organize our teaching stations for the wetlands festival with a local elementary school. These preservice teachers would be teaching wetlands-related activities to elementary school children in a few weeks. A non-profit wetlands educator came to class that day as a guest to help explain and organize the teaching stations. She described each station's activity, showed the students the equipment and described the science journal activity at each one that tied the activities together with the overall theme.

One of the stations involved demonstrating a ground water flow model. It consisted of a sealed thin clear plastic container that allowed a view of the underground "soil" and several moving parts such as a water pump. The idea was to pour water into the model and then work the pump, which made the water flow through the ground water system. The model demonstrated the movement of groundwater.

"Who wants to teach the ground water flow model?" the wetlands educator asked.

The pause grew longer.

"It really is very easy. All you have to do is pump the hose."

No hands went up.

Finally, one woman raised her hand, "OK, I'll do it. But I shouldn't do this station. I'm such a klutz when it comes to mechanical things. I usually break stuff."

The class giggled nervously.

I have observed that a high percentage of preservice elementary teachers exhibit low self-confidence and aptitude for manipulating physical objects. They demonstrate reluctance to teach a lesson that requires any sort of mechanical manipulation. There are many factors at work that predispose preservice teachers toward these attitudes and behaviors, which inhibit their confidence leading EE activities. Traditional gender role-based upbringings, media exposure, poor prior experiences, lack of role models, or a lack of opportunity to successfully engage in mechanical manipulations are just a few. Regardless of how these attitudes were formed, this characteristic prevents teachers from feeling confident with physical materials that often accompany EE activities.

One successful strategy I have used involves forming partnerships with local schools to set up structured EE field experiences. Many different possibilities exist for teacher educators to organize actual teaching with small groups of children that result in positive EE instructional experiences. One example of this I found very helpful was an elementary school/science methods course collaboration in which preservice teachers taught wetlands ecology activities at teaching stations during a wetlands festival for elementary school students. The preservice teachers researched their environmental content for the stations and planned their teaching activities. Many of the activities required manipulating nets, buckets, microscopes, magnifiers, models, and other apparatus. The preservice teachers were given opportunities in class through modeling, guided practice, and independent practice to become proficient with their equipment. A local wetlands environmental educator provided extra support to assist preservice teachers with their preparation. Careful coordination with practicing teachers and administrators aligned the activities to state standards. Attention to the logistics permitted the preservice teachers to experience small groups of 8–10 students for their wetland lessons and repeat their teaching with different groups multiple times over the day ensuring success.

On the day of the wetland festival, the nervous preservice teachers began their first lesson and immediately connected with the students. The elementary students displayed much energy and joy typical of children who find themselves outside.

The preservice teachers tentatively began their first lesson and soon were caught up in the children's enthusiasm. As they taught the same lesson again, their confidence improved and their comfort with the equipment grew stronger. By the end of the wetland festival day, the preservice teachers gained much skill and confidence with their equipment. Preservice teachers expressed comments about their increased confidence and understanding of the wetlands content by teaching the same lesson multiple times allowing them to make adjustments. Overall, partnering with schools to create positive EE teaching experiences has substantial benefits. Preservice teachers can grow tremendously in their ability to reduce their mechanical disinclination with prior preparation, modeling, mentoring, and monitoring.

Need for Highly Structured Assignments

> The class had been wonderful. We were outside demonstrating how to make observations using nature journaling techniques. The instructional task was to share ideas and resources for creating a science/nature journal with elementary children. Each preservice teacher had participated in the nature journaling activities, received an 8-page handout and participated in examples of nature journal projects (Leslie and Roth 2003). Formal scientific systematic observation journal formats as well as free-form journal formats were discussed and practiced. I explained that their course assignment was to spend some time outside and create one journal entry. A detailed scoring rubric was provided for the assignment. The students had some questions:
> How many pages does it have to be?
> How much time do I have to spend outside?
> How many sketches do I need?
> Can I draw animals or do you want us to draw plants?
> Does it have to be in a notebook or can I staple paper together?
> What if I can't draw very well, will it affect my grade?

Their unspoken question was, "What do I have to do to get an A?" Their questions revealed concern for the trivial format of the assignment not the instructional intent. These students were focused on getting the assignment done rather than focusing on the learning. They asked questions, very important to them, but sounded to me more like, "What color do you want me to paint my tree?" I would have preferred them to ask, "What drawing activities help elementary children increase their observation skills?" When preservice teachers ask these questions, I patiently explain my expectations, refocus them on the intent of the task, and promise support through the process. Their questions may be a result of a need, accumulated over years of schooling experiences, for knowing specific details for meeting course requirements.

There are several strategies I have used to address this need for structure. Over the years, I have revised assignments over and over to better address their need for detailed, quantified assignment structure. In effect, I increased the specificity and quantifiable aspects of the assignment. In some cases, I provided assignment descriptions that contained so much detail that the assignment essentially became a prescription for them to follow. While some students feel comfortable with this level of structure, I have come to feel that in the long run, this does not provide the best learning experience. One strategy that successfully challenges the students' need for structure is to provide little, if any, of the detailed quantitative procedures they request. Providing less structure, not more, becomes an effective discrepant event if the teacher educator can provide a supportive environment from which to negotiate through the anxieties of the preservice teachers. They have to trust that their instructor will not hold them accountable to unannounced criteria for their grade.

If we hope to educate teachers who can creatively solve unique curricular dilemmas without clear direction from authority figures, then it follows that we should set up similar situations in their university coursework. Similar to using open inquiry during instruction, preservice teachers initially feel confused and even angry at the lack of specific procedures to follow for their assignment. They say things such as, "What do you want me to do?" My goal of providing practice in self-directed learning conflicts with their image of professor as the knower of all answers and authoritarian prescriber of learning tasks. As an environmental educator, I feel it is an essential skill for citizens to be able to look at their community, assess its needs, and develop multiple approaches for resolving environmental issues. Teachers, in particular, need to experience learning tasks where there are not highly structured pathways to find the correct answers. Such tasks model environmental social decision-making.

Summary

This chapter provides a discussion of my journey as a teacher educator seeking to understand the nature of preservice elementary teachers and the pedagogical content knowledge involved with integrating EE into my elementary science teaching methods courses. Through these phenomenological vignettes, I hope to have illuminated key preservice teacher characteristics in a way that resonates with the elementary science teacher educator. The discussion of EE instructional strategies provides teacher educators with the opportunity to reflect on their own practices that address preservice teacher characteristics. The interdisciplinary nature of EE lends itself well to inclusion in elementary science teaching methods courses. While this chapter focuses on science teacher education, it should be noted that EE can be integrated into other disciplinary areas in the teacher education curriculum.

The challenges faced by teacher educators are substantial. Environmental educators, who are not science teacher educators, often discuss integrating EE into science teacher education as "outsiders," unaware of the constraints and pressures involved in a relatively short 45 contact hour course. As a science teacher educator and an environmental educator, I hope to have provided some insight into several of the dilemmas derived from the nature of the learner. The goal of helping preservice elementary teachers to learn to implement EE in their future classrooms through integrating EE into a science teaching methods course is an important one. The task requires teacher educators to become dedicated champions of EE, persevere through substantial challenges, and refine instructional strategies that take into account the nature of the preservice elementary teacher.

References

Abell, S. K., Rogers, M. A. P., Hanuscin, D. L., Lee, M. H., & Gagnon, M. L. (2009). Preparing the next generation of science teacher educators: A model of developing PCK for teaching science teachers. *Journal of Science Teacher Education, 20*(1), 77–93.

Bacon, S. (1983). *The conscious use of metaphor in outward bound.* Denver, CO: Colorado Outward Bound School.

Bleicher, R. E. (2006). Nurturing confidence in preservice elementary science teachers. *Journal of Science Teaching, 17*(2), 165–187.

Coyle, K. (2005). *Environmental literacy in America: What ten years of NEETF/Roper research and related studies say about environmental literacy in the U.S.* Washington, DC: The National Environmental Education and Training Foundation. Retrieved April 14, 2008, from http://www.neefusa.org/pdf/ELR2005.pdf

Disinger, J. F. (1983). *Environmental education's definitional problem* (Information Bulletin No. 2). Columbus, OH: ERIC Clearinghouse for Science, Mathematics and Environmental Education.

Driver, R. (1990). *Constructivist approaches to science teaching* (Paper presented at the Seminar Series: Constructivism in Education). Athens, GA: University of Georgia Mathematics Education Department.

Driver, R., Guesne, E., & Tiberghien, A. (Eds.). (1985). *Children's ideas in science.* Philadelphia, PA: Open University Press.

Driver, R., Squire, A., Rushworth, P., & Wood-Robinson, V. (1994). Research into children's ideas. In *Making sense of secondary science* (pp. 1–13). New York: Routledge.

Driver, R., Leach, J., Millar, R., & Scott, P. (1996). *Young people's images of science.* Philadelphia, PA: Open University Press.

Ernst, J. (2007). Factors associated with K-12 teachers' use of environment-based education. *Journal of Environmental Education, 38*(3), 15–32.

Gass, M. A. (1993). *Adventure therapy: Therapeutic applications of adventure programming.* Dubuque, IA: Kendall/Hunt Publishing Company.

Hungerford, H. R., & Volk, T. L. (1990). Changing learner behavior through environmental education. *The Journal of Environmental Education, 21*(3), 8–20.

Krall, R. M., Lott, K. H., & Wymer, C. L. (2009). Inservice elementary and middle school teachers' conceptions of photosynthesis and respiration. *Journal of Science Teacher Education, 20*(1), 41–55.

Leslie, C. W., & Roth, C. E. (2003). *Keeping a nature journal: Discover a whole new way of seeing the world around you.* North Adams, MA: Storey Publishing, LLC.

McKeown-Ice, R. (2000). Environmental education in the United States: A survey of preservice teacher education programs. *Journal of Environmental Education, 32*(1), 4–11.

National Council for the Accreditation of Teacher Education. (2007). *North American association for environmental education standards for the initial preparation of environmental educators.* Washington, DC: National Council for the Accreditation of Teacher Education.

National Environmental Education and Training Foundation (NEETF), & Roper ASW. (2002). (2001 NEETF/Roper Report Card) *Americans' low "Energy IQ:" A risk to our energy future; The tenth annual national report card: Energy knowledge, attitudes, and behavior.* Washington, DC. Retrieved May 20, 2008 from http://www.neefusa.org/pdf/roper/Roper2002.pdf

National Research Council. (2007). *Taking science to school: Learning and teaching science in grades K-8.* Washington, DC: National Academy Press.

Powers, A. (2002). Teacher preparation for environmental education: Faculty perspectives on the infusion of EE into preservice methods courses (Unpublished master's thesis). Keene, NH: Antioch University New England.

Schoel, J., Prouty, D., & Radcliffe, P. (1988). *Islands of healing: A guide to adventure based counseling*. Hamilton, MA: Project Adventure.

Schön, D. (1983). *The reflective practitioner: How professionals think in action*. New York: Basic Books.

Schön, D. A. (1987). *Educating the reflective practitioner*. San Francisco, NY: Jossey-Bass.

Schön, D. A. (1991). *The reflective turn: Case studies in and on education practice*. New York: Teachers College Press.

Shulman, L. S. (1987). Knowledge and teaching: Foundations of the new reform. *Harvard Educational Review, 57*(1), 1–22.

Sobel, D. (1999). *Beyond ecophobia: Reclaiming the heart in nature education* (Vol. 1). Great Barrington, MA: Orion Society.

Stapp, W. B. (1969). The concept of environmental education. *Journal of Environmental Education, 1*(1), 30–31.

Trundle, K. C., Atwood, R. K., & Christopher, J. (2006). Preservice elementary teachers' knowledge of observable moon phases and pattern of change in phases. *Journal of Science Teacher Education, 17*(2), 87–101.

United Nations Scientific and Cultural Organization (UNESCO). (1977, October). *First intergovernmental conference on environmental education final report*. Tbilisi, Georgia, USSR: Author.

van Manen, M. (1990). *Researching lived experience: Human science for an action sensitive pedagogy*. London: The State University of New York.

von Glasersfeld, E. (1989). Cognition, construction of knowledge, and teaching. *Synthese, 80*, 121–140.

von Glasersfeld, E. (1995). *Radical constructivism: A way of knowing and learning*. Washington, DC: The Falmer Press.

von Glasersfeld, E. (2008). *An exposition of constructivism: Why some like it radical*. Retrieved March 21, 2009 from http://www.oikos.org/constructivism.htm

Wilson, E. O. (1984). *Biophilia*. Cambridge: Harvard University Press.

Yolen, J. (1987). *Owl moon*. New York: Philomel Books.

Name That Plant! Overcoming Plant Blindness and Developing a Sense of Place Using Science and Environmental Education

Jennifer Kreps Frisch, Matthew M. Unwin, and Gerald W. Saunders

Introduction: Part of the Problem

Carrying her cup of coffee, Donna walked into her daughter's living room. "Oh!" she enthused, "you got a new plant!"

Donna leaned close to the plant, inspected the leaves and examined the stems closely, then frowned. "Huh. What's the name of this one?"

"Um..." her daughter stalled. "It's...a..." There is a long pause. She smirked and shrugged, "I call it 'Fred'."

The mother shook her head in mock dismay. "All those fancy college biology courses, and you still don't know anything..."

I admit it. I have been part of the problem. When I began teaching high school about 15 years ago, I often sped through the plant unit in my biology classes so that we would have time for "important" topics like animals and ecology. In the years since, however, I have come to admire the importance of plants, their diversity, and the people (like Matt, Gerry, and my mom) who know their names and how they are special. Noticing and understanding the plants around us helps us know the place where we live: the community not only in biological terms but also in sociological terms. Once our students are able to notice and learn about plants, they will have new anchors upon which to build connections that foster ecological awareness, knowledge, and action. The purpose of this chapter is to describe plant blindness (Wandersee and Schussler 1999) and some potential methods of preventing it by helping our students learn science and develop a sense of place.

J.K. Frisch (✉) and M.M. Unwin
Department of Biology and Physics, Kennesaw State University, Kennesaw, GA 30144, USA
e-mails: jfrisch1@kennesaw.edu; munwin@kennesaw.edu

G.W. Saunders
Center for Experiential and Environmental Education at Unity College,
Unity, Maine 04988-9502, USA
e-mail: gsaunders@unity.edu

A.M. Bodzin et al. (eds.), *The Inclusion of Environmental Education in Science Teacher Education*, DOI 10.1007/978-90-481-9222-9_10,

Plant Neglect and Plant Blindness

In 1993 and 1996, Hershey wrote about plant neglect: the idea that science teachers spend little time teaching about plants in the classroom. As he explains, plant neglect is not a new problem (e.g., Nichols 1919). In fact, the phenomenon has been around so long that, like many other educational issues, the concept has been revisited and renamed several times: plant neglect, zoochauvinism, zoocentrism. The general cycle seems to be that biology teachers get little training in botany, and as a result, they do not teach much in the way of botany themselves.

Wandersee (1986) gave 136 rural junior-high students a "science interest query" to identify the science topics in which they were most interested. His evidence matched that of previous studies: students had a significant preference for animals over plants as a science topic ($\chi^2 = 10.9$, df =1, $p < .02$). Wandersee suggested that this preference may have been due to students' tendency to anthropomorphize and therefore relate more to animals and to cultural influences, including cartoons, books, and television shows, which tend to focus on animals rather than plants. This study set the stage for Wandersee's later work with plant blindness.

Kinchin (1999) did a "head to head" comparison of plants vs. animals by showing a sample of 162 girls specimens of *Arabidopsis thaliana* (a flowering plant) and *Mellitoba* (an insect). Kinchin then asked open-ended questions including which organism they found most interesting and why. Of the 162 participants, 144 chose *Mellitoba* as more interesting, most frequently citing the fact that *Mellitoba* can move as the reason they found the organism interesting.

Barman et al. (2003) investigated over 2,400 students' ideas about plants. When shown images of different organisms and one nonliving thing (a telephone pole), students were asked which of these items were plants. The data indicated that young students saw plants as green things with stems and leaves that grew in the soil. This conception of plants led students to be unsure about whether trees, grass, and Venus flytraps were plants, since trees have brown wooden trunks, grass has no obvious leaves, and Venus flytraps move and eat insects. The authors suggested that children begin with a narrow definition of the word "plant" and then broaden the definition as they get older, based on examples they gather in school and their daily lives. However, if students do not get the chance to learn about plants from teachers and/or parents, they may continue to hold a fairly narrow definition (Gatt et al. 2007).

Plant blindness, rather than introducing yet another term to describe the same phenomenon, is an attempt to get at the root cause for our tendency to overlook the plant kingdom. We pay little attention to plants because of the way our visual information-processing systems work (Wandersee and Schussler 1999, 2001; Wandersee and Clary 2006). Essentially, human vision is capable of "seeing" a great deal, but what we "attend to" is much more limited; "visual consciousness is like a spotlight, not a floodlight" (Wandersee and Clary 2006, p. 3). Humans do not attend to things unless they hold meaning, survival value, or both. As Mack (2003) explains, this is inattentional blindness, and some studies indicate that our brains process meaningful stimuli while "ignoring" meaningless stimuli.

The observation that biology teachers (and their students) often neglect plants is a symptom of plant blindness (Wandersee and Schussler 2001). People who exhibit plant blindness show "an inability to see or notice the plants in one's own environment" (p. 3) which leads to a lack of awareness and knowledge of plants' role and importance in ecosystems. One common misconception among students, the idea that the bulk of a growing plant's mass comes from the soil, is based on a lack of understanding about plants' role in the cycling of matter and transfer of energy on Earth (Barman et al. 2003; Harvard-Smithsonian Center for Astrophysics 1995; Wood-Robinson 1991). Orr (1994) wrote that "we routinely produce economists who lack the most rudimentary understanding of ecology or thermodynamics" (p. 11) and that may help explain why our natural areas' values are so underestimated. Perhaps instilling understanding of the value (economic, ecological, and aesthetic) of biological communities needs to begin earlier, and continue throughout our education.

Moon's (1921) textbook, *Biology for Beginners*, devoted about 25% of its content to plants, but Glencoe's *Biology: The Dynamics of Life* (Biggs 2004) has less than 10% plant content, though it includes about 20% genetics and biochemistry. Certainly, these branches of science are necessary and important for a broad understanding of the way the world works. However, as textbooks devote more space to emerging fields of biology, plants receive less coverage; this may lead teachers to believe that plants are less important and so devote less time to naturalistic studies outside the classroom. With limited opportunities to encounter and learn about plants, children are not given the opportunity to appreciate "real" nature. In fact, plant blindness may itself be a symptom of what Louv (2005) called "nature-deficit disorder." Children are wired into their worlds, surrounded by technologies that make it unnecessary to venture out and play in nature. Some authors suggest that students know more about rainforest plants and animals than they do the biology in their own back yards (Ashworth et al. 1995; Brewer 2002). Because our students are learning organisms only from their textbooks and the Internet, rather than their local environments, they have become disconnected from the living things actually found in their communities (Bebbington 2005; Brasher 2006; Lock 1995; Paraskevopoulos et al. 1998).

One of the main predictors of students' awareness, knowledge, and appreciation of plants is the presence of a plant mentor, or "someone who help[s] the mentee observe, plant, grow, and tend living plants" (Wandersee and Clary 2006, p. 3; see also Gatt et al. 2007). Teachers need not look far to find plant mentors; ask students' parents if they enjoy gardening, or contact a local master gardeners' association to find enthusiastic helpers. If teachers can learn to "see" plants themselves, they are better equipped to help their students see plants – this is the first step toward a greater understanding, awareness, and appreciation of the community and ecosystem around them.

I'll Sign a Petition, But I'm Not Going In There

An upper-level plant course in a midsized suburban university is full of students who care about the environment. They are biology majors, and they want to change the world. They passionately discuss An Inconvenient Truth and the plight of the polar bears. When their instructor brings up the possibility that part of the university's forest may be cut down to provide more parking, they are irate. The students suggest writing letters, circulating petitions, and a few more extreme solutions. A week later, the instructor walks them out to that same forest. The instructor enters the forest to point out some trees and plants, but the students hover at the periphery of the forest as if there was some invisible fence keeping them out. "Come on in here," shouts the instructor. Many of the students demur. "We can see and hear fine from over here," one student yells back. Another student whispers, "but there are bugs in there!"

On a positive note, there is increased environmental awareness percolating in schools. Students are ready to save the rainforest, the whales, and the polar bears. Science in schools has focused on those exotic locations and organisms because they are interesting and great exemplars of certain ecological principles.

However, despite having strong emotional attachments to the environment as an abstract concept, our students seem to be deeply disconnected to the concrete environment outside their windows. What should be familiar has become alien. There is a reason that the environmental aphorism asks us to "Think Globally, Act Locally." We must start with where we are.

Teachers tend to consider going on field trips to distant areas when trying to construct lessons that help students learn about ecology and the environment (Simmons 1996). Although a trip to a botanical garden or state park could be a wonderful way to help students explore plants and nature (Bowker 2004), the logistics and costs that are involved can be daunting, particularly for beginning teachers. The schoolyard itself can be just as useful as a field trip for teaching about ecology, while eliminating most costs and logistical concerns.

Brewer (2002) reported that when teachers are asked what they need in order to start including outdoor inquiry in their classes, one thing the teachers ask for is resources that are specific to their region and schoolyards, including "the names and general natural-history traits of common organisms in their schoolyards" (p. 578). Brewer suggested that scientists could work with graduate students to develop field guides for teachers, but another option might be for teachers and students to develop their own schoolyard field guides with the assistance of nearby science and/or education faculty. When the teachers and students are given ownership over the process, it will help them develop or enhance their sense of schoolyard natural areas as a part of their community.

Lindemann-Matthies (2005, 2006) examined the results of a Swiss conservation organization's educational program, "Nature on the Way to School." The program combined structured ecology-related information with more inquiry-oriented activities, and participating teachers reported back on which parts of the program they used, and which parts of the program they found to be most effective. One of the activities in the program that teachers and students rated most highly was one that gave students the chance to investigate different plants and animals that they found on their walk to school. Students were encouraged to observe and identify species that they found to be interesting. These investigations often included what the program called the "Nature Gallery," in which students created picture frames, and then "framed" their favorite plant or animal that they found along their walk. Students could investigate their plant and stand next to their frame and tell passers-by all about their organism. Lindemann-Matthies (2005) used multiple regression analyses to show that the biggest predictors of children's learning gains about common local plants and animals was time spent on investigations on the way to school, and time spent depicting plants and animals. In addition, Lindemann-Matthies (2006) examined how children's appreciation for wild plants and animals changed over the course of the program, and found that the more plant and animal taxa the children noticed in their environment and could identify, the more likely they were to appreciate these organisms.

Sense of Place

> "If you don't know where you are, you don't know who you are." – Wendell Barry in *The Sense of Place*
>
> (Stegner 1992)

The next time you have a meeting that requires an "icebreaker," consider the activity in Project Learning Tree's (2006) *Places We Live*: "Personal Places." The activity asks the group to think back to when they were 10 years old, and draw a map showing their house and features of their neighborhood. Afterwards, individuals can share their drawings in small and/or large groups. Drawings often include nearby creeks or special trees, parks or open areas where children might have converged. More often than not, an animated discussion ensues about how "things were different then." Often, this discussion leads participants to wonder what today's 10-year old would draw.

Sense of place is defined as having ecological knowledge, social knowledge, and attachment to community (human and nonhuman) about and in a particular place (Worster and Abrams 2005). The concept is multifaceted, much like environmental education (EE) itself; in order to have a sense of place, one must have acquired knowledge about it, positive affect toward it, and skills to be a part of it. Worster and Abrams' qualitative examination of farmers and fishermen in New England revealed that these people felt a connection to the land. One farmer commented that, "[t]he ultimate goal has been that you know the place so well, that somehow … you all of a sudden become a part of it" (p. 531). The participants felt closely connected not just to the land and the sea, but also to the people in their place. Worster and Abrams wrote that the fishermen's "perceived relationship with the local social and ecological context led to a heightened ecological knowledge of the ocean" and a desire to learn skills in order to treat that environment responsibly (p. 532).

Giving students the chance to strengthen their sense of place in the classroom also has benefits for multicultural classrooms. As Derr (2002) found, children's sense of place is often deeply influenced by extended family and direct experience. Giving students a chance to share their "place" with others can help them connect the classroom with their real lives. For example, one of the participants in Derr's study was 11-year-old Teresa, a girl who felt deeply connected to a park across from the Boys' and Girls' Club. She found quiet and solace there, and took time to help keep the river that ran through the park clean. Teresa seemed to take great joy in sharing her park with Derr. Other participants, Leo and Marcos, both 10 years old, had deep place attachments to natural areas around their homes, and had plant mentors from their families that allowed them to grow to know and love the land, including plants, in their areas. This kind of place attachment is an important predictor of environmental concern and action (Peterson, 1982; Vorkinn and Riese 2001). If children like Derr's participants are given an opportunity to share their sense of place and their ways of knowing about a place with other children in their classrooms, the classroom itself can start to "feel" more like part of a community. In addition, giving children a chance to share the knowledge of plants and trees that they have gathered from their extended family validates their cultural knowledge.

Not all students will have the kind of place knowledge that is found in Derr's (2002) participants, but they can learn from each other and from their teachers. By allowing students to discover their place, teachers can connect classrooms with communities (Ebersole and Worster 2007). In fact, Ebersole and Worster asserted that "place-based education strives to contextualize curriculum into the local culture and ecology and bridge the gap between schools and communities" (p. 20). Further, they suggest that place is an excellent way to integrate all aspects of a curriculum and meet the standards. They describe two education methods courses (a science/math course and a language arts/social studies course) that were designed to increase teacher candidates' knowledge, skills, and attitudes about facilitating children's exploration of their place. Ebersole and Worster found that teachers who integrated local ecology and culture into their unit planning could target state standards effectively in all of these disciplines.

Knowing about the natural history of one's local environment, including the local flora and fauna, can contribute to a place-conscious education (Gruenewald 2003), which may boost student achievement (Lieberman and Hoody 1998; Theobald and Curtiss 2000). In learning about place, students can discover connections between ecology, social science, and the arts.

What's in a Name?

By encouraging teachers (and thus, their future students) to use taxonomic keys and learn the names of common trees and plants in their area, we are teaching them to combat their own plant blindness (Wandersee and Clary 2006). When students learn a name and the associated characteristics for something, they are attending to it. The process of comparing and contrasting flower and leaf structures is a higher-level skill (Bloom 1956, revised in Anderson et al. 2001), and allows students to form schema and connections between schema that will help them better understand both the nature of science and nature itself.

However, the external constraints on teachers can make it difficult to convince them about the importance of teaching about plants. Time and space constraints may prevent teachers from growing plants in the classroom, and budget constraints or liability concerns may keep teachers from taking students outside. While trying to gauge interest in a teacher workshop on plant taxonomy and nature-deficit disorder, we came into close contact with these constraints. One district science coordinator scolded that our objective of having teachers learn the names of twenty common local plants was unreasonable, because "we're trying to get away from memorizing now." She went on to suggest we develop a workshop on molecular genetics or something else teachers "needed."

While memorization is part of the process of learning about the land around us, it is hardly the main learning outcome. When we learn the names of plants, we are learning part of their history. Every plant has a story, and their names, both common and scientific, can help tell that story. For example, the genus name *Dracaena* comes from the Ancient Greek δρ καινα, meaning female dragon. Plants in this genus include the so-called Dragon Trees, which, according to legend, sprouted from the blood of Landon, the hundred-headed dragon slain by Hercules. The genus includes the Canary Island Dragon Tree, *Dracaena draco*, which produces a red resin known as dragon's blood, often used for ancient Roman magic and alchemy. If this story inspires students to want to get a look at a dragon tree, they might have a relative in their home. Relatives of the dragon trees include shrubby plants often sold in the USA as ornamental plants, like the *Dracaena fragrans*. *D. fragrans* is commonly called a "corn plant" because its leaves resemble the leaves of a corn plant, but the common name is deceptive since it is not closely related to corn plants. Instead, *D. fragrans* is a shrubby dracaena, or sort of a domesticated dragon tree, with honeysuckle–lilac scented, fragrant flowers that led to the specific epithet "fragrans."

Like us, plants have names that show their relationships with each other and their own unique characteristics. Learning and sharing plants' names help make the plants themselves more visible, more real, and more important. However, learning names alone may not eliminate plant blindness. Schussler and Olzak (2008) found that college students shown a slideshow including both plant and animal images recalled significantly more of the animals than plants shown, even if the students had taken a botany course. However, we suggest that giving students the opportunity to learn not only local plants' names but also their stories will be a good strategy to combat plant blindness; even though teachers and students may still have a preference for animals, their appreciation and knowledge of plants will increase. The goal is not rote memorization, it is appreciation and understanding, and a good plant mentor (which can be any teacher that helps students learn about plants) is important to achieving that goal.

There is a great deal of science content for teachers to "cover" in a short amount of time, and "learn the names of the plants in your area" is not stated explicitly in any state's standards. However, giving students the opportunity to learn the names, structures, and functions of the plants in their own communities can fit well into any science curriculum. Thoughtful design and some of the ideas included here and elsewhere can allow teachers to use plants to teach about evolution, ecology, and classification.

Teaching Plants and Meeting Standards

In elementary classrooms, science is often overlooked, in part because of schools' and standardized tests' emphasis on mathematics and reading. However, place-based teaching is one way to thoughtfully integrate reading and science. In middle and secondary science classrooms, reading and writing across the curriculum is at least encouraged, if not mandated, and activities that help students develop and investigate the ecological community around them can often help meet these goals.

Wandersee et al. (2006) developed "a writing template for probing students' botanical sense of place (BSP)." The template found in this article is designed to be used with students in phases: the first phase allows students to access their memories of plants from their childhood; the second phase allows them to choose two of the memories and write mini-essays based on them. The third phase is designed for use at the end of a plant-based unit, and helps the student to connect plant biology concepts they have learned with the memories they have described. Wandersee et al. (2006) report that their BSP helped students reconnect to their sense of wonder toward plants, motivated students to learn plant biology, inspired new botanical awareness and appreciation, and helped establish a "plant-centered community of learners" (p. 421). The article includes detailed instructions for using the template in the classroom.

Sanger (1997) described how storytelling is an important way that students may connect to their places. He wrote: "[I]ndividuals have stories that represent their personal histories, just as the land has a story of its own that includes people and their stories" (p. 5). Teachers can take students to a natural area, encourage them to tell stories about this or similar natural areas, and then guide them in an exploration of a new question related to the area. Through this guided inquiry activity, the teacher is reaffirming the students' lives and stories, while encouraging them to add to their knowledge by uncovering some of the stories the land has to tell.

As an example of this storied–guided inquiry approach, let us use the Ginkgo tree. Students in an urban forest might be curious about how an exotic ornamental tree like *Ginkgo biloba* came to be found in the urban area. The history of the tree reveals numerous interesting tidbits, including the fact that *Ginkgo* is considered the oldest genus of living trees, and was around at the time of the dinosaurs. Global learning objectives could be met by showing students where *Ginkgo* came from in China, and how it was cultivated on temple grounds for many years. A group of students might then discover the natural history of the tree, its unique taxonomic position, when it became introduced to the USA, and the properties that make it tolerant to pollutants. Another group of students might decide to design an experiment to investigate how *Ginkgo* seedlings react to various other environmental conditions. As a culmination of this activity, students could do expressive writing activities about the stories the tree might tell if it could. A unit such as this incorporates reading and writing across the curriculum, global learning, science inquiry, and life science standards.

For elementary- and middle-school students, the "go-to" EE curricula, Project Learning Tree, and Project WET, include useful activities that teach about plants and place. Project Learning Tree includes several activities ("To Be a Tree," for example) that allow younger students to examine, compare, and contrast different types of leaves, seeds, and flowers, and begin to learn the names of the trees to which they belong. "Adopt a Tree" from PLT can extend students' understanding by giving them a chance to study and get to know a tree over time. Students can even do a phenology of their tree, improving their science process skills by measuring and recording the tree's bud and leaf growth or senescence over a semester or a year. Students might even compare the dates for "first flowers" or "first leaf" on their tree with data from other places. Project WET, for its part, includes an international supplement, River of Words, which encourages young children to develop and explore their senses of place, and express their place in poetry or artwork. These products can be evaluated statewide and internationally (http://www.riverofwords.org/about/index.html).

Teaching about plant taxonomy aligns well with classification standards; for example, National Science Education Standard (NRC 1996) middle-school life science standard C: diversity and adaptations of organisms includes facilitating students' understanding of the multitude of different species of living things and how their similarities and differences are analyzed, how evolution accounts for the diversity of species, and extinction of species. Although teachers tend to spend a great deal of curricular time teaching the animal kingdom, these principles can be illustrated just as effectively with plants (Uno 1994), and students' tendency to be less interested in plants can be overcome with enthusiasm and interest of a teacher (Strgar 2007). It could be particularly effective to teach about names of local plants while helping students learn about these plants and their place. Later, the local plants studied could be placed in the broader context of known plant taxonomy, in order to show the breadth of the kingdom. Students could be taught the skills to use field guides, not just for plants but also for many other organisms. Taxonomy could lead into lessons in ecology, to demonstrate the importance of plants and other producers in the energy dynamics of an ecosystem.

Evolution is a major standard in high school, as well. One tool that biologists use to illustrate and model evolutionary relationships is a phylogenetic tree. Interestingly, the lay public's misconceptions about how to interpret phylogenetic trees can contribute to general misunderstanding of evolution in general and human evolution in particular. Baum et al. (2005) article describes how and why phylogenetic trees are often misread, and includes some suggestions for how scientists and science educators can help people read them correctly. The article also includes a Web link to quizzes that can be used to assess students' developing understanding of phylogenetic trees. To integrate the understanding of local plants' classification and evolution, we suggest that a lesson on taxonomy be followed by examination of a phylogenetic tree that includes the plants just discussed (such as the sample tree in Fig. 1). Students could then study the phylogenetic tree and discuss how it illustrates similarities and differences between the trees studied.

With resources that are available, why not consider creating a children's garden? Children can be allowed to make choices about the types of plants that will be kept, and they can be given responsibilities commensurate with their ages that will allow them to feel more connected to their place. Wake (2007) describes a case study of a children's garden in New Zealand, including how it can serve as a model for other

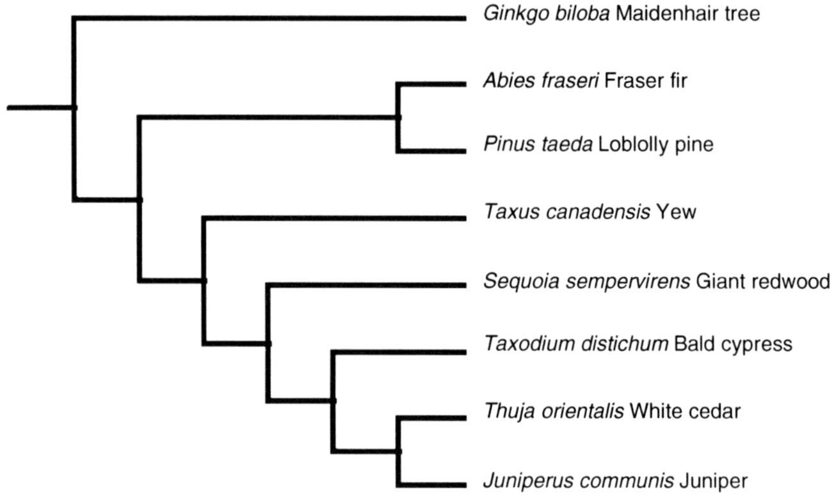

Fig. 1 Sample phylogenetic tree showing relationships of *Ginkgo biloba* to other gymnosperm trees and shrubs based on molecular data

learning gardens. If properly maintained, gardens can be versatile teaching tools. Vegetable gardens are an excellent project for students. In fact, Learning Gate Community School in Florida has structured much of its interdisciplinary curriculum around an organic garden, with impressive gains in both students' knowledge and their sense of place (Howes et al. 2007).

Environmental Education, Sense of Place, and Finally Seeing Plants

One of the foundational documents of environmental education, the Tbilisi Declaration (UNESCO 1977), asserts five goals of EE: awareness, knowledge, attitudes, skills, and action. By teaching our students (and the teachers of our students) to see their environment, including the essential details of plants, we begin at the awareness level. From there, we can begin to address the knowledge of our students by introducing them to ecology and evolution through plants. As students and teachers become aware of and more deeply connected to their own senses of place, attitudes toward their environment cannot help but change. We also seek to give our students the skills to continue developing that sense, by teaching them how to use field guides and how to take care of the land.

If we want students to finally "see" plants, *our* action piece is the most important part. Science teacher educators can (and should) act as a bridge between the research on plant blindness and sense of place, and the practice of science teachers in the classroom. We know that science teachers are faced with constraints that keep them from teaching about plants, and if we expect them to overcome these constraints we need to overcome them ourselves when we teach methods courses. We can all enhance our plant skills. Take a methods class on a walk, pointing out the distinguishing features of the trees on your campus. Have your students transfer their plant understanding to their neighborhoods or the school where they student-teach, by asking them to draw a map that includes trees they can identify. Include plant curricula in life science methods courses whenever feasible; there are well-designed curricula (including PLT and FOSS kits) that preservice teachers may not be exposed to once they are teaching, so give them a chance to explore them when you teach methods courses. A list of other potential lesson sources is included in Appendix 1. We can solve this problem, now that our eyes have been opened.

Appendix 1: Botanical Lessons and Lesson Source Ideas

Hopefully, we have convinced you to start teaching about and with plants. In order to capitalize on your enthusiasm, here are some lessons and resources. The resources meet some or all of the objectives described in the chapter. Our intention is to provide teachers and teacher educators with many ideas and options; however, this is by no means an all-inclusive list. The Botanical Society of America (http://www.botany.org) maintains a list of web links and resources, and that is another great place to look.

Plant Curricula (Aligns to National Science Education Standards)

- American Museum of Natural History (AMNH) "Biodiversity Counts" http://www.amnh.org/education/resources/biocounts/ Has plant identification lessons at http://www.amnh.org/education/resources/biocounts/plant_id.php Plant ecology curriculum materials for grades 6–12 online at: http://www.amnh.org/education/resources/biocounts/ecology.php
- Botanical Society of America's PlantingScience http://www.plantingscience.org This project connects teachers, students, and plant scientists with a venue to connect and develop relationships. Includes inquiry units ("The wonder of seeds" and "The Power of Sunlight") for middle- and high-school classrooms where students can follow the science process by engaging in real research.
- C-Fern®: Using *Ceratopteris richardii* to teach about plants. http://c-fern.org Lessons and ordering information.
- FOSS modules http://www.delta-education.com/science/foss/scopesequence.shtml "Tree" (K); "New Plants" (Grades 1–2); "Plants and Animals" (1–2); "Insects and Plants" (1–2)
- Literature in the Garden http://www.jmgkids.us/index.k2?did=11882&sectionID=2013 Elementary curriculum that incorporates children's literature with garden activities designed by Junior Master Gardener program
- Project Learning Tree ® http://www.plt.org *Pre K-8 Environmental Education Activity Guide.*

- The Private Eye http://www.the-private-eye.com Activities use loupes, observation, and "thinking by analogy" to help students get out, increase their sense of wonder, and explore nature.
- Wisconsin Fast Plants ® http://www.fastplants.org *Brassica rapa* plants that are especially fast cycling, for use in the classroom. Lessons and ordering information.

Plant Classification Guides and Phylogeny Resources

- Tree World http://www.domtar.com/ARBRE/english/index.asp (Designed for 4–6 grade students. Includes a simple online tree identification dichotomous key, and quizzes to test your students' tree ID skill.
- The Missouri Botanical Garden (MBG): The Unseen Garden. http://www.mobot.org/MOBOT/Research/unseengarden/unseengarden1.shtml Includes information on systematics and how scientists study plants.
- Missouri Botanical Garden and University of Missouri's Angiosperm phylogeny Web site. http://www.mobot.org/MOBOT/research/APweb/welcome.html Very extensive taxonomy resource.
- Tree of Life web project http://www.tolweb.org/tree/ Phylogeny and phylogenetic trees.

Latin and Legends: Plant Names and How They Got Them

- Barnette, M. (2005). *A Garden of Words.* Lincoln, NE: ASJA Press.
- Virginia Tech (dendrology): The meanings of Latin names http://www.cnr.vt.edu/dendro/dendrology/syllabus/meanings.cfm
- Wells, D. & Patterson, I. (1997). *100 Flowers and how they got their names.* New York, NY: Algonquin.
- Wildflower name origins http://www.wildflowerinformation.org/WildflowerNames.asp Includes information on common names, botanical names, and brief description of systematics.

Other Good Resources for Teaching Outside and About Plants

- Photographic Atlas of Plant Anatomy http://botweb.uwsp.edu/anatomy/
- Exploratorium: Science of Gardening. http://www.exploratorium.edu/gardening/feed/index.html Interactive videos and information on composting, carnivorous plants, garden vegetables, and soil science.
- Grissino-Mayer, H. D. (University of TN-Knoxville) "Ultimate Tree-Ring Web Pages" http://web.utk.edu/~grissino/ Information on dendrochronology.
- Plants in Motion http://plantsinmotion.bio.indiana.edu/plantmotion/starthere.html Indiana University-based web site includes many good time-lapse photography movies (QuickTime) so that your students can see that plants really do move.
- Avoid misconceptions when teaching about plants with this D.R. Hershey (2004) article available online at: http://www.actionbioscience.org/education/hershey.html

- More misconceptions to avoid when teaching about with this D.R. Hershey (2005) article available online at: http://www.actionbioscience.org/education/hershey3.html Junior
- Master Gardener Program http://www.jmgkids.us/ Part of 4-H, information, lessons, and instructions on starting your own JMG group (includes ordering information for several curricula, including Wildlife gardener, literature in the garden, health and nutrition from the garden).

References

Anderson, L. W., Krathwohl, D. R., Airasian, P., Cruikshank, K. A., Mayer, R., Pintrich, P. R., et al. (Eds.). (2001). *A taxonomy for learning, teaching, and assessing — a revision of Bloom's taxonomy of educational objectives*. New York, NY: Addison Wesley Longman.

Ashworth, S., Boyes, E., Paton, R., & Stanisstreet, M. (1995). Conservation of endangered species: what do children think? *Journal of Environmental Education and Information, 14*, 229–244.

Barman, C. R., Stein, M., Barman, N. S., & McNair, S. (2003). Students' ideas about plants: Results from a national study. *Science and Children, 41*, 46–51.

Baum, D. A., Smith, S. D., & Donovan, S. S. S. (2005). The tree-thinking challenge. *Science, 310*, 979–980 (supporting online material available at http://www.sciencemag.org/cgi/content/full/310/5750/979/DC1)

Bebbington, A. (2005). The ability of A-level students to name plants. *Journal of Biological Education, 39*, 62–67.

Bloom, B. S. (Ed.). (1956). *Taxonomy of educational objectives: The classification of educational goals*. White Plains, NY: Longman.

Biggs, A. (2004). *Biology: the dynamics of life*. New York, NY: McGraw-Hill/Glencoe.

Bowker, R. (2004). Children's perceptions of plants following their visit to the Eden Project. *Research in Science and Technological Education, 22*, 227–243.

Brasher, J. (2006). The southern rocky mountain interactive flora (SRMIF) and factors correlated with recognition of plants and mammals (Unpublished Dissertation). Greeley, CO: University of Northern Colorado.

Brewer, C. (2002). Conservation education partnerships in schoolyard laboratories: A call back to action. *Conservation Biology, 16*, 577–579.

Derr, V. (2002). Children's sense of place in northern New Mexico. *Journal of Environmental Psychology, 22*, 125–137.

Ebersole, M. M., & Worster, A. M. (2007). Sense of place in teacher preparation courses: Place-based and standards-based education. *Delta Kappa Gamma Bulletin, 73*(2), 19–24.

Gatt, S., Tunnicliffe, S. D., Borg, K., & Lautier, K. (2007). Young Maltese children's ideas about plants. *Journal of Biological Education, 41*, 117–121.

Gruenewald, D. A. (2003). Foundations of place: A multidisciplinary framework for place-conscious education. *American Educational Research Journal, 40*, 619–654.

Harvard-Smithsonian Center for Astrophysics. (1995). *Minds of their own*. Burlington, VT: Annenberg Media.

Hershey, D.R. (1993). Plant neglect in biology education. *Bioscience, 43*.

Hershey, D. R. (1996). A historical perspective on problems in botany teaching. *The American Biology Teacher, 58*, 340–347.

Howes, E. V., Lim, M., & Solomon, S. (2007, October). *Developing an Environmental "Sense of Place" for Science Teacher Education* (Paper presented at the annual meeting of the Southeastern Association of Science Teacher Education) Valdosta, GA.

Kinchin, I. M. (1999). Investigating secondary-school girls' preferences for animals or plants: a simple "head-to-head' comparison using two unfamiliar organisms. *Journal of Biological Education, 33*, 95–99.

Lieberman, G., & Hoody, L. (1998). *Closing the achievement gap: Using the environment as an integrated context for learning*. Poway, CA: Science Wizards.

Lindemann-Matthies, P. (2005). 'Lovable' mammals and 'lifeless' plants: How children's interest in common local organisms can be enhanced through observation of nature. *International Journal of Science Education, 27*, 655–677.

Lindemann-Matthies, P. (2006). Investigating nature on the way to school: Responses to an education programme by teachers and their pupils. *International Journal of Science Education, 28*, 895–918.

Lock, R. (1995). Biology and the environment- A changing perspective? Or "there's wolves in them there woods!". *Journal of Biological Education, 29*, 3–4.

Louv, R. (2005). *Last child in the woods: Saving our children from nature-deficit disorder*. Chapel Hill, NC: Algonquin.

Mack, A. (2003). Inattentional blindness: Looking without seeing. *Current Directions in Psychological Science, 12*, 180–184.

Moon, T. J. (1921). *Biology for beginners*. New York: Henry Holt.

National Research Council (NRC). (1996). *National Science Education Standards*. Washington, DC: National Academy Press.

Nichols, G. E. (1919). The general biology course and the teaching of elementary botany and zoology in American colleges and universities. *Science, 50*, 509–517.

Orr, D. W. (1994). *Earth in mind: On education, environment, and the human prospect*. Washington, DC: Island Press.

Paraskevopoulos, S., Padeliadu, S., & Zafiropoulos, K. (1998). Environmental knowledge of elementary school students in Greece. *The Journal of Environmental Education, 29*, 55–60.

Peterson, N. (1982) *Developmental variables affecting environmental sensitivity in professional environmental educators*. (Unpublished master's thesis). Carbondale, IL: Southern Illinois University.

Project Learning Tree. (2006). *Exploring environmental issues: Places we live*. Washington, DC: American Forest Foundation.

Sanger, M. (1997). Sense of place and education. *Journal of environmental education, 29*, 4–9.

Schussler, E. E., & Olzak, L. A. (2008). It's not easy being green: Student recall of plant and animal images. *Journal of Biological Education, 42*, 112–118.

Simmons, D. (1996). Teaching in natural areas: What urban teachers feel is most appropriate. *Environmental Education Research, 2*, 149–157.

Stegner, W. (1992). *The sense of place*. New York: Random House.

Strgar, J. (2007). Increasing the interest of students in plants. *Journal of Biological Education, 42*, 19–23.

Theobald, P., & Curtiss, J. (2000). Communities as curricula. *Forum for applied research and public policy, 15*(1), 106–111.

UNESCO. (1977). The Tbilisi declaration. Available: http://www.gdrc.org/uem/ee/tbilisi.html. Accessed 12 April 2008.

Uno, G. E. (1994). The state of precollege botanical education. *The American Biology Teacher, 56*, 263–267.

Vorkinn, M., & Riese, H. (2001). Environmental concern in a local context: the significance of place attachment. *Environment and Behavior, 33*, 249–263.

Wake, S. J. (2007). Designed for learning: Applying "learning-informed design" for children's gardens. *Applied Environmental Education and Communication, 6*, 31–38.

Wandersee, J. H. (1986). Plants or animals – which do junior high school students prefer to study? *Journal of Research in Science Teaching, 25*, 415–426.

Wandersee, J. H., & Schussler, E. E. (1999). Preventing plant blindness. *The American Biology Teacher, 61*, 84–86.

Wandersee, J. H., & Schussler, E. E. (2001). Toward a theory of plant blindness. *Plant Science Bulletin, 47*, 2–9.
Wandersee, J. H., & Clary, R. M. (2006). *Advances in research towards a theory in plant blindness*. Chapter in the Proceedings of the 6th International Congress on Education in Botanic Gardens at Oxford University. London, England: Botanic Gardens Conservation International.
Wandersee, J. H., Clary, R. M., & Guzman, S. M. (2006). A writing template for probing students' botanical sense of place. *The American Biology Teacher, 68*, 419–422.
Wood-Robinson, C. (1991). Young people's ideas about plants. *Studies in Science Education, 19*, 119–135.
Worster, A. M., & Abrams, E. (2005). Sense of place among New England commercial fishermen and organic farmers: implications for socially constructed environmental education. *Environmental Education Research, 11*, 525.

Place-based Inquiry: Advancing Environmental Education in Science Teacher Preparation

Somnath Sarkar and Richard Frazier

Introduction

Future generations face unprecedented environmental challenges such as global climate change, worldwide food crises, species extinctions, and increasing demands for energy. Teachers play a vital role in preparing students to address such complex and interconnected problems. Place-based inquiry provides a rich setting to educate our science teachers so that they can help their students deal with these challenges. This teacher education strategy emphasizes the authentic practice of science, develops deeper content knowledge in an interdisciplinary context, and focuses on questions of environmental importance.

The Guidelines for Excellence by the North American Association for Environmental Education emphasize that, "Environmental education is learner-centered, providing students with opportunities to construct their own understandings through hands-on, minds-on investigations. … Environmental education provides real-world contexts and issues from which concepts and skills can be learned" (NAAEE 2004, p. 1). A comparable recommendation from the National Research Council has identified inquiry and direct experience with scientific phenomena as best practices in science teaching (National Research Council 1996). Place-based inquiry makes it possible to put these recommendations into action in the teaching and learning of science and to broaden the applicability of environmental education. Our approach defines inquiry as the practice of science (Duschl et al. 2007) and integrates environmental education through place-based pedagogy (Sobel 2004; Swope 2005).

S. Sarkar (✉)
Department of Biochemistry Chemistry and Physics, University of Central Missouri, Warrensburg, MO 64093, USA
e-mail: Sarkar@ucmo.edu

R. Frazier
Department of Elementary and Early Childhood Education, University of Central Missouri, Warrensburg, MO 64093, USA
e-mail: frazier@ucmo.edu

A.M. Bodzin et al. (eds.), *The Inclusion of Environmental Education in Science Teacher Education*, DOI 10.1007/978-90-481-9222-9_11,

Place-based pedagogy uses a particular place for the context of investigation where the integration of a variety of scientific and environmental concepts occurs. Students make meaningful connections with the physical and natural world and seek solutions to environmental problems through a multidisciplinary approach. Investigations focus on places from a range of environments, such as a free-flowing stream, an untouched section of the schoolyard, a local park, or even the school building itself. Bodzin (2008), Cronin-Jones (2000), and Martin (2003) all discuss the potential of using the schoolyard itself as an important focus for science and environmental education.

Focusing on questions about a place and embedding the questions in an environmental context enable teachers and students to employ a wide variety of scientific protocols. Investigations can include field surveys, field experiments, naturalistic observations, and controlled laboratory experiments. When questions are tied to a particular place, students are likely to be familiar with the characteristics of the place. When the students are connected to the place, they take ownership of the investigations, and thus the investigation becomes personally meaningful. The combination of caring and science often leads to thoughtful action on the part of students. Placed-based pedagogy extends beyond the boundaries of traditional science lessons, and places science learning in a larger social and environmental context.

When a teacher knows the "right" answer to a question or problem and then directs students toward that answer, those students may never consider the scientific evidence and argument that lead to the best possible answer. On the other hand, when an answer is not predetermined for teachers and students, they have the opportunity to work together toward solutions that are grounded in evidence and reasoning and that are born of an authentic (rather than an authoritarian) practice of science.

This chapter describes a 3-year professional development project for inservice science teachers (Sarkar & Frazier 2008). The goal of the project was to develop teachers' facility with content-rich inquiry in the context of a particular place. The chapter is divided into three sections. "Summer Learning Experience" details a summer workshop where teachers learned how to use place-based pedagogy effectively. "Application in the Classroom" presents three cases in detail and makes reference to additional studies. "Discussion" includes challenges, reflections on successes, and some recommendations for preservice science teacher preparation.

Summer Learning Experience

A 60-h summer workshop was offered annually for 3 years to provide the tools to implement place-based pedagogy and environmental education in science classrooms. More than 80 inservice teachers participated in the PD experience. Most taught the middle grades but the range extended from grades 5–12. Participants attended the summer workshops and conducted inquiry-based projects in their classes. One-third of the total participants continued for 3 years, and we accepted new teachers each year over the 3-year period.

We used examples that focused on air, water, and soil, three primary categories which incorporate big ideas in life science, physical science, and environmental education. We used a variety of data-gathering measures to help us design the activities. Ongoing formative data measures were employed to modify the activities on a continual basis. We assessed teachers' content knowledge and skills throughout the summer workshop with pretests, performance tasks, observations, and discussions. We were particularly interested in teachers' depth of knowledge in the sciences as well as their facility with skills central to inquiry.

In one of the initial place-based activities, teachers in small groups surveyed a local recreational area that included woods, lakes, heavily used grassy areas, small wetlands, streams, and small hills. The study area is surrounded by agricultural land, a golf course, and residential neighborhoods. We provided directions for surveys that focused on plant diversity along a trail; bird diversity in woods, grassy area, and lakeside; soil characteristics along the slope of a hill; humidity near the lake and at various distances from the lake; water quality (pH, dissolved oxygen, nitrate) and temperature of water at various locations along the lake; and the environmental impact of the proposed development of a conference center in this area.

From the surveys, teachers gleaned a long list of relevant concepts that corresponded to those in their schools' science curriculum. In focus groups, teachers considered a big idea such as "water quality" or "air quality" and developed concept maps completed with specific details from the survey activities. They also generated more focused questions for subsequent investigations.

The investigation of place-based problems requires understanding multiple dynamic and interacting factors in the environment and having the necessary skills to isolate variables for measurement. Before conducting subsequent investigations in the field, we helped teachers perform controlled experiments in the laboratory to learn about the importance of isolation, manipulation, and measurement of specific variables. The experience of conducting investigations under controlled conditions prepared them for deeper and more meaningful place-based inquiry. In one example, we chose to investigate the rate of evaporation of water under controlled conditions because during discussion teachers developed questions about the effects of various physical parameters on the lakes at the survey site. Some of these questions involved variables that required careful isolation and accurate measurement. For example, one group, who proposed that the temperatures of water in sunny areas should be higher compared to shady areas, did not take account of currents in the lakes or the "movement" of shade as the day progressed.

In response to a question about how wind speed might affect the rate of evaporation and humidity at the lake, we guided teachers to perform an experiment in the laboratory with fans set at different speeds at fixed distances from equal amounts of water placed in containers of the same size, shape, and composition. This experiment isolated the wind-speed variable from all other variables such as light intensity, humidity, temperature, atmospheric pressure, and properties of the container. In another experiment, teachers set up hot plates at different temperature settings and compared the effect of temperature on the rate of evaporation. In this case, the temperature was the isolated variable and the rest of the variables that could have affected the rate of evaporation were kept constant throughout the experiment. In all experiments, groups gathered and analyzed data and then presented their conclusions to the entire class.

From the initial surveys, controlled experiments, and continual discussion, teachers in groups modified their questions pertaining to the survey site. We provided equipment and taught them the necessary skills of measurement before the teachers returned to the site to investigate their specific place-based questions. Since the lakes are located near agricultural land and a golf course, one of the questions a group of teachers investigated had to do with the water quality of one lake and whether this was affected by runoff that entered the lake through several inlet streams. Teachers tested for nitrates, dissolved oxygen, phosphates, and pH at various locations of the lake and also gathered water-quality data from the inlet and outlet streams. Teachers found that there were no appreciable differences in their results from the lake and the streams and inferred that the streams were not influencing the level of these pollutants in the lake. As they compiled their results, they asked what parameters were used by the state water-quality monitoring agencies. They knew that nitrates and phosphates are commonly used fertilizers and that an excess of these pollutants in water would result in less dissolved oxygen. They wondered if the city had taken any measures to control runoff and what those measures might be. These follow-up questions were not investigated during the workshop because of time constraints but illustrate how place-based studies lead to deeper inquiry in an interdisciplinary context.

Teachers employed technology in appropriate and effective ways during their summer investigations. They used topographic maps, GPS systems, field microscopes, digital cameras, and a number of other measuring devices. Teachers also built instruments to measure variables defined in their specific experiments. This design activity helped them to understand the nature of measurement including concepts of accuracy and precision. In a representative soil study, one of the teachers designed and built a compaction tester. The device consisted of a cylinder in which a sliding rod could be dropped from a standard height. To use the compaction tester, the foot of the cylinder was placed over the spot to be sampled and then the rod was raised to a fixed height and dropped. The rod penetrated the soil to some measurable depth and provided nonstandard, but controlled values for soil compaction. These values were correlated with types of soil, plant diversity, and plant health across various areas.

During the last phase of the summer workshop, teachers developed a "task analysis plan" focused on a place in their schoolyard or local environment. The guidelines for the task analysis plan are provided in Appendix 1. Teachers developed questions associated with air, water, soil, or other science areas connected with their curriculum. Their plans included testable place-based questions, investigation design, assessment items, concepts and standards covered, time management, organization, and safety. The plan served as a flexible tool to guide investigations.

Application in the Classroom

In this section we describe three cases where teachers employed place-based pedagogy in their schools. Additional projects are mentioned briefly. There were a variety of ways in which projects were conceived and developed. With information from various assessments (such as pretests, concept maps, and performance tasks), teachers refined their initial question(s) and modified their task analysis plans. Three retired university science faculty served as mentors and provided necessary help to the teachers throughout the project. Mentors played an important role right from the beginning by visiting teachers and their students regularly to offer guidance. Teachers and mentors communicated electronically with each other extensively throughout the project. Mentors answered queries, directed teachers to relevant resources, and provided assistance in organization, evaluation, and analysis of student-generated data. Sometimes mentors even facilitated discussion among students and the teacher to help them draw reasonable conclusions and generate new questions for further inquiry. Teachers occasionally asked experts from external agencies to provide guidance in relevant areas. Whenever findings suggested any potential impact on human health, experts were consulted.

The projects varied in terms of organization, instructional design, management style, and resources available to the teachers. All projects followed school safety protocols aligned with NSTA recommendations. In most projects, groups of students gathered data for different parts of an investigation and then contributed their results to a combined database for the entire class or classes. Students recorded their observations and data in journals and periodically discussed them in class. The journals contributed to students' growing science literacy and provided a tool for both reflection and assessment.

Many of the projects showed a transition from teacher-led investigation to student-directed inquiry. Teachers observed students' actions, discussions, and journal writing throughout the investigations, and they used students' reports and posttests for summative evaluation. At the end of the school year, students attended a science symposium at our university. They displayed and discussed posters and made oral presentations to an audience of their peers from other schools, teachers, administrators, parents, and university students and faculty.

Case Study I: Mystery Water

In a rural school in the Missouri Ozarks, teachers and students had wondered about the source of a small stream that seemed to originate from a seep in the playground. Ms. S. and her sixth-grade class of 20 students decided to investigate the nature of the source. Although this seep was in the schoolyard, students found out that nobody in the school really knew the source of the water. Students observed that the flow of the stream was fairly constant and concluded that storm water was not a primary contributor to the seepage. Students then suspected that the source was a sewer leak. With the help of their mentor, the teacher and students designed a study to gather evidence to test their suspicion. They measured the depth of the water for several weeks at different sites on Mondays and Fridays to rule out the possibility that the use of water from school facilities contributed to the flow. If it were from the school facilities, students reasoned that there would be more water on Friday compared to Monday because of very little water use during the weekend.

Students tested the seep water and a control (tap water) for dissolved oxygen, nitrates, phosphates, pH, chlorine, and bacteria, all indicators of water quality. Students had considered possible sources as storm water drainage, a sewer leak, a leak from the public water supply, and a spring. The set of tests in combination with observations of rainfall and flow rate were designed to rule out any three out of the four possibilities. Results indicated that the seep water in the playground was spring water. Enthusiastic about the finding, the students named the previously undocumented spring. Afterwards the Missouri Department of Natural Resources was contacted and experts from the agency recommended a plan to preserve the character of the spring.

Over the next 2 years the project evolved from a teacher-led investigation to inquiries more centered on students' questions. Students asked what organisms they could find in the stream originating from the spring, and if the kinds and number of organisms would change with distance downstream. In the last year of the project, students traced changes in water quality, vegetation, and faunal diversity along the water course that eventually led to a large lake.

This project began with a scientific question, and the investigation proceeded in accord with accepted scientific practice. Students used concepts from physics, such as the flow rate and velocity, from chemistry such as solubility and physical change, and from biology such as the diversity of bacteria, macro-invertebrates, and plants. Students responded to their conclusions by naming the spring and initiating plans for its preservation and enhancement as a school resource.

Case Study 2: They're Buggin' Me

In a large middle school in a suburban district in central Missouri, Ms. H., a sixth-grade science teacher, noticed an area in the schoolyard, previously developed as an outdoor classroom but neglected for years. Ms. H. realized that the site could provide engaging learning opportunities for her 120 students. During the initial exploration of the site just after the school had opened in late summer, Ms. H. and her students identified three distinct areas (called nature centers) with characteristics of prairie, wetland, and woodland habitats. During the initial exploration, Ms. H. observed that her students were intrigued with the insects they found. Ms. H. helped the students develop a survey question on how the types of insect might vary across these three habitats. She and her mentor designed a protocol to count the types of insects in a consistent and scientific manner.

Several times a week for 6 weeks, students surveyed the number and the types of insects from the three areas, described and photographed the specimens, and identified them from field guides. During the surveys students noticed that the number of insects they found seemed to vary from the beginning of the study in late summer to the end of the study in early fall. Ms. H. also noticed that the number of insects depended on the time of the day. While analyzing the data during a class discussion, students studied the varying numbers over the course of a day as well as over the duration of the 6-week study. They wondered how temperature related to the number and the kinds of insect found. The original question, which had been introduced by Ms. H., involved insect populations across habitat. New questions from the students retained this interest but added the effect of temperature on insect numbers and on insect diversity.

In the second year of the study students measured the air temperature with a handheld thermometer and used graphs to correlate different types of insects with temperature. The large set of data gathered by her students from different periods provided the opportunity for making the study a sound field-based investigation. In the third year, students kept the same questions but refined the technique by directly measuring the temperature of the insects using an infrared (IR) thermometer. The IR thermometer provided accurate and precise information about insect body temperature.

Ms. H. reflected on several aspects of the project. She noted that the use of her SMART Board greatly enhanced the collaborative nature of data entry and analysis on an ongoing basis. Students could see the emerging patterns and would point out data that appeared discrepant. She remarked that her own attempts to make sense of class data led her to pay more attention to students' ideas and thinking.

The size and complexity of the project involved such a variety of tasks that all students found areas in which to be successful. In addition, the project provided a rich set of examples that enabled students to measure physical properties and understand concepts in ecology. Students referred to their project experience throughout their science curriculum. In the case of food chains and food webs in the study of ecology, textbooks often use dramatic examples like a mountain lion killing a deer. Children who were involved in the place-based investigation referred to their observations of spiders and praying mantises taking insects as prey. According to the teacher, her students developed a richer understanding of the pervasive importance of food chains and predator–prey relationships due to their experiences in the outdoor classroom study of insect diversity.

In addition to the refined questions and techniques, the project grew in other ways, especially in response to students' concerns and interests. Ms. H. invited her companion special-education teacher and her students to participate in the project. They embarked upon restoration and expansion of the nature centers, another example of a place-based investigation resulting in caring for the environment. The enthusiasm of these teachers and their students established the project as an integral and recognized part of the school science curriculum.

Case Study 3: Structures in the School Environment

In an urban school, Ms. M wanted to engage her seventh-grade students in a place-based problem to explore concepts in physical science. She proposed to her students that they design and conduct an experiment on the effect of temperature on cracks in the schoolyard concrete. Initially, students used calipers to measure the width of the cracks in different parts of the schoolyard at different times of the day for a week. Ms. M's mentor suspected that calipers would not adequately register changes because of the irregularity of the cracks and the magnitude of the change that would occur during heating and cooling. With suggestions from her mentor, Ms. M. developed a technique to measure the changes of the width in the cracks in the concrete. The method involved taking a digital image of a ruler laid across the crack and marking the position for subsequent photos. The pictures were blown up and a scale was devised to convert the width in the magnified view to that of the actual crack. Students used graphs to correlate widths with air temperatures for the dates and times and found that the width decreased with increasing temperature.

In the following semester, students took up a different project, but one that also involved the human-made environment of the school and other buildings. During a class discussion on the inclined plane, Ms. M's students wanted to know more about the design of access ramps. Because of the students' interest in this topic and Ms. M's previous positive experience with place-based pedagogy, she advised students on a procedure to address their question. They determined the slopes of the ramps and also carried out force measurements on several ramps using a wheelchair, spring scale, and a student as a load. During the investigation, Ms. M presented information to students about building codes for accessibility. The students reported their findings in a letter to the principal, stating that while all ramps met the legal requirements, new construction should follow guidelines that would provide easier access. This investigation led to a thoughtful action on the part of the student showing care and concern for their school environment.

Ms. M. became so convinced of the value of a place-based approach for student learning that when she was assigned to teach life sciences, she applied the approach to another schoolyard project. She led a group of students and teachers in selecting an area in the schoolyard in order to design and construct a rain garden to solve an erosion problem. In the beginning, students analyzed storm water drainage patterns and conducted soil percolation tests in different parts of the schoolyard to find the best place to build a rain garden. While the size, position, and the physical parameters of the rain garden were being established, students investigated native plants in order to identify those best suited to the particular habitat. A number of classes joined in the planting and began to use the rain garden for their own place-based investigations. The project brought greater educational value by involving additional students and enhancing the ecological resources of the school.

Other Participant-Place-Based Projects

Other projects carried out by the participant teachers and their students suggest the range of possibilities for place-based inquiry. These investigations, like those described in more detail above, involved a variety of scientific protocols and included surveys, experiments, measurements, tests, and counts. Water was the topic for several studies. Students from one school investigated water quality in a favorite swimming stream. At a different school, students wondered about the quality of drinking water from wells in their homes. In another location, a schoolyard pond was rejuvenated and changes in flora and fauna were monitored. In a school near a popular recreation lake, students were concerned about the effect of old and inadequate septic systems. They studied the difference in levels of nitrogen, potassium, phosphorus, and pH in soil collected from both populated and non-populated areas near the lake. Weather captured the attention of a class who wondered about the extent that weather indicators would vary across the schoolyard from data reported in weather broadcasts. Grasshoppers appeared on walls at one school, and students investigated the effect of solar heating on the temperature of the bricks and possible correlations with the number of insects seen. At a school near heavily travelled streets, students designed collection devices and sampled particulate matter from the air at a number of locations on the school site. These projects developed in a variety of ways, but all were shaped by students' concerns about the local environment, their questions and ideas, their teachers' instructional interests, and mentors' suggestions and assistance.

Discussion

One of the most important questions in science is: "How do we know what we know?" Practicing scientists study background literature, gather initial exploratory data, formulate testable questions, design and conduct experiments, and analyze and interpret results in order to reach reasonable conclusions. The knowledge becomes the foundation for asking new questions. A scientist makes decisions about a course of action at different stages of the process and revisits issues. Scientific inquiry methods are sometimes perceived to be linear but rather they are cyclic and dynamic in nature. Our approach to integrating environmental education in science teacher preparation involves taking the teachers through learning experiences where they function like practicing scientists. They use their experiences to develop strategies for teaching science in an environmental context.

National organizations such as AAAS, NSTA, NRC, and NAAEE advocate hands-on, minds-on scientific investigations as the best way to teach science, but the lists of recommended concepts remain lengthy. When teachers think that they have to cover these concepts separately, they face the problem of not having enough time to engage students in deeper inquiry. The case studies described above show that each placed-based investigation incorporates several science concepts and skills within a unit. These concepts are covered in an integrated rather than in an isolated fashion, and the skills are developed within a context. Place-based pedagogy provides a solution to the challenges related to the coverage of a broad range of topics in science curriculum through deeper and more meaningful inquiry.

Successful implementation of the place-based approach required us to individualize instruction because teachers were at different stages of professional growth. These differences included their depth and breadth of content knowledge and their ability to engage their students in inquiry science practices. In particular, teachers' skills varied in devising testable questions, designing experiments, acquiring data, conducting analysis, and drawing conclusions. We assisted individual teachers in areas such as learning a new instrument, analyzing data to recognize patterns, designing instruction assessments, understanding particular science content, and finding additional resources. Mentors played an important role in this aspect. For example, in the case of the insect diversity study, the mentor assisted the teacher in organizing a vast amount of data in order to examine patterns and draw conclusions. In the study of the schoolyard spring, the mentor provided the teacher with additional content knowledge about macro-invertebrates and vegetation that related to healthy water. Even though this individualized instruction was time-consuming, we felt that the successes in professional growth and student achievement made it worth pursuing.

We noticed that in successful projects teachers were flexible and prepared. They exhibited clear learning goals, provided opportunities for students to take ownership, used assessment to tailor instruction, and made use of assistance from mentors. A common characteristic of successful investigations included extensive dialogue between teachers and their students, teachers and the mentors, and the teachers and the project investigators. Most importantly, in all projects the dialogue centered on data.

The participant teachers grew in several ways throughout the project. They took more action for themselves and became more self-directed. They addressed questions where they did not know the answers in advance, and they learned the necessary skills to develop possible answers. They felt confident in pursuing new directions and learning new teaching methods when they felt that their students would learn better. They also started to integrate their curriculum and delved into extended in-depth inquiry instead of small segregated units.

Participant teachers also demonstrated professional leadership by working with their administrators and fellow teachers in reshaping school curriculum and programs. They often recruited other teachers into school-wide place-based projects. Some teachers presented their work at regional and national meetings to share their successes. They also reflected on how place-based pedagogy enabled them to foster deeper inquiry within the existing demands of school curriculum (Frazier et al. 2008).

Teachers brought their students to annual science symposiums to disseminate findings from their investigations. Students consistently demonstrated high levels of enthusiasm and excitement during their oral and poster presentations. They explained their findings to an audience consisting of their peers, teachers, school administrators, and others. Students responded to questions from the audience with confidence and clarity. The level of exchange reflected the depth of engagement of students in their place-based investigations. The communication that occurred among teachers and their students during the symposiums displayed the characteristics of a scientific community.

Implications for Preservice Teacher Education

The project described in this chapter was designed for the professional development of inservice teachers. However, several elements are applicable to preservice teacher preparation. The investigation of questions tied to a familiar place can have a great impact on preservice teachers' understanding of science. When preservice teachers view science as a body of "fixed and furnished" information, they encounter great difficulty in addressing new questions and challenges. Experiences with place-based investigation can mitigate the effects of such preconceptions about science and can move a preservice teacher's view toward an understanding of science as a way of knowing. Learning how to investigate new questions, using the practices and tools of science, promotes the realization that teachers do not need to know answers to all questions in advance in order to teach science effectively.

From our multiple assessments, we realized that some inservice teachers in our project had difficulty in recognizing significant variables, and some would overestimate the effect of a suspected variable. We also observed that many teachers found it challenging to recognize patterns in data and to develop testable hypotheses. In one example, a group of teachers predicted that air and soil temperature would decrease as they ascended a hill with an elevation of only 50 ft. They measured the land and the air temperature at different points along the trail to the top and found that the temperature did not follow any correlation with altitude. They had not realized that what might be true for a change in altitude of 6,000 ft in the Colorado Rockies would not be observed over a much smaller change in elevation. Place-based investigations would provide preservice teachers with greater experience in the meaningful interpretation of data and the generation of reasonable explanations. Such an experience will help them respond to students when they encounter anomalous results.

It is important for preservice science teachers to have opportunities to work with K-12 students in schools as part of their field experience. Some of the projects and ideas described in this chapter can form the basis of appropriately scaled projects conducted by interns and students under the supervision of host teachers and university faculty. It is equally important for preservice science teachers to have the opportunity to conduct their own place-based investigations so that they gain experience in the practice of science. Such opportunities could be a part of science content courses designed for teachers as well as courses on science teaching methods. Any university or college campus could serve as the setting for projects similar to those conducted by the teachers and students described in this chapter.

The application of the content knowledge in a place-based setting encourages the development of deeper content knowledge. For example, in the first case study on the schoolyard spring, the investigation of water quality through chemical tests led to further inquiry into organisms that live in healthy water. Place-based investigations provide the opportunity for preservice teachers to apply their content knowledge in ways that go beyond their regular science courses. Setting the context of inquiry in familiar places brings immediate relevance and provides multiple opportunities to connect a variety of disciplines, concepts, and concerns. Place-based inquiry equips teachers for the vital role in preparing students to solve multifaceted environmental problems of the present and the future.

Acknowledgements This work was funded by Missouri Department of Higher Education Title II Teacher Quality Improvement grant. Pseudonyms have been used for all teachers. We would like to acknowledge the contributions of Dr. Mike Powers, Dr. Steve Mills, and Dr. Terry Berkland for their outstanding service as mentors. We also acknowledge Mr. Vance Shook who handled many logistical details during the course of the project.

Appendix 1: Task Analysis Plan Guidelines

Design Phase: Curricular details

- Question(s), problem(s), and the general topic that the project will be based on
- Rationale for why the question(s) are worthy of investigation
- A detailed concept list, map, or web that will be part of this field project (relate to national science standards and environmental education standards for the specific grade)
- Experimental design (tentative procedure). An indication of the types of measurement students will make
- A list of individual skills that a student needs in order to carry out the specific investigation

Resource Evaluation: Opportunities and Constraints

- Place of study, time frame of the project, space, materials including approximate budget, people, which class, number of students, school schedules, school policies, safety considerations and precautions, travel arrangements including cost

Implementation Phase: Questions to Consider

- How will you introduce the investigation?
- How will you assess students' prior knowledge (e.g., pre-tests, concept maps, performance events)?
- How will you solicit input from students regarding the question being investigated and the design of the investigation?
- What in-class activities will lead to the field project and what in-class activities will follow the field project?

- What will students produce?
- How will students be assessed (e.g., post-test, concept maps, student generated reports)?
- How will the students be encouraged to evaluate the research findings and their explanations?
- How will the individual differences among students be handled so that everybody is included (students of different abilities and different interests)?
- How will students disseminate their results (e.g., report, posters, video documentaries, web sites)?

References

Bodzin, A. (2008). Integrating instructional technologies in a local watershed investigation with urban elementary learners. *Journal of Environmental Education, 39*(2), 47–57.

Cronin-Jones, L. L. (2000). The effectiveness of schoolyards as sites for elementary science instruction. *School Science and Mathematics, 100*(4), 203–211.

Duschl, R. A., Schweingruber, H. A., & Shouse, A. W. (Eds.). (2007). *Taking science to school: Learning and teaching science in grades K-8*. Washington, DC: National Academies Press.

Frazier, R., Hegger, P., Grotzinger, T., & Sarkar, S. (2008). *"They're buggin' me" – real data, real sixth graders, and real insects in the school yard.* Boston, MA: Presentation at the Annual Conference of the National Science Teachers Association.

Martin, S. C. (2003). The influence of outdoor schoolyard experiences on students' environmental knowledge, attitudes, behaviors, and comfort levels. *Journal of Elementary Science Education, 15*(2), 51–63.

National Research Council. (1996). *National science education standards*. Washington, DC: National Academies Press.

North American Association for Environmental Education. (2004). *Education materials: Guidelines for excellence.* Retrieved April 3, 2009 from http://naaee.org/npeee/materials_guidelines/guidelines.pdf

Sarkar, S., & Frazier, R. (2008). Place-based investigations and authentic inquiry. *The Science Teacher, 75*(2), 29–33.

Sobel, D. (2004). *Place-based education: Connecting classrooms and communities*. Great Barrington, MA: The Orion Society.

Swope, S. (2005). A sense of place. *Teacher Magazine, 16*(6), 42–45.

Summer Methods in Summer Camps: Teaching Projects WILD, WET, and Learning Tree at an Outdoor Environmental Education Center

Charles J. Eick, Sarah Carrier, Karni Perez, and Doyle E. Keasal

Introduction

Our preservice science teacher education programs at Auburn University emphasize the teaching of environmental science curricula in grades K-12 in our science methods and curriculum courses (Powers 2004). Teachers often feel that they do not have the ability to teach environmental sciences because they lack the training (Smith-Sebasto and Smith 1997). We address this need programmatically. Content preparation in environmental science for undergraduate elementary preservice teachers comes in their second required biology course, where half of it is devoted to environmental science and ecology. Secondary alternative masters preservice teachers in biology education who are seeking initial certification take advanced courses in ecology and other natural science courses. This knowledge is put into practice teaching with young students as part of the methods and curriculum course requirements.

As in many large teacher education programs, we utilize the summer term for practice teaching in our methods courses in informal science settings (Kelly 2000). In our case, we have developed an ongoing relationship with our Forest Ecology Preserve, an outdoor classroom covering 110 acres on the edge of Auburn and run by the School of Forestry and Wildlife Sciences. The mission of the Forest Ecology Preserve is: *To provide programs, experiences, nature trails, and natural habitats for education, study, and relaxation for students and citizens of all ages throughout the area while creating an atmosphere of discovery and stewardship toward our natural world.* Outdoor settings like this one hold tremendous value for providing authentic learning experiences in environmental education (EE) for preservice teachers and their students (Bouillon and Gomez 2001; Hammerman et al. 1985; Negra and Manning 1997). As a form of service learning, our preservice teachers work as camp counselors and prepare lessons from Project WILD/Aquatic WILD,

C.J. Eick, S. Carrier, K. Perez (✉), and D.E. Keasal
Department of Curriculum and Teaching, Auburn University, AL 36849, USA
e-mails: eickcha@auburn.edu; k.r.perez@charter.net

A.M. Bodzin et al. (eds.), *The Inclusion of Environmental Education in Science Teacher Education*, DOI 10.1007/978-90-481-9222-9_12,

Project Learning Tree (PLT), and Project WET (henceforth *Projects* curricula) to teach to small groups of elementary and middle grades summer campers at the Preserve. This teaching opportunity occurs before practice teaching in schools.

Our chapter begins with a discussion of the role of EE in our science teacher education programs. We theoretically support the importance of EE and its interdisciplinary nature from the Science-Technology-Environment-Society (STES) approach to learning science. We review the development and nature of the Project WILD, Project Learning Tree, and Project WET curricula as well as their format for teacher use. We present our collaborations with the Forest Ecology Preserve and the Alabama Cooperative Extension System in preparing preservice teachers for teaching in these summer courses. We follow descriptions of our collaborations with a typical week at summer camp. We then highlight our impact on elementary candidate thinking about utilizing these environmental curricula in outdoor settings from their reflections on the experience before concluding our chapter.

Role of EE in Our Methods Courses

The state of Alabama plays a prominent role in supporting the teaching of environmental curricula, particularly because of its abundant water resources and species diversity (Eick et al. 2008). Through hunting, fishing, and car licenses, the State supports the dissemination of a number of environmental curricula including Project WILD/Aquatic WILD, Project Learning Tree, and Project WET. The Alabama Cooperative Extension System supports an EE specialist who works with preservice and inservice teacher training on these specific curricula. In both our elementary and secondary teacher education programs, we invite Cooperative Extension to work with our preservice teachers within their first methods or curriculum course after we have spent time preparing them for planning and teaching science within a learning cycle model (Karplus 1977). Our approach to the curiosities and wonderments of students is to embed inquiry within this basic framework for teaching.

Early in pedagogical development, environmental science and its related inquiries often become the content that we use in modeling effective practice and in preparing science lessons. Environmental science builds on preservice teachers' content strengths because of greater preparation in biology, ecology, and environmental science. With EE as our context, we emphasize the need to experience concepts and principles first-hand through activities that explore them before more direct forms of instruction. We also emphasize formative assessment as a critical component of student learning, even within one simple lesson. Furthermore, we make connections in lessons with other nonscience disciplines demonstrating the importance of environmental science in society, utilizing the STES framework in addressing national standards and scientific literacy (American Association for the Advancement of Science 1993; Hofstein et al. 1997; Kumar and Chubin 1999). EE as both content and context for teaching is strongly interdisciplinary in nature and meaningfully integrated into the broader context of society for the improvement of quality of life for our citizenry (Roth and Lee 2002). The interdisciplinary nature of environmental science also supports teaching across scientific disciplines as well as other subjects like social studies, mathematics, language arts, art, and music. This integrated approach is especially important to our elementary preservice teachers who have limited time to teach science along with other mandated subjects.

We also try to create a positive "first teaching" experience for our preservice teachers. The outdoor, informal setting seems to support the "fun" nature of camp and camp learning. As a first teaching experience for preservice teachers, the Forest Ecology Preserve summer camps are a low-risk setting to begin learning about how to manage and teach small groups of students (Luehmann 2007). The *Project* curricula also support preservice teachers in early practice because lessons are highly organized in a lesson plan format that is ready-to-use. Each lesson also includes extensions and authentic assessment options. These nationally known programs focus as much on effective teaching methods as on the content included in activities (Irvin 2007).

The Nature of the *Project* Curricula

During the late 1960s and early 1970s, the environment took center stage due to events like the Cuyahoga River fire in Cleveland, Ohio, and also with the birth of Earth Day. These events and others helped several organizations and agencies come to the realization that educating the public about our natural resources and their management was of great importance. This realization led to the creation of several award winning, multidisciplinary, activity-based EE programs that facilitate and promote awareness, appreciation, knowledge, and stewardship of our natural resources for PreK-grade 12 in an unbiased fashion.

Project Learning Tree

The first such program was Project Learning Tree (PLT) (PLT, 2007), which was created in 1976 to ensure that future generations of Americans understand the importance of our country's publicly and privately owned forests and that young people learn the skills to be good stewards of the environment. PLT was created through a partnership between the American Forest Institute (now, the American Forest Foundation) and Western Regional EE Council (WREEC), which is now the Council of Environmental Education (CEE). Since its inception, PLT has expanded from the original 13 western partnership states to all 50 states as well as gone international in the 1980s. The PLT curriculum went through limited revisions during the 1990s and a major revision again in 2005 to address education reform and today's most pressing environmental issues. PLT is reprinted yearly with updates and revisions to meet current demands. In addition, PLT has developed several modules for the secondary level on topics such as forest ecology, solid waste, and risk and has also published an Energy and Society kit for PreK-8 grade. The mission and goals of PLT are to increase students' understanding of our environment by encouraging critical and creative thinking, developing the skills with which to make informed decisions regarding environmental issues, and to be good stewards of our environment by using the forest as a "window" to the world.

Project WILD

The second program was Project WILD and its companion, Aquatic WILD (Project WILD, 2006). The idea for the development of Project WILD was conceived in 1979 by a group of representatives of state-level departments of education and natural resource-related agencies from 13 western states. These representatives identified a need for the development of a wildlife-based conservation and environmental program. Once again, the WREEC was instrumental in developing this program, which premiered after several years of development and testing in 1983 with an initial introduction in 20 western states. Since its inception, Project WILD is now available in all 50 states as well as internationally. During its existence, Project WILD and Aquatic WILD have been revised, with the latest major revisions being made in 2000, which included several new activities and a new conceptual learning framework that aligns to national education standards. The mission and goals of Project WILD focus on promoting responsible actions towards wildlife and its related natural resources through wildlife-based conservation and EE. This is accomplished by helping learners of all ages develop the skills, knowledge, and commitment to make informed decisions while displaying responsible and constructive behavior towards wildlife and the environment.

Project WET

The last conservation and EE program was Project WET (Project WET, 1998). This program was established by the North Dakota State Water Commission in the late 1980s as a means to educate the public about water resources and their management. From 1989 to 2005, the Project WET program received funding from the U.S. Department of the Interior, Bureau of Reclamation, and was moved to Montana State University where additional development and research resulted in the publication of the Project WET Curriculum and Activity Guide in 1995. Since then, the Project WET network has made the program available in all 50 states and has also been expanded internationally. In 2005, Project WET left Montana State to establish an independent Foundation called the Project WET International Foundation. Through Project WET, over 40 diverse water education guides and books for children and educators have been published since 1995. The mission and goals of Project WET are to educate people of all ages and communities of the world about water by facilitating and promoting the awareness, appreciation, knowledge, and stewardship of water resources through the development and dissemination of classroom-ready educational materials.

Instructional Design

During the development of these programs, an extensive multilayered process was employed that included research, surveys, writing workshops, and reviews by educators and resource professionals. In addition, revisions to the various materials have been made based on pilot and field-testing in classroom settings as well as by recommendations by external program evaluators. To assure that the activities being created meet specific objectives, each program has developed a conceptual framework that serves as a guide to help facilitate students' active participation in the learning process.

The instructional materials developed for all three programs are designed specifically to support state and national science standards for grades K-12. In fact, in Alabama and many other states, these programs and their activities are also correlated with the state's science, social studies, language arts, and mathematics learning objectives. During the development of these activities, various teaching methods and strategies were employed to guide the learners through the process of awareness, understanding, challenge, motivation, and action using active involvement and hands-on experiences. The activities also contain background information for the educator, which includes current statistical and factual information so that they remain relevant to the learners. Educators are also free to choose from a number of activities to enhance the teaching of a concept or skill as well as integrate them into an existing course of study.

Educators will find that all three programs are laid out in an easy-to-use format, giving educators valuable information that will help them teach the lesson with minimal preparation time. To help educators, each lesson contains an educators' box with information such as intended grade levels, subject areas, concepts, process skills, duration, materials list, and vocabulary. In addition, each activity follows a user-friendly format that includes objectives, background information, preparation instructions, procedural instructions, learner's pages, evaluation ideas, extensions, and in PLT's case, literature connections. All three EE programs are of great value to educators who want to prepare their students to learn *how to think, not what to think* in regard to environmental issues that are facing us today and in the future.

Preparing for Camp

In an effort to provide summer preservice teachers with experiences teaching elementary and middle grades students, a partnership was developed with the local Forest Ecology Preserve to help teach environmental science lessons in their week-long, outdoor camps. The Forest Ecology Preserve summer day-camp program is located approximately five miles from the university and is a community outreach program of the School of Forestry and Wildlife Sciences. The Preserve's education director was eager to coordinate a partnership between the university and their summer camp program for children in the community because our preservice teachers were needed to act as both counselors and teachers. Approximately 24–30 children per week attend the summer day-camps. Their parents pay a small fee to defray the costs of snacks and materials. Elementary preservice teachers (in rotation with another informal science site) would be present during 2 weeks of the camp (for children in grades 1–2 and 3–4), while secondary alternative masters preservice teachers would be present during 1 week of the camp (for children in grades 5–6).

Prior to the field experience at the Forest Ecology Preserve, the preservice teachers were introduced to the learning cycle instructional model and participated in the modeling of age-appropriate environmental curricula utilizing the learning cycle. These lessons included teaching outdoors, logistical considerations for smooth materials distribution, as well as provided opportunities for student inquiry. Discussion on effective questioning skills included strategies for eliciting student questions, encouraging higher-order thinking, and using wait time. These discussions presented the role of teacher as facilitator rather than lecturer. The preservice teachers were able to discuss presentation strategies with the professor and with the entire class. These opportunities for group reflection served as a valuable learning method. The potential to extend the activities based on students' questions and ideas helped the preservice teachers make connections between the courses' discussions and readings on inquiry to field work with students. These modeled lessons also allowed the preservice teachers to experience lessons from the students' point of view, which opened up opportunities for reflection and discussions that helped them as they planned their lessons for implementation with actual children.

Afterwards, preservice teachers completed a day-long training on Project WILD, Project Learning Tree, and Project WET curricula with the Environmental Education Specialist from the Alabama Cooperative Extension System. The training was held at the University's arboretum and outdoor classroom located on campus. The Specialist reviewed the layout of the curriculum guides and how to use the teacher-friendly format. Most of the time was used for preservice teachers to practice selected lessons from each curriculum guide (see Table 1) and later discuss how they would integrate it into their science teaching and assessment practice.

On the following day, class time was provided to assist the preservice teachers in their planning and preparation of select lessons suggested by the Preserve's education director. Many potential lessons were targeted by the director as meeting the

Table 1 Sample activities practiced in preservice teacher training workshop from the *Project* curricula

Project curriculum	Activity title	Description
Project WILD	Grasshopper gravity	The students carefully examine a grasshopper or cricket and then record their observations by drawing a sketch and/or writing a description of what they see
	Wildlife is everywhere!	The students go on a walk outdoors as they look for signs of life. The students quickly discover that living things do not have to be large. They also learn that a chewed leaf or a footprint in the mud is an indicator that a living organism is around
	The thicket game	The students play a form of hide-and-seek in which one person is the predator and everyone else is the prey. The predator is placed in a stationary location and hunts only with his or her eyes. The prey is allowed to hide while the predator is blindfolded. The only rule for the prey is that once they hide, they must be able to see the predator at all times. If the predator sees a part of them, they are caught. This activity emphasizes the importance of camouflage
	Learning to look, looking to see	This activity has the students observe their natural surrounding using their senses. While observing in the outdoors, the students record what they see, smell, hear, and feel over a period of time
	Oh deer!	Through physical movement, the students role-play deer and their habitat. Through this activity, the students experience the cyclical increasing and decreasing of animal populations due to a shortage of basic needs (food, water, and shelter) in the environment
Aquatic WILD	Migration headache	The students portray migrating water birds that are traveling between nesting habitats in one area and wintering grounds in another area. The students learn how changing land usage can directly impact the migration of waterfowl as well as other wildlife species

(continued)

Table 1 (continued)

Project curriculum	Activity title	Description
	Water canaries	The students investigate the organisms that live in a stream or pond by using simple sampling techniques. Once they collect the organisms, they discover how the quality of the water can be determined based on the availability of the aquatic macroinvertebrates found in the stream or pond
	Aqua words	The students observe a body of water such as a stream or pond and record water-related vocabulary. Once they have created a word bank of approximately 50 words, they work in groups to write about water
Project Learning Tree	Birds and worms	The students learn that camouflage is an important survival strategy. Through this activity, the students discover the importance of protective coloration as they role-play birds in search of colored worms
	The closer you look	To begin this activity, the students are asked to draw a picture of a tree from memory. Once they have completed their drawing, the drawings are put away while the students take a walking field trip to get a closer look at trees. Once they finish looking at trees using their senses, they are asked to draw another picture and then compare their two drawings
	The fallen log	In this activity, the students investigate a fallen log and become familiar with some of the organisms that call the log home as it decays. Through this activity, they will gain a greater understanding of decomposition and its importance
	Name that tree	The students learn to identify trees by looking at several different features: leaves, bark, twigs, flowers, fruit, and seeds
	Tree cookies	The students learn to tell the age of a tree by counting its annual rings and also discover how environmental changes are recorded through the ring growth in a "tree cookie"
Project WET	Poison pump	Through a series of clues just like in a CSI television show, students unravel a mystery that resulted in the death of hundreds of people in London during 1854. The students discover that water can have a negative impact on living things if it becomes contaminated

(continued)

Table 1 (continued)

Project curriculum	Activity title	Description
	The incredible journey	The students learn about the water cycle in a simulation as they become a water molecule that moves through the water cycle
	Stream sense	Students use their senses to observe a local stream and record their observations
	Wetland soils in living color	Students create a wetland soils color key as they learn about soil properties and classify soil types
	Macroinvertebrate mayhem	Students take on the role of a variety of aquatic macroinvertebrates as they play a game of tag to simulate the effects of environmental stressors

camp theme and age-level of the students attending camp during the given week. The Preserve also provided many of the materials needed for the lessons. The preservice teachers worked in pairs to coplan lessons from the *Project* curricula that they would coteach to campers later during the camp week. Also, preservice teachers received the scheduled camp program for their assigned week prior to the first session. The first few days of camp would be spent shadowing their assigned group of campers to better learn their students and the camp program. Before going out to the week-long camp, the preservice teachers attended a walking tour and orientation of the Forest Ecology Preserve and outdoor classroom areas.

The Camp Experience

Camp and Teaching Context

Forest Ecology Preserve summer camps consist of 1-week sessions with camp days from 8:00 a.m. to 11:30 a.m. Each of the three sessions is designed for 24–30 children entering either grades 1–2, 3–4, or 5–6. Within each session, campers are divided into four groups that rotate in varying combinations between activities lasting from one-half to one hour, with a mid-morning break for a snack.

The camp curriculum is organized around an ecological theme (e.g., water and life, nocturnal ecology, food webs). Six different themes are cycled over 6 years to avoid repeating themes and activities for children attending camp all six summers of their elementary years. Maintaining a certain amount of flexibility allows taking advantage of opportunities to have unexpected guest presenters or other ecology-related experiences.

Activities are selected, and adapted when necessary, from Project WILD and WET, PLT, and other types of resources or are created and designed specifically to meet the needs and goals of the camp session. The activities are grade-appropriate, incorporate state standards as guidelines, and offer an interdisciplinary experience involving science, mathematics, art, and language skills, among others. The week includes daily theme hikes and, often, physically active games or movement activities as part of a lesson or as a break between rotations. Theme-related art or craft activities are offered two to three times a week, and special presentations by guests often involve some type of live animal presentation. Special presenters may be asked to include theme-related information in their presentation, as they see fit and possible. Music is in the curriculum usually through songs.

To ensure that the preservice teachers can participate as daily counselors and teachers during each week-long camp, their class schedules have to be taken into account. The amount of time preservice teachers are present can vary from day to day because of other obligatory class commitments. An orientation meeting provides a necessary introduction to the camp. The preservice teachers need to know what is expected of them as camp staff, what they need to expect of the campers, and what they can expect from the regular teachers, guest teachers, and the camp experience in general. The orientation sets an anticipatory and relaxed mood and provides them with information and talking points that will help them be a positive addition to the campers' experience as well as have a positive experience themselves (see Appendix 1: Orientation Points).

The preservice teachers participate in ecology camp as teachers and counselors responsible for their camper groups. As counselors, they are viewed as camp staff and are responsible for camp tasks as their schedules allow. Initially, their role as counselors (and teaching assistants) allows both them and the campers time to familiarize themselves with camp routine and expectations and with each other. The preservice teachers typically start out oriented toward classroom teaching and management of children, and many show varying degrees of unease in a wilderness setting. Having the first two or three days in which to become more comfortable with the campers and with the wilderness of the camp setting promotes a higher level of ease and ultimately greater success in their planned teaching experience at camp.

Accompanying their groups of six to eight campers to the various rotations through the first two or three mornings provides preservice teachers the opportunity to observe a variety of teaching styles and methods and to assist the regular teachers informally in various learning activities before they are expected to be prepared to teach their assigned activity. These lead teachers come from the university community and include scientists, artists, extension agents, and the like. Preservice teachers typically coteach their assigned lesson with a fellow student to at least two different groups (rotations) of campers. Some pairs choose to share presenter roles equally, while others choose to have one main presenter per lesson with the second teacher responsible for materials distribution and working individually with students as they participate in their activities. The latter choice was acceptable because some student groups had members requiring individual attention. Preservice teachers are assured that the regular camp teachers are present at each rotation and are there to support and help them in whatever way is necessary. Preservice teachers will typically coteach two different lessons per week of camp.

Typical Camp Day

On a typical camp day, preservice teachers arrive shortly before campers are received. They sign in, put on name-tags, and possibly assist with last-minute setup tasks if there is time. Otherwise, they head for the parking lot to help receive campers as parents drop them off. They are encouraged not only to watch over the campers during the quarter-hour of waiting before camp begins but also to interact with them and to promote campers' interest in the natural surroundings before everyone gathers at a central meeting area (the "Canopy" amphitheater).

In mid-morning, the groups return from their early rotations to the Canopy for a snack and bathroom break. The preservice teachers take turns through the week helping with the setup and distributing food and drink. During the break, campers may rest, socialize, and investigate the "book tent," where theme-related books, science artifacts, specimens, and magnifiers are available. Inevitably, the campers use some of the break time for independent exploration of the environment around the Canopy. Items they find may be set out for display at the book tent to share with others. The preservice teachers are encouraged to participate with the campers in free-time explorations as well as to encourage curiosity and inquiry by engaging them with book table items.

After snack, another round of rotations takes the preservice teachers and campers again to that day's activity sites. On Wednesday or Thursday, the day of their own teachings, preservice teachers are expected to have prepared their lessons, help set up the requisite materials, and are declared in charge during their 30 or more minutes of teaching (see Fig. 1). During their lessons, one of the teachers observes and evaluates the preservice teachers and is also available to provide support if needed (see Appendix 2: Candidate Evaluation Form). After their teaching rotation is completed, the preservice teachers resume their counselor/assistant roles, enriched by their experience and better able to take advantage of informal teaching opportunities during the remaining camp week.

Elementary Candidate Reflections in Teaching at the Camps

During the training and planning for their lessons at the Forest Ecology Preserve, many of the elementary preservice teachers express hesitation about the upcoming field experience. In class meetings, they describe a lack of comfort in the outdoor setting as a location for learning, most having no experience with learning outside of a traditional classroom. This is not the case with secondary alternative masters preservice teachers who have had many "out in the field" experiences in their undergraduate biology and wildlife course work. In a written journal assignment on their experience, elementary preservice teachers express initial apprehension with working with students in an outdoor field setting. They express doubts in their abilities to teach science lessons in an outdoor setting where they

Fig. 1 Preservice teacher leaders teach their assigned lesson activities to the campers

particularly fear lack of classroom organization and structure. However, most leave the experience of teaching in the outdoors with newfound learning about the value of utilizing the outdoor classroom and a greater self-efficacy in their science teaching abilities:

> At first I was apprehensive to go to the Forestry Preserve because I did not think that it would be a good experience because it is not an 'in classroom' experience. I felt that the Forestry Preserve would just be a waste of time and I could not have been more mistaken. I came away from the Forestry Preserve with a feeling that there are teachable events that occur outside of the classroom.
>
> When I found out that we were doing it I was kind of apprehensive about it because I hadn't been exposed to that before. But once we got out there everything seemed to flow and having the director that was there, she really made everything run very smoothly...

The summer camp experience is typically exciting to the K-6 students who attend. This excitement and enthusiasm for the outdoors and for learning about it contributes to preservice teachers' positive attitude and feelings about the camp experience:

> ....how excited they [students] were all the time. It didn't matter that it was summer and they were learning, but they came everyday and they were all excited. I heard students say, "Oh my gosh this is so cool."
>
> My best memory with the students would have to be getting to see them excited about science.... and being outside.

The outdoor setting was a unique and stimulating location that offered a new experience to the elementary preservice teachers. They soon recognized the opportunities for teaching and learning in the outdoors. Many described the experience as contributing to their plans to conduct lessons with their future students in the outdoors.

> It was definitely the most fun experience that I've had. Sometimes it just makes a difference for them to be outside. I'm a big believer in the eight intelligences...Being outside is also visual, there's auditory, you can hear, you can see. So I really realized being outside is a way to get all of those intelligences.
>
> I think it was great and I would definitely do it again. It gave me a whole new perspective on science and how you can take it outside rather than being confined to that classroom.
>
> I think the students loved having lessons outdoors just because I mean you're outside. It's a lot of fun. You're in a different environment. You're not just cooped up inside the classroom all day... some of the things are not going to clearly be available in a normal classroom. But I feel like I could take my students outside now and it's exciting to plan.

Preservice teachers describe the benefits of students learning environmental science and related interdisciplinary studies in the outdoors particularly in experiencing their learning from a constructivist perspective:

> If I hadn't had that experience I don't think I would be more likely to teach outside. And I can see such a big difference in the student[s] and how it just helps them. You can talk about it all day in the classroom but I think it helps to experience it. If they have that experience then it just opens up, you know. Then they go do different things. That's an experience they have and then that's their prior knowledge.

Conclusion

Our elementary and secondary preservice teachers learn that EE is an important and easily accessible, experiential means of addressing various science standards as well as standards in other subject disciplines. As preservice teachers with little teaching experience, they find value in EE, in our approach to learning how to teach EE utilizing the outdoors, and in the *Project* curricula that support them (Moseley et al. 2002). Their first teaching experience in the outdoors is one that bolsters their confidence and positive attitude about teaching science to children before they begin more formal teaching in the classroom (Meredith et al. 1997).

Appendix 1: Orientation Points for Student Preservice Teachers

What We Ask of You

1. Wear closed-toe shoes and socks, not sandals; the provided name tags; and your camp tee.
2. Be at camp as early as you can. The more time you spend here, the sooner you'll feel comfortable and know how camp works.
3. Check the Camp Task List on the office door to find out what needs to be done for setup before camp and what needs to be done after camp.
4. We encourage you to actively assist the teachers – look or ask for things that need to be done to help. You'll learn more that way.
5. As a group "counselor," you are one of the responsible adults on whom campers can rely for guidance, help, information, and encouragement. (Your campers will be in the group you teach.)
6. Be a role model for the campers, especially with regard to curiosity about and appreciation for (exploring) nature and the science we do. Try not to show squeamishness or fear.

Tips on Teaching at Camp

1. This is hands-on, experiential science. The campers are supposed to have an opportunity to observe and explore. This isn't the classroom. This is informal education. The campers will be expected to participate with the group and behave appropriately, but the learning is active and often noisy, and that's good. It means they're enjoying what they're doing. We are at camp, and the most important thing is that the children have fun while they're discovering nature.
2. Watch the teachers. Learn from them. Don't hesitate to ask us questions, like "Why did you do this?" etc. You'll have your own way of teaching, but use this opportunity to learn, to develop your repertoire of teaching skills and techniques. I learn new things about teaching (and about nature and science) every day at camp.

3. You aren't expected to know the answers to all or necessarily even most of the campers' questions. It's okay to say things like, "Gosh, I don't know," or "That's a great question. I wonder about that too."
4. If you see that some children aren't understanding what needs to be done, help explain it to them or, if needed, help them do it as best as you can.

Miscellaneous

1. Location of First Aid supplies.
2. Chigger bites can be discouraged by insect repellent and not sitting on the ground except in designated places or on a tarp.
3. Special health information about any of the campers.
4. What does poison ivy look like (an important reason we urge campers to stay on the trails on hikes)? (Show examples.)

ABOVE ALL: *Enjoy camp yourselves*! Let yourself be curious and respond to the campers' curiosity. When adults are interested in what interests the children, it adds to their motivation to learn.

Appendix 2: FEP Outdoor Environmental Education Teaching Rubric

Directions: Complete this evaluation of your university student teacher with *specific written comments* (REQUIRED) and a rubric score for each performance indicator. This is a formative evaluation and will NOT count as a grade for the FEP teachings, so be honest in your appraisal and feedback.

Rubric		
4 = Excellent demonstration of meeting this indicator with *no need for improvement at all* 3 = Very good demonstration of meeting this indicator with little need for improvement 2 = Almost meeting this indicator with room for some improvement 1 = Not meeting this indicator with much room for improvement needed		
Performance indicator	Written comments	Rubric Score
Preparation and planning: The instructor was very well organized having the lesson on hand, hands-on materials, handouts, and other items readily available when needed		
Opening: The instructor opened the lesson with a point of engagement, question, or link to students' prior knowledge or previous learning/ lessons		
Instructions: The instructor gave clear and concise directions for what the students were to do and adequately answered any procedural questions before beginning activity		

(continued)

(continued)

Performance indicator	Written comments	Rubric Score
Activity: The instructor guided the students' actions and learning throughout the activity including what they were to do, what they were NOT to do, pacing the students, answering questions, and the general welfare and safety of the students		
Closing: The instructor promptly assembled the students at the end of the activity in order to facilitate what they did/found/learned through knowledgeable questioning and guiding students during student sharing or presentations		
Disposition: The instructor demonstrated a strong teacher presence through multiple indicators including enthusiasm, positive attitude, interaction with students, clear commands, self-assuredness, a "take charge" approach, and/or other indications that a "teacher is present"		

References

American Association for the Advancement of Science (AAAS). (1993). *Benchmarks for scientific literacy*. New York: Oxford University Press.

Bouillon, L., & Gomez, L. (2001). Connecting school and community with science learning: Real world problems and school-community partnerships as contextual scaffolds. *Journal of Research in Science Teaching, 38*, 878–889.

Eick, C., Deutsch, W., Fuller, J., & Scott, F. (2008). Making science relevant. *The Science Teacher, 75*(4), 26–29.

Hammerman, D., Hammerman, W., & Hammerman, E. (1985). *Teaching in the outdoors*. Danville, IL: Interstate Printers & Publishers.

Hofstein, A., Bybee, R. W., & Legro, P. L. (1997). Linking formal and informal science education through science education standards. *Science Education International, 8*, 31–37.

Irvin, T. (2007). Nature lessons. *Educational Leadership, 64*(8), 54–56.

Karplus, R. (1977). Science teaching and the development of reasoning. *Journal of Research in Science Teaching, 14*(2), 169–175.

Kelly, J. (2000). Rethinking the elementary science methods course: A case for content, pedagogy, and informal science education. *International Journal of Science Education, 22*, 755–777.

Kumar, D. D., & Chubin, D. E. (Eds.). (1999). *Science, technology, and society: A sourcebook on research and practice*. New York: Kluwer/Plenum.

Luehmann, A. L. (2007). Identity development as a lens to science teacher preparation. *Science Education, 91*, 822–839.

Meredith, J., Fortner, R., & Mullins, G. (1997). Model of affective learning for nonformal science education facilities. *Journal of Research in Science Teaching, 34*, 805–818.

Moseley, C., Reinke, K., & Bookout, V. (2002). The effect of teaching outdoor environmental education on preservice teachers' attitudes toward self-efficacy and outcome expectancy. *The Journal of Environmental Education, 34*, 9–15.

Negra, C., & Manning, R. (1997). Incorporating environmental behavior, ethics, and values into nonformal environmental education programs. *The Journal of Environmental Education, 28*(2), 10–21.

Powers, A. L. (2004). Teacher preparation for environmental education: Faculty perspectives on the infusion of environmental education into preservice methods courses. *The Journal of Environmental Education, 35*, 3–11.

Project Learning Tree. (2007). Retrieved September 24, 2008, from http://www.plt.org/index.cfm
Project WET: Water Education for Teachers. (1998). Retrieved September 24, 2008, from www.projectWET.org/aboutus.htm
Project WILD. (2006). Retrieved September 24, 2008 from http://www.projectWILD.org/index.htm
Roth, W. M., & Lee, S. (2002). Breaking the spell: Science education for a free society. In W. M. Roth & J. Desautels (Eds.), *Science education as/for sociopolitical action* (pp. 67–95). New York: Peter Lang.
Smith-Sebasto, N., & Smith, T. (1997). Environmental education in Illinois and Wisconsin: A tale of two states. *The Journal of Environmental Education, 29*, 23–31.

Teachers Connecting Urban Students to Their Environment*

Sherri L. Brown, Nikki L. Votaw, and Thomas R. Tretter

"Why do environmentalists ignore a third of the U.S. population?" Oladipo, a local environmentalist, asked in an *Orion* article titled *Global Warming is Colorblind.* In her 2 years of volunteering and working at an urban nature preserve, Oladipo had "never seen another face like [hers] come through our doors. At least, I've not seen another black woman come for a morning hike or native-wildlife program" (Oladipo 2007, p. 11). She elaborated on her observations by stating

> nobody benefits from the perception that enjoying and caring for the environment is an exclusively white lifestyle. The truth is that brown, yellow, red, and black people like to go backpacking, too. Those of us with the means are buying organic, local, and hybrid. If environmentalism continues to appear mostly white and well-off, it will continue to be mostly white and well-off, even as racial and economic demographics change. The environmental movement will continue to overlook the nuances, found in diversity of experience, that reveal multiple facets of environmental problems - and their solutions. (p. 11)

As a science education community preparing science teachers, we must ask ourselves if we are meeting the needs of preservice and inservice science teachers in presenting experiences that provide a foundation to teach all learners. And if not, what are we doing to address this deficit? It has been widely reported from science achievement scores on national and state standardized exams that minority students perform lower than white students. The 2005 National Report Card showed that minority students had lower scores than their white counterparts at all grade levels, despite the fact that at fourth and eighth grades the gap had lessened slightly since 1995 (National Assessment of Educational Programs [NAEP] 2005). Not only do minority students have lower scores on standardized tests, but they also tend to lose interest in science and develop negative attitudes toward science through middle school (e.g., Atwater et al. 1995; Hill et al. 1995). One reason for this may be that

S.L. Brown (✉), N.L. Votaw, and T.R. Tretter
Johnson Bible College, Teacher Education, 7900 Johnson Drive, Knoxville, TN 37998, USA
e-mail: s.brown@louisville.edu

A.M. Bodzin et al. (eds.), *The Inclusion of Environmental Education in Science Teacher Education*, DOI 10.1007/978-90-481-9222-9_13,

the urban school environment often tends to have few resources, large enrollments, little equipment, and less experienced teachers. Because urban schools often tend to have larger percentages of minority students than nonurban schools, the potential disconnect of urban students with science would disproportionately impact minority students. For the purposes of this chapter, references to urban students imply a relatively large percentage of nonwhite students included in this population.

To impact urban students' attitudes and understandings of science, school science needs to connect to the real world. Educators must emphasize the relevance of science to the lives of all students, not just the white-middle class experiences that are typical to school science (Atwater 1996). This requires that science teachers be informed about processes of "knowing, doing, and communicating science that are not mirrored in traditional school science" (Barton 1998, p. 528). Research in the field of science education suggests three underlying ideas that need to be considered when encouraging students to connect science to their own lives. First, students must be exposed to culturally relevant teaching; hence, school science should value the ways of knowing that are reflective of the students (Atwater 1996). Second, science must be a social practice (O'Neill and Barton 2005). Learning takes place in a social environment and the content learned cannot be separated from the context in which it was learned (Rahm 2001). Third, students must have ownership over their learning (O'Neill and Barton 2005).

Thus, the purpose of designing the graduate course, *Science Beyond the Classroom (SBC)*, was to provide preservice and inservice K-12 science teachers content and pedagogical experiences with multiple authentic opportunities that engage high achieving, low socioeconomic status (SES), urban students in connecting science to their community. The science teachers and urban students learn concurrently about scientific concepts that are applied in nonschool settings from actual visits to community sites (e.g., sewage treatment plant, water treatment facility, forest and arboretum, power plant, etc.). The course outcomes provide evidence for supporting the collective interaction of all of these groups (e.g., urban students, teachers, and site visit guides) to create a positive sociocultural context of this course. This chapter summarizes the relevant literature and theoretical framework underlying the course, the implementation and expectations of the course, and the content and pedagogical outcomes from the teachers who participated in the course. Readers interested in student outcomes should see Votaw (2008) or Brown, Votaw, and Tretter (2009).

Relevant Literature

Although numerous studies have focused on student learning during outdoor and place-based environments, there has been limited research examining the impact of site visits on teachers' content knowledge and pedagogical practices. Limited research exists in spite of the fact that there are over 180,000 teachers nationwide who participate in professional development events provided by informal science education institutions each year (Association of Science-Technology Centers 1996). This number of participating teachers suggests that teachers find these experiences valuable. The National Science Teachers Association (1998) highlighted the importance of informal science institutions in a position statement on informal science education where they stated that "informal science learning experiences offer teachers a powerful means to enhance both professional and personal development in science content knowledge and accessibility to unique resources" (p. 17). Research on teachers' experiences from informal learning environments involving professional development workshops or programs are discussed within three domains: content and pedagogy, confidence and enjoyment, and creating learning opportunities.

Content and Pedagogy

Although the research to support these claims is limited, there are several studies which suggest that informal learning environments positively affect teacher content knowledge and pedagogical skills (Boykie 1986; Melber and Cox-Peterson 2005; Neathery et al. 1998). After using Star Lab™ and participating in hands-on experiences, teachers displayed an increased understanding of basic astronomy principles (Boykie 1986). After a museum workshop, teachers stated that the "hands-on activities and interactions with museum artifacts and specimens" were the most valuable components of the workshop and they were able to apply the content and instructional strategies into their science lessons (Melber and Cox-Peterson 2005, p. 111). Following a professional development at either a science center, wildlife refuge, or zoo, teachers said they had greatly enhanced their content knowledge and understanding (Neathery et al. 1998).

Confidence and Enjoyment

Informal learning experiences develop enjoyment and increased confidence in teaching science as well as content and pedagogical benefits (Kyle et al. 1990; Seidman 1989; Sukow 1990). After a museum-based inservice program, teachers reported feeling less anxious and had increased confidence and competence about teaching science (Seidman). Reporting similar findings with teachers who participated in physical science workshops in a science museum, Sukow found that teachers had an increased confidence in understanding science and in using inquiry-based science instruction. In addition to confidence in teaching science, Kyle et al. found that teachers in their informal learning program now thought of science as fun and interesting, and they displayed an excitement for teaching science.

Creating Learning Opportunities

In providing teachers experiences to increase their science content knowledge, pedagogy, confidence, and enjoyment, Freeman et al. (2004) also noted necessary factors for creating effective learning opportunities for teachers. First, teachers must be engaged in hands-on experiences similar to what they will be providing for their students. Second, the learning opportunity must provide teachers opportunities to work with experts in the field. Third, teachers must have leadership opportunities. Lastly, the learning opportunity must provide teachers practical applications and then create occasions for follow-up discussions of these applications in the K-12 school setting. Informal learning institutions are ideal learning environments to offer all of these experiences to teachers. Grinell (1988) summarized that science museums are well positioned to address the needs of teachers as museums have the resources, skilled staff, and knowledge that would benefit teachers and schools, in turn benefiting the students.

Summary

Despite the recognition that informal learning institutions are providing professional development resources for teachers, a review of the relevant research literature points to few studies that have been conducted to determine the impact of site visits on teaching content knowledge and pedagogical methods. The *SBC* course is unique in that it allows teachers to experience six site visits in a 10-day time span, and does so with course participants in the role of teacher to middle school students, closely simulating their professional responsibilities in the classroom rather than situating them exclusively in the role of a learner. Rather than repeated experiences with the same site as recommended by Falk (1983), this multiple-site structure offers opportunities for the teacher to have myriad experiences that may have a cumulative effect as recommended in the NARST policy statement of the Informal Science Education Committee (Dierking et al. 2003).

Theoretical Framework

The underlying theoretical basis for the *SBC* course aligns with the Informal Science Education Ad Hoc Committee's policy statement (Dierking et al. 2003) that states

> learning rarely if ever occurs and develops from a single experience. Rather, learning in general and science learning in particular, is cumulative, emerging over time through myriad human experiences. ... The experiences children and adults have in various situations dynamically interact to influence the ways individuals construct scientific knowledge, attitudes, behaviors, and understanding. (p. 109)

Within the *SBC* course experience, environmental education (EE) content learning occurred within a community of practice (Lave and Wenger 1991) among teachers, urban middle school students, course instructors, and experts in the field. This community's shared domain of interest included the observing and learning of science practices within authentic contexts (i.e. site visits to nearby community venues). All community members engaged in similar experiences, which required relationships in which they learned from one another. These shared experiences and use of tools included the tour of facilities, reflections on experiences, completion of supplemental learning activities, and components of the digital narrative (e.g., taking pictures, interviewing experts, etc.). All members of this community of practice were involved in some aspect of the shared experiences and use of tools.

Camp and Course Background

The *SBC* course provides teaching and learning experiences for preservice and inservice science teachers and urban youth concurrently during a *Hands-on, Minds-on Summer Science Camp* experience.

Hands-On, Minds-On Summer Science Camp

Conducted each summer since 2006, the *Hands-on, Minds-on Summer Science Camp* has received funding support for camp participants' recruitment, travel, food, tickets, cameras, and instructional supplies from General Electric and E.ON U.S. Foundations. Recruiting for the camp participants was done in collaboration with the Lincoln Foundation (n.d.), which is a "premier nontraditional educational programs provider for disadvantaged youth" (p. 1). The camp participants are Lincoln Foundation's "Whitney M. YOUNG (Youth Organized to Understand New Goals) Scholars (WYS) program members who are academically talented, economically disadvantaged seventh grade students" (Whitney M. Young Scholars, n.d., p. 1). The rationale for recruiting the *WYS* was that those students from low socioeconomic backgrounds often lack opportunities to participate in educational programs outside of school and are often not exposed to scientific careers in the community. These same students also frequently belong to underrepresented groups in science and technological professions.

The 10-day *Camp* focused on site visits to community-based venues where environmental science issues were addressed on a daily basis. The community sites (e.g. local cave system, water treatment facility, sewage treatment facility, zoo, forest and arboretum, and power plant) were purposefully selected because environmental science concepts such as conservation and recycling were routinely practiced. For example, during the visit to the power plant, participants learned about the plant's conservational efforts: (a) scrubbers were used in the stacks to remove sulfur dioxide gases; (b) fly ash was a by-product recycled for use in gypsum board and concrete; and (c) water from the cooling towers was tested daily before returning to the river to avoid thermal or chemical pollution to the river. The participants later observed the use of the fly ash product at the forest and arboretum visitor center (the first LEED™ Platinum building in the state); the visitor center's concrete floor was constructed from recycled fly ash.

The selected sites were also very applicable and relevant to the lives of the urban students as personal daily consumers of electricity, water, and natural resources. Participants examined their personal conversational efforts at each site, such as preserving electricity by turning off the lights, preserving water by using less (i.e., taking shorter showers, turning off water while brushing teeth), and reducing amount of waste water (i.e., urban runoff from car washing, lawn care). From each site visit, the participants were able to weave the conservation content together. For example, the conservation of water was addressed through multiple examples during the site visits. The participants designed a town based on limited water supply at the water treatment plant and interacted with a city model to show runoff activity to storm drains at the waste treatment facility. In addition, they toured the visitor center at the forest and arboretum to learn about specific building constructions that allow for reuse of rain water for toilet flushing and parking lot purification methods that funnel run-off water through a peat-based treatment. All site visits included elements to promote awareness of environmental issues, personal consumption habits, and appreciation of natural settings (e.g., forest and arboretum and cave system).

They also completed activities on the university campus that underscored the environmental science concepts they observed during the site visits (see Table 1). Examples of the camp activities included the design and construction of (a) a filter from various materials (sand, pebbles, cotton, screen, etc.) to "clean" a sample of "dirty" water; (b) a karst model from gypsum board and leaf litter to simulate acid rain (vinegar) effects on limestone rock (gypsum board); and (c) a food web model based on Kentucky animals from the All Wild about Kentucky's Environment (n.d.) to visually demonstrate the interconnectedness of all life. To assess impact of these camp experiences and activities on the student participants' environmental science learning and attitudes, each participant completed a pre- and postcontent assessment and modified environmental attitude survey (Wojtowicz 1995).

A detailed account of the sewage treatment facility visit and previsit supplemental activities is as follows. Prior to the site visit, camp participants completed a homework assignment to investigate the inner-workings of their own bathroom toilet (see Appendix 1; Tretter 2004). The following day, the teachers guided the students in completing a demonstration using "mock toilet" toilet stations, which display the physical science concepts used in flushing toilets (see Fig. 1).

Before arriving at the treatment facility, the camp participants predicted in their notebooks what they thought occurred at a sewage treatment facility and what they expected to see. At the sewage treatment facility, the tour guides provided a brief overview and displayed a short video to explain the inner workings and design of their facility. This particular sewage treatment plant was purposefully selected

Table 1 Description of sites visited and supporting camp activities

Site	Science content	Supporting camp activities
Large cave system	Formation of caves	Karst model
Power plant	Energy transfer Recycling	Building a motor Steam engine demonstration Pollution control (thermal and air)
Water company	Water filtration and conservation	Building a water filter Calculating daily water usage Designing a "mock" town
Zoo	Interdependence of living things Carrying capacity Animal classification Endangered species conservation	Physical food web *Oh Deer*[a] Animal identification
Sewage treatment facility	Cleaning of sewage water	Run-off activity How a toilet works[b]
Local forest and arboretum	Stream restoration Recycling, green building techniques	Discussion of connections to all sites visited

[a] Council for Environmental Education 2002
[b] Tretter 2004

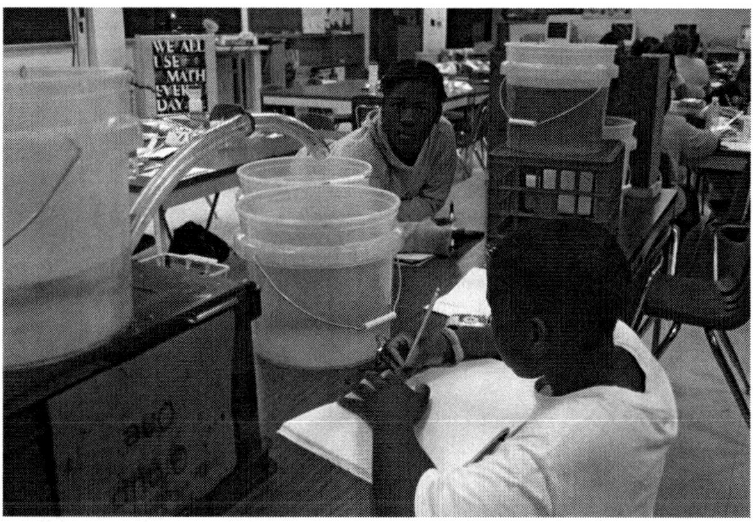

Fig. 1 Camp participants completing activity at "mock toilet" station

because it was architecturally designed to use the natural landscape of the land (i.e. utilizing gravity) and minimally impact the natural landscape. The students were divided into two groups: one group toured the entire facility (see Fig. 2), while the other group completed a waste water runoff activity (see Fig. 3).

Fig. 2 Sewer treatment facility tour

Fig. 3 City model run-off activity

Then, the groups alternated roles so that all experienced each event. After returning to campus, the student participants reflected on their learning by using their notebooks and instructional worksheets about the sewer treatment process. After the students had witnessed the process of cleaning sewage water, they were able to apply EE water conservational efforts to the inner workings of the "mock toilet". They were able to discuss strategies such as purchasing a high-efficiency toilet (HET), installing an adjustable flush flapper valve, or adding an object to the toilet tank (brick, etc.). They were also able to discuss strategies to reduce city wastewater run-off.

During the camp for all site visits and activities, participants documented their experiences using a science notebook, which included pre- and postsite reflections to engage students in synthesizing their understandings. Groups were assigned a particular site to showcase their learning and each student in the group received a disposable camera to document his/her assigned site visit. Under the guidance of their group teacher, each group prepared and presented a digital narrative of a site visit during a culminating event in a public forum to their parents, university faculty, and staff, Lincoln Foundation personnel, and community members. To construct this narrative, camp participants revisited their notebooks to ensure accurate site visit details, wrote a storyboard (i.e. the narration) about their learning, and then selected specific digital photographs to support their narration. They utilized Windows Moviemaker™ software at a university campus computer lab and microphone headsets to construct their digital narratives. This digital narrative process was an authentic learning task for the camp participants because it showcased their learning in a story context rather than a traditional test-type of assessment.

Science Beyond the Classroom (SBC) Course

The *SBC* course addressed multiple course goals by providing the teachers methods (pedagogical strategies) that engaged and connected urban adolescent learners with "real-life" environmental science applications in the community. The course provided teachers the opportunities to address the disconnect adolescent urban students may have between school science and real science. It is common for science teachers to address student queries such as "why am I learning this?" and "when will I ever use this again?" Learning about scientific concepts that are applied in nonschool settings (e.g., power plants, water treatment facilities, etc.) was an important goal.

Specifically, the *SBC* course addressed the following seven major goals: (1) help teachers increase their awareness of environmental science learning opportunities within everyday contexts and learn how to plan and coordinate the use of informal learning centers in teaching K-12 science; (2) nurture positive attitudes of urban students toward environmental science learning by increasing awareness of science in the community; (3) plan student-centered instructional activities/lessons that support science learning for all students regardless of race, gender, socioeconomic status; (4) enhance science instruction with integrated technology; (5) evaluate and reflect on instructional choices, classroom management techniques, and diverse student needs; (6) integrate physical, earth/space, and life science with other academic disciplines; and (7) foster collaborative relationships with colleagues and community resource personnel.

Course population. The target population for the *SBC* course included preservice and inservice science teachers who wanted to experience multiple authentic opportunities that engage high achieving, low socioeconomic status, urban students in connecting science to their community. The 23 "teachers" (5 males; 18 females) who completed the course had various years of teaching experience (0–25), certification levels (six elementary; three high school, six middle school; one special education; two primary/K; five none/preservice) and subject area expertise (e.g., physics, mathematics, biology, anatomy, chemistry, English language learner, environmental).

Course structure and rationale. To address the course goals, the teachers were provided opportunities to enhance their own teaching skills during the community site visits by teaching the camp participants within informal learning contexts. Each teacher worked closely with a small group (four to five) of low-socioeconomic status middle school urban youth (camp participants) for a 10-day *Hands-on, Minds-on Summer Science Camp*. The teachers prepared supplemental activities for site visits, instructed small groups of camp participants during each site visit, guided assessment reflections (i.e. notebook entries) of camp participants, and directed the digital narrative process.

To underscore that the informal learning site visits should not be isolated occurrences but interwoven into the classroom learning context, the *SBC* course instructors encouraged the use of specific pre- and postsite visit supplemental activities during the camp. To collaborate on the design of these supportive activities, the teachers convened on the university campus for 2–3 days prior to the camp. They developed or modified the existing environmental science content-based lessons from Project WILD (Council for Environmental Education 2002), Project Learning Tree (American Forest Foundation 2007), Pure Tap Water Adventures in Water Curriculum (Dearing-Smith 2002), etc. to supplement each site visit (see Table 1). The teachers also constructed a water filter, built a motor, and constructed a digital narrative prior to camp participants' arrival. In doing so, they were able to assist the camp participants in meaningful ways to troubleshoot any problems and address any content misunderstandings.

Although the *SBC* course included multiple applications to "real-world" contexts, it cannot adequately address all the opportunities that may be available to teachers (i.e., science centers, museums, aquariums, laboratories, etc.). Therefore, the ability to continue learning about informal learning contexts and how to connect this real-world application to students' lives was an invaluable goal for the course. To provide teachers the skills and confidence in using informal venues for students' science learning, course instructors introduced them to several site-specific personnel whom they would be able to contact for planning future site visits. Not only did teachers receive personal contact information, but they also received instructional materials for use in their classroom (e.g., recyclable fly ash samples from power plant, curriculum book from water company, etc.). The teachers had direct learning experiences in facilitating site visits to various venues in diverse conditions (outside/inside, hot/cold, rainy/sunny) and they developed a logistical awareness of how to prepare student groups for a site visit. The teachers were highly involved in the daily organization of each camp participant's experience and his/her materials used during site visits; they were actively engaged in all activities at the site visits. Even though the site visits were led by tour guides, the teachers were not peripheral participants because they needed to encourage and facilitate students' questions, model appropriate behavior, model appropriate science notebook use, etc. They monitored group logistical issues (e.g., attendance, participation, restroom visits, arrival, and departure) and students' materials (e.g., clipboards, cameras, science notebooks, water bottles, pens, sunscreen, bug spray, coats, etc.).

To address science teachers' abilities to connect fundamental science principles and concepts to applications in everyday life, the *SBC* course engaged teachers in the teaching and learning of environmental, physical, earth, and life science content knowledge. Prior to, during, and after each site visit, teachers facilitated camp participants in making connections among the science content, community professions, and conservational efforts. The teachers experienced the learning of science content in a variety of ways, which included inquiry-based learning, team planning, team teaching, individual content reflections, large and small group discussions, demonstrations, and the tours during the site visits. These experiences provided multiple approaches to enhance the teachers' own content knowledge as well as their approaches to teaching the content to their future students.

To prepare the teachers in guiding camp participants in using technology to demonstrate science conceptual understanding, the *SBC* course provided teachers a meaningful context to implement the use of disposable and digital cameras, Windows Movie Maker™ software, and various audio files in guiding camp participants to communicate their understandings from their community site visits. Rather than learning *about* technology applications in an abstract context, students and their teacher learned *with technology* by grounding the application in the completion of a culminating student product, a digital narrative.

To determine content knowledge and pedagogical methods gained from the course, the teachers maintained a science notebook throughout the entire course and camp experience in which they wrote about various aspects from each site visit (i.e., pedagogical ideas from tour provider, site visit connections). From their notes, diagrams, and handouts, the teachers wrote specific content and pedagogical reflections regarding their impressions of the camp and the camp participants' learning that occurred. To provide information regarding the use of content and pedagogical knowledge gained from the course in the teachers' classroom, the course instructors interviewed teachers at the end of the upcoming school year (approximately 9 months after the completion of the course).

Impact of Course Experiences on Teachers

The results of reflections and interviews with teachers revealed ways in which this course impacted them in their content knowledge and pedagogical methods. In spite of a substantial time span (9 months after participation) between course experiences and interviews, each teacher remembered his/her experiences and could speak to the impact that those experiences had on him/her. This suggests that the impacts reported from teachers are most likely to be enduring rather than transient.

Cultural Awareness

Samantha became aware of the different background experiences of people of different cultures. She was surprised that

> A lot of kids were scared of the cave, but even when we were walking around [the forest] just in the woods, they acted freaked out. For me, it wasn't that big a deal because I grew up around that stuff.

This awareness helped Samantha learn about "how to work with kids from other cultures and what it is like for them to experience something that is normal for [her], but not for them." Samantha taught at a school with a diverse population of students; however, seeing a diverse group in an environment other than school brought about an awareness of the different experiences that they have outside of school.

Based on her limited experience with middle-grade students, Erica expressed skepticism both in verbal conservations and in reflective writings about what urban kids would enjoy at a forest and arboretum. She stated

> When Dr. Brown told us that a [silent walk] would be a great part of the trip, I was doubtful. How much can these kids get out of walking around silently in the forest? I was proven wrong. I think the students gained much from the experience. Some of them were a bit frustrated with silent communication, but otherwise they participated extraordinarily willingly. They were joyful, relaxed, contemplative, and some actually became withdrawn. I think it was a shock to the system of some students. I think students learned something about themselves and how they can experience things without verbal communication.

Erica expressed dismay that urban students were "extraordinarily willing" and seemingly "joyful, relaxed, contemplative, and withdrawn" while participating in a silent walk in the forest.

Erica's reflection regarding previously held assumptions to what actually occurred allowed her to experience an increased awareness of previously held biases about urban students to surface. Erica's main goal for enrolling in the *SBC* course was to gain experience in working with urban middle school students. Again, similar to Samantha's experience, Erica was able to observe and interact with a diverse group of students in an environment other than school, and in doing so she had an increased awareness of her own personal biases.

Increased Environmental Awareness and Action Implementation

Through the course experiences, several teachers elaborated on how they utilized the newly learned content knowledge in their own classroom or home settings.

After visiting the power plant and observing the steps that were taken to protect the environment (i.e. using by-products to create new materials in gypsum board and concrete), Caroline created an activity in which her own classroom students had to apply that concept to another product. Her students used something that was a "left-over" and applied it in a different way. For example, her elementary students brought various left-over items from home (e.g., metal objects, aluminum cans, etc.), which they decorated and arranged to construct a wind chime. In doing this, Caroline helped her students experience how "leftovers" can be recycled and used to make new things.

Another teacher, Ashley, became more aware of ways that she could conserve energy and water. Her awareness led her to action. She said that she attempted to "use less water and I've gotten energy efficient light bulbs throughout the house. I try to keep the thermostat down or off when I don't need it." Similar to Ashley, Emily said that her family "doesn't buy bottled water and they recycle more." Another teacher stated that she "had no idea how electricity really worked at a power plant and the different things that they had to do to get power to my own house and how to get water to my house." This awareness provided her with background knowledge, which she used in teaching her own children at home. Since her family lives near the power plant, she discussed it with her own children, discussions such as "what the smoke was [that was] coming out of the smoke stacks."

Robert said he included the concepts from the course to teach his students about the cyclical nature of water. He specifically referred to the sewage treatment facility and recent problems the facility was having due to flooding. He connected the flooding event to the student families' water bill by making students aware that their water bill includes sewage and drainage. He stated "you pay more to treat the water than you do for the water itself." He referred to this in class when they discussed the cost of running a household and how that connects to science. Yet another teacher noted "I'd been to the forest and arboretum, but didn't realize all of the environmental ways that they had designed it, the welcome center and everything to help out the environment. I didn't realize that even though I had been there before." Another teacher had no idea that the "forest and arboretum existed. It opened my eyes that there are ecologically sensitive places that I need to seek out and use as examples for my kids. It made me more sensitive that there are more options out there."

Roger's increased awareness of what occurred at the water and sewage treatment companies prompted him to create a rain garden to conserve water and drink more tap water. After visiting the power plant, he implemented "the air condition saving device thing" [agreeing to have the compressor to his air conditioner remotely turned off by the power company for brief periods during peak demand hours] that helped reduce the amount of energy used by the air conditioner. Sarah reported that the site visits made her more personally aware of small things that she could do to help conservation efforts. To promote energy efficiency, she "changed all of the light bulbs in the house" and she uses "cold water to do laundry because we now know how much energy the hot water takes." Along with energy efficiency, she also tried to conserve through recycling and drinking tap water. She said that she never really felt like she overused resources, but now is "more conscious of it."

Summary

Results of teachers' reflections and interviews emerged around two themes that repeatedly became evident from the teachers' data. The particulars of these themes varied, in part due to the unique personal context for each teacher. First, the different aspects of the course and camp experiences brought about new areas of awareness for the teachers. From this awareness, teachers developed knowledge, attitudes, and skills to imbed EE tenets within their current school curriculum. Therefore, they gained additional resources and possibilities for their students. Additionally, teachers became more aware of the impact that they can have on the environment and they used their newly acquired knowledge to discuss some of the concepts from these community sites in their classrooms with their students. The data supported that the teachers moved along the EE goal continuum (UNESCO/UNEP 1978) from "awareness and sensitivity to the total environment" to "knowledge gained via experiences" (i.e. supporting learning of camp participants before and after through community tours). The teachers developed "attitudes" and "skills" necessary to identify and ameliorate environmental problems as indicated by personal changes within their own practices.

Implications

Results from the *SBC* course implementation support the effective learning opportunities elucidated by Freeman et al. (2004) in that teachers must (a) be engaged in hands-on experiences similar to what they will provide for their students, (b) have opportunities to work with experts in the field, and (c) make practical applications to the K-12 school setting. Results from the *SBC* course also demonstrated the impact that teaching urban students in multiple informal learning environments can have on K-12 science teachers. Teacher learning from this experience occurred in multiple intersecting dimensions: learning about middle school students' lack of connection to environmental concerns; enhancing their own personal learning of related content and interconnections; and coming to understand the power of the site-based pedagogy and how that can impact students. One of the main factors that affected all of these dimensions of learning was sensory interactions with the environment. Ideally, the best way for teachers or students to learn is to be directly engaged in the site themselves. Teachers need the direct experience both for their own learning and for strengthening their abilities to assist students in making a personal connection to the environment. Pragmatically, teachers will not be able to visit every site with all of their own students during the school year, but providing teachers themselves with at least one of these site-based experiences would be beneficial to their own learning and would provide them with a perspective (and possibly physical artifacts such as photos or samples) to share with their students.

Teachers expressed a growing awareness that urban middle school students may not fully understand the interconnections between the natural environment and their urban infrastructure. This enhanced awareness coupled with a strengthened understanding themselves of the interconnectedness of science with their lives both motivated and enabled teachers to enact pedagogical change in their own classrooms the year following the *SBC* course. Each teacher used his/her experiences at these sites in unique ways within his/her classrooms. These personal experiences at the site provided a context which teachers can use to enhance their instruction.

Through these direct connections, teachers showed examples of how these facilities are relevant to students' own lives. The feedback also demonstrated that teachers change their habits and personal choices when they have an increased understanding in how their actions affect the environment. Through the use of local environmental facilities, teachers can connect the products and resources that they use everyday to science. When teachers assist students' connections of these facilities to their daily lives, their students can relate their lives to their environmental impact. The EE implications include both the acquisition of resources (e.g., power plant, water treatment plant) and the disposal of waste (e.g., sewage, urban runoff, power plant refuse). Most importantly, these sites are venues that directly deal with their houses, communities, schools (not some generic power plant somewhere else) and are places where they could feel empowered to understand the underlying science. This understanding removes the mysterious, magical quality of household items used daily, such as clean water from a faucet, or electricity from a switch, or toilets flushing water "away." After students make these connections to their own lives (i.e. increased awareness), they need to be provided with opportunities to share their knowledge with others (i.e. knowledge to action). This can be done through formal presentations to parents and the community. Students can create a school display in the school hallway or media center to convey important environmental information learned to the rest of the school. Another effective outlet could be a student-created web page posting information that they learned from sites visited. The goal for the teachers is to provide experiences for their students to learn how science and the environment connects to their lives and in turn share that new knowledge with others.

Appendix 1: Toilet Homework

1. Clean toilet thoroughly after checking with your parents about how to do so safely and thoroughly.
2. Pour one cup (approximately 250 mL) of water into the toilet bowl and carefully observe the results. Record your observations on a piece of paper.
3. Pour ten cups of water, one at a time, into the toilet bowl and carefully observe and record the results.
4. Pour a large container of water (approximately 4–8 L) slowly into the toilet bowl, pouring a stream of water no thicker than about your thumb until the container is empty. Observe and record the results.

5. Pour the same large container of water quickly into the toilet bowl, emptying the container all at once. Observe and record the results.
6. Write down any questions that arise during your investigation to share with the class.

---

*Pseudonyms are used to preserve teachers' anonymity.

References

All Wild About Kentucky's Environment (AWAKE). (n.d.). Retrieved July 1, 2005, from http://www.kentuckyawake.org

American Forest Foundation. (2007). *Project Learning Tree: Pre K-8 environmental education activity guide*. Washington, DC: American Forest Foundation.

Association of Science-Technology Centers (ASCT). (1996). *Survey of informal science education institutions: Connections with schools*. Retrieved on January 31, 2008, from www.aste.org/resource/education/infrastructure_findings.htm

Atwater, M. (1996). Social constructivism: Infusion into the multicultural science education research agenda. *Journal of Research in Science Teaching, 33*, 821–838.

Atwater, M., Wiggins, J., & Gardner, C. (1995). A study of urban middle school students with high and low attitudes toward science. *Journal of Research in Science Teaching, 32*, 665–667.

Barton, A. C. (1998). Reframing 'science for all' through the politics of poverty. *Educational Policy, 12*(5), 525–541.

Boykie, T. S. (1986). Stars in the schools: Teacher training at the New York Hall of Science. *Journal of Museum Education, 11*(2), 10–11.

Brown, S. L., Votaw, N. L., & Tretter, T. R. (2009, April). *Impact of summer science institute on urban, low socioeconomic status middle school students' science attitudes and content knowledge*. Paper presented at the annual American Educational Research Association, San Diego, CA.

Council for Environmental Education. (2002). *Project Wild K-12 curriculum and activity guide*. Houston, TX: Council for Environmental Education.

Dearing Smith, K. (2002, May). *Pure tap adventures in water: A learning curriculum*. Louisville, KY: Louisville Water Company.

Dierking, L. D., Falk, J. H., Rennie, L. J., Anderson, D., & Ellenbogen, K. (2003). Policy statement of the "informal science education" ad hoc committee. *Journal of Research in Science Teaching, 40*(2), 108–111.

Falk, J. H. (1983). Field trips: A look at environmental effects on learning. *Journal of Biological Education, 17*(2), 137–142.

Freeman, J. G., Marx, R. W., & Cimellaro, L. (2004). Emerging considerations for professional development institutes for teachers. *Journal of Science Teacher Education, 15*(2), 111–131.

Grinell, S. (1988). Science centers come of age. *Issues in Science and Technology, 4*(3), 70–75.

Hill, G., Atwater, M., & Wiggins, J. (1995). Attitudes toward science of urban seventh-grade life science students over time, and the relationship to future plans, family, teacher, curriculum, and school. *Urban Education, 30*, 71–92.

Kyle, W. C., Bonnstetter, R. J., Sedotti, M. A., & Dvarskas, D. (1990). Sciencequest: A program that works. *Science and Children, 27*(8), 20–21.

Lave, J., & Wenger, E. (1991). *Situated learning: Legitimate peripheral participation*. Cambridge, UK: University of Cambridge Press.

Lincoln Foundation. (n.d.). *Lincoln Foundation: Overcoming adversity through education since 1910*. Retrieved September 28, 2008, from http://www.lincolnfdn.org

Melber, L. M., & Cox-Peterson, A. M. (2005). Teacher professional development and informal learning environments: Investigating partnerships and possibilities. *Journal of Science Teacher Education, 16*, 103–120.

National Assessment of Educational Programs. (2005). *The nation's report card: Science 2005* (NCES 2007455). Retrieved November 1, 2007, from http://nces.ed.gov/pubsearch/pubsinfo.asp?pubid=2007455

National Science Teachers Association. (1998). An NSTA position statement: Informal science education. *Journal of College Science Teaching, 28*, 17–18.

Neathery, M. F., Glynn, J., & Long, K. (1998). Professional development for elementary teachers: A collaborative effort involving a university, forest learning center, industry, and state learning agencies. *Electronic Journal of Science Education, 2*(4), 1–15.

Oladipo, J. (2007, November/December). Global warming is colorblind: Can we say as much for environmentalism? [Electronic Version]. *Orion, 26*(6), 11.

O'Neill, T., & Barton, A. C. (2005). Uncovering students' ownership in science learning: The making of a student created mini-documentary. *School Science and Mathematics, 105*(6), 292–301.

Rahm, J. (2001). Emergent learning opportunities in an inner-city youth gardening program. *Journal of Research in Science Teaching, 39*(2), 164–184.

Seidman, C. M. (1989). A community centered on learning. *Science and Children, 26*(5), 25–26.

Sukow, W. W. (1990). Physical science workshops for teachers using interactive science exhibits. *School Science and Mathematics, 90*(1), 42–47.

Tretter, T. R. (2004). Science in the toilet: The flush of learning. *Science Scope, 27*(5), 30–33.

United Nations Education, Scientific and Cultural Organization (UNESCO)/United Nations Environment Program (UNEP). (1978). The Tbilisi declaration. *Connect, 3*(1), 1–8.

Votaw, N. L. (2008). Impact of an informal learning science camp on urban, low socioeconomic status middle school students and participating teacher-leaders (Doctoral dissertation, University of Louisville, 2008). *Dissertation Abstracts International, 69,* 10.

Whitney M. Young Scholars. (n.d.). *Lincoln Foundation educational programs*. Retrieved September 28, 2008, from http://lfdn2008.xpressions.com/programs/whitney.asp

Wojtowicz, G. G. (1995). *Health and environmental protection: A survey of student attitudes*. (ERIC Document Reproduction Service No. ED 386447)

Exploring Preservice Teachers' Mental Models of the Environment

Christine Moseley, Blanche Desjean-Perrotta, and Courtney Crim

Several decades of research on teacher education shows that preservice teachers come into the profession with strong mental models of teaching and learning in the various content areas (Calderhead and Robson 1991; Pajares 1992; Thomas and Pederson 2003). These mental models or conceptualizations are individually, socially, and culturally derived (Rickinson 2001). Teachers bring to the classroom implicit theories and understandings that impact their teaching. If we accept this premise as true, it is reasonable to assume that preservice teachers bring to the classroom their conceptualizations or mental models of the environment. Their prior experiences with the environment provide the framework from which preservice teachers develop their mental models regarding the teaching and learning about this subject. It would seem logical, therefore, that teacher preparation programs should investigate preservice teachers' mental models of the environment, in order to identify, and if necessary, shift and refine any misconceived mental models before they become practitioners.

In 1943, Kenneth Craik suggested that people rely on mental models and that "the mind constructs 'small-scale models' of reality that it uses to anticipate events" (as cited in Johnson-Laird and Byrne 2003). Elaborating on Craik's work, Johnson-Laird and Byrne (2003) describe mental models as the representations in the mind of real-life or imaginary situations. Senge (1990) further explains mental models as "deeply held internal images of how the world works" (p. 174). Preskill and Torres (1999) tell us that mental models can be thought of as the "values, beliefs, assumptions, and knowledge that have been developed over time, are thought of as 'truths', and are what guide people in their everyday lives" (p. 66). From these definitions, we can assume that mental models are powerful and can influence how we act upon the world. Research indicates that mental models are never complete, but continue

C. Moseley (✉) and B. Desjean-Perrotta
University of Texas at San Antonio, One UTSA Circle, San Antonio, TX 78249, USA
e-mail: Christine.moseley@utsa.edu; Blanche.perrotta@utsa.edu

C. Crim
Education Department, Trinity University, One Trinity Place, San Antonio, TX 78212-7200
e-mail: courtney.crim@trinity.edu

A.M. Bodzin et al. (eds.), *The Inclusion of Environmental Education in Science Teacher Education*, DOI 10.1007/978-90-481-9222-9_14,

to expand as new information is assimilated. However, Senge (1990) and Rogers and Dunn (1997) suggest that sometimes these mental models may be so deeply buried below the surface, that they are difficult to alter and change.

Shepardson, Wee, Priddy, and Harbor (2007) argue that "students' conceptualizations of the environment or their mental models of the environment shape the ways in which they understand an environmental issue and guides their environmental behaviors" (p. 328). To further extend this thought, we believe that preservice teachers' mental models of the environment influence how they will teach about environmental education and related issues. If preservice teachers do not have a clear understanding of the natural environment, its systems and processes, they will not be able to help their students develop accurate mental models of the environment. In the interest of environmental education, it would seem, therefore, that teacher preparation programs should assist preservice teachers in identifying their mental models of the environment, and facilitate the reframing of their experience-based conceptions about the environment before they enter a classroom. Replacing preservice teachers' misconceptions about the environment with scientifically correct conceptions seems a logical first step in ensuring quality environmental education programs for all students.

Furthermore, researchers argue that a person's mental model of the environment shapes the way an individual will respond to environmental issues (Shepardson et al. 2007; Loughland, Reid, and Petocz 2002). For example, if an individual's mental model of the environment does not include human beings as an integral part of the environmental system, the impact that humans can have on the environment may not be fully appreciated by that individual. Although people do not readily change their mental models, their mental models usually become more pliable when presented with incontrovertible evidence that their current mental model is faulty (Duffy 2003). A major premise underlying the study presented in this chapter is that if preservice teachers are aware of their mental models of the environment, they will want to develop more accurate mental models that inform how they will teach about the environment in the future. Following is a description of how we used drawings as a survey tool to uncover preservice teachers' mental models of the environment, our disturbing findings, and programmatic changes that we implemented as a response to our study.

Use of Drawings as a Research Tool

The use of drawings or illustrations is one method of research that can be used to explore mental models or images. According to Knight and Cunningham, "[i]mages are a powerful form of communication, thus exploring and understanding images has important theoretical and practical implications" (2004, p. 2). Drawings have been used successfully in art therapy focusing on "how images and their expression reflect emotional experiences and how the emotional experiences affect thought and behavior" (Lusebrink 2004, p. 129). Art and thinking have been closely linked (Eisner 1997; Vygotsky 1971) and the visual arts provide an outlet for visual thinking (Arnheim 1969). According to Van Manen (1990), objects of art, or drawings, can be viewed "as lived experiences that are transformed into transcended configurations" (as cited in Alerby 2000, p. 209).

Research in science education has used drawings as a methodological tool for over 50 years. In the mid-1950s, Mead and Metraux (1957) initiated a major study investigating images of the scientist held by thousands of American high-school students. Drawing on this study and those of Goodenough's (1926) Draw-A-Man Test, Chambers (1983) developed the Draw-A-Scientist Test (DAST) to provide information regarding children's perceptions of scientists. In studies using the DAST, the drawings of scientists revealed issues related to motivation and self-efficacy in learning and teaching science (Kahle 1988), stereotypical images of scientists (Barman 1996), precollege students' ideas about science (Flick 1990; Schibeci and Sorensen 1993), preservice teachers' stereotypical views of scientists (Moseley and Norris 1999), and choice of science as a career (Finson 2002).

Finson et al. (1995) developed the Draw-A-Scientist Test Checklist (DAST-C) to facilitate ease of assessment of the DAST. The DAST-C was modified by Thomas et al. (2001) to include elements judged to be characteristic of science classrooms and teachers, calling the revised instrument the Draw-A-Science-Teacher Test Checklist (DASTT-C). Thomas and Hairston (2003) also modified the DAST-C to analyze students' perceptions of an environmental scientist, and developed the Draw-an-Environmental-Scientist Test Checklist (DAEST-C). Finally, Farland (2006) modified the DAST-C to include a rubric rather than a checklist to analyze drawings of scientists by school-age children.

Building on previous science education research using drawings that describe preservice teachers' beliefs about the subject of science, we adapted an instrument developed by Shepardson (2005) to create the Draw-An-Environment Test (DAET). We then developed a rubric for scoring the drawings, the Draw-An-Environment Test Rubric (DAET-R), to quantitatively analyze the data collected using the DAET.

Development of the Draw-an-Environment Test and Rubic (DAET-R)

The Draw-an-Environment Test (DAET) uses the draw-and-explain protocol and consists of a single page with two prompts (Appendix A). The rubric used to assess the DAET was developed from the *NAAEE Guidelines for the Preparation and Professional Development of Environmental Educators* (2004), a set of recommendations about the basic knowledge and abilities educators need in order to provide high-quality environmental education. The guidelines are designed to apply:

- Within the context of preservice teacher education programs and environmental education courses offered to students
- To the professional development of educators who will work in both formal and nonformal educational settings
- To full-time environmental educators as well as for those for whom environmental education is just one of their responsibilities

The *Guidelines* state that preservice teachers should be able to "describe the broad view that environmental education takes of environment, incorporating concepts such as systems, interdependence, and interactions among humans, other living organisms, the physical environment, and the built or designed environment" (p. 9). This statement was used to develop the DAET-R using each of the four factors in the definition of environment – humans, other living organisms (biotic), physical environment (abiotic), and the built or designed environment – as rubric categories for scoring the drawings (Appendix B).

The DAET-R is divided into four sections that focus on the degree of evidence in the drawings of interactions of the four environmental factors with each other. Degrees of evidence of these factors were assigned using a score of 0–3: Factor Not Present (0), Factor Present (1), Factor Interacting with other Factors (2), and Two or More Factors Interacting within a Systems Approach (3). The range of possible total scores on an individual rubric is 0–12. The higher the score, the more evidence there is of the participant's understanding of the environment, as defined by the *Guidelines.*

The DAET-R was used in a pilot study to test reliability and validity of the instrument and rubric. The subjects for the pilot study included a convenient sample of 390 ethnically diverse undergraduate early childhood preservice teachers (defined as preK–4th grade) from a large urban university. Data were used to determine frequency of factors drawn by individual participants and interactions of those factors. Results of descriptive statistics are reported in Tables 1 and 2. Table 1 suggests that preservice teachers do not consider humans to be an integral component of the environmental system. Sixty-nine percent of the drawings do not contain any drawn humans. Twenty-two percent of the participants drew humans with no obvious interaction with other factors in the environment. Only 6% of the sample population drew humans interacting with other factors and only 2.6% actually indicated any kind of systems approach in their drawings of humans. Results of the DAET-R also indicate that generally where preservice teachers drew one or more factors, they merely drew and labeled the factors. Very few of the participants' drawings evidenced an understanding of a systems approach to the environment with interactions among factors. The factor drawn the most showing no evidence of interaction was Abiotic (72.3%), followed closely with Biotic (64.6%).

The total scores from each individual drawing were then collapsed into three broad categories: One or more Factors Present, One or more Factors Interacting with Another Factor, and Two or more Factors Interacting within a Systems Approach (Table 2). Seventy-nine percent of the drawings scored a 4 or less, indicating the lack of one or more factors in the drawings. Only 0.3% of the drawings scored 9–12, indicating factors depicting interactions within a systems approach. In fact, only one drawing out of the total 390 drawings scored a 12, indicating interactions of all four factors within a systems approach.

Results of this pilot study using the DAET-R for scoring early childhood preservice teachers' drawings of the environment suggest that these participants' initial mental models of the environment are incomplete when compared to the *NAAEE Guidelines* (2004). It is assumed in the *Guidelines* that teachers have the mental

Table 1 Participants and percent of factors included in drawings

	Human		Living		Abiotic		Built	
Points on rubric	# of participants	%	# of participants	%	# of participants	%	# of participants	%
0	270	69.2	39	10.0	63	16.2	170	43.6
1	85	21.8	252	64.6	282	72.3	147	37.7
2	25	6.4	90	23.1	32	8.2	44	11.3
3	10	2.6	9	2.3	13	3.3	29	7.4

Table 2 Percent of total scores

Total points	Category	# of participants (n = 390)	% of total (n = 390)
0–4	Factor present	306	78.5
5–8	Factor interacting with one other factor	83	21.3
9–12	Factor interacting with one or more factors with systems approach	1	0.3

model or image of the environment that incorporates the concepts of systems, interdependence, and interactions among humans, other living organisms, the physical environment, and the built or designed environment. However, almost 70% of the participants did not even include humans in their drawings and only 2.6% drew humans interacting with other factors in a systems approach. In contrast, 66% of the preservice teachers included items of the built or designed factor in their drawings such as drawings of their homes or personal bedrooms, a school or specific classroom, or urban neighborhoods with streets, commercial buildings, cars, and residences. For preservice teachers, the word "environment" produced mental images that did not depict the more naturalistic images of the environment as proposed by the *Guidelines*.

We next utilized the DAET-R as a research tool to analyze the impact of participation in an environmental education professional development experience on preservice teachers' mental models of the environment. Approximately 100 early childhood preservice teachers participated in a full day of an environmental education (EE) workshop at a local natural area. The EE workshop used the *Project WILD K-12 Curriculum and Activity Guides* (2004) as the framework for the teacher training. Results of the pre- and post-data analyzed using the DAET-R are recorded in Tables 3 and 4. Participation in the EE workshop slightly influenced the preservice teachers' inclusion of humans in their drawings: 71% initially did not include any humans in their drawings which decreased to 65% after participation. The same analysis was true for the built factor, decreasing from 62% who did not initially include built factors in their drawings to 54% who did not include them after participation. The percent of total scores in the three categories did not substantially change as seen in Table 4.

Impact of Workshops on Mental Models

Reflecting upon the results of the study that indicated little substantial impact of the EE workshop on the preservice teachers' mental models of the environment, it was determined that our data support other studies regarding the impact of professional development. As teachers acquire new information, they will slowly incorporate it

Table 3 Participants and percent of factors included in drawings

Points on	Human		Living		Abiotic		Built	
rubric	Pre %	Post %	Pre %	Post %	Pre %	Post %	Pre %	Post %
0	71.4	64.7	4.1	4.4	7.1	11.8	62.2	54.4
1	16.3	20.6	54.1	52.9	75.5	70.6	24.5	32.4
2	10.2	11.8	36.7	38.2	13.3	10.3	9.2	10.3
3	2.0	2.9	5.1	4.4	4.1	7.4	4.1	2.9

Table 4 Percent of total scores

Total points	Category	% of total PRE (n = 98)	% of total POST (n = 68)
0–4	Factor present	75.5	75.0
5–8	Factor interacting with one other factor	24.5	23.5
9–12	Factor interacting with one or more factors with systems approach	0.0	1.5

into their existing understanding and practice. Rapid changes will rarely occur. Most significant changes will result after participants have engaged in long term professional development programs that integrate innovative materials, opportunities to practice new teaching ideas, time for reflection on practical experiences, and discussions in a supportive environment (Van Driel et al. 2001). As reported by Garet et al. (2001), a longer time span of professional development is linked to more opportunities for active learning, which eventually leads to implementing the new information within the classroom. Additionally, professional development that spans a longer time frame supports a deeper understanding of content, which results in greater alignment with content standards and increased communication with other professionals. These findings and our initial data support that even though we expose preservice teachers to new information about environmental education, they need multiple opportunities to incorporate this information into their own mental models if it is to have any lasting effect on their practice.

Realizing the benefits of extended professional development that occurs over time, we recognized the need for programmatic changes and began to infuse EE curricula into our teacher education program. Rather than one environmental education experience, we have initiated a sequence of experiences for the preservice teachers throughout the program. The first step was to train faculty as facilitators in a range of exemplary EE curricula. Next, all preservice teachers in the early childhood certification program receive professional development in EE curricula that is integrated into their coursework at specific intervals in the teacher preparation program. These EE curricula provide the preservice teachers with inquiry-based activity guides, materials, and resources to use in their future classrooms. The curricula, in the order that they are offered in the preservice teacher preparation program, are outlined in Table 5.

Our teacher education program's commitment to EE has led to the development of an on-going partnership called *S*trengthening *A*wareness and *V*aluing the *E*nvironment (SAVE) that brings together the University's College of Education and Human Development, Texas Parks and Wildlife Department, Texas Forest Service, and San Antonio Parks and Recreation Department. The purpose of SAVE is to increase preservice teachers' knowledge of, and interaction with, the city's natural areas and to influence their perceptions of the environment. This is accomplished by using systematic, authentic professional development experiences in environmental education that includes opportunities for practice and reflection.

Table 5 EE curricula

Curricula	Description	Course connection
Global Learning and Observations to Benefit the Environment (GLOBE) (2008)	Developed by NASA, this integrated curriculum targets mathematics, science, and technology as they pertain to the study of earth systems science in an outdoor setting	Earth Systems Science lecture and lab (sophomore year)
Wildlife in Learning Design (Project WILD) (2004)	Through connections to wildlife, this curriculum was developed by the Council for Environmental Education (CEE)	Approaches to Teaching Science (junior year)
Project Learning Tree (PLT) (2008)	Developed by the American Forest Foundation, PLT uses a conceptual framework based on the forest ecosystem	Approaches to Teaching Social Studies (senior year, prior to student teaching)

Implications for Preservice Environmental Education Professional Development

The *NAAEE Guidelines* (2004) challenge teacher education programs to support preservice teachers' development of a conceptual model of the environment that integrates humans and the abiotic and biotic factors. The results of our research support the findings of studies by Loughland et al. (2002), Shepardson et al. (2007), and Rickinson (2001) in which the majority of young children viewed the environment as an object, with little or no human interference. In our study, the preservice teachers did not hold even a minimally accurate perception of the environment as described in the NAAEE *Guidelines*. Yet, these existing mental models, if not addressed, will be what teachers incorporate into their practice. Only by changing these inaccurate mental models can we ensure that we are infusing the public school system with transformative leaders in the area of environmental education.

As Loughland et al. (2002) suggest, students that perceive an *object* view of the environment may not see the need to take any responsibility for the environment. They propose that, for environmental education to be more meaningful and purposeful, students' own personal experiences must be explored and challenged. Environmental education should start at home, using local resources and environmental issues, relating one's actions to one's own experiences. We think that the same approach needs to be taken with preservice teachers as we work to strengthen their mental models of the environment. The question is how can initial beliefs and mental images of the environment held by early childhood preservice teachers be changed?

Perceptions about the environment are not innate: they are learned (Rickinson 2001). It is from this foundation that we made conscious changes to our preservice teacher education program. Negative perceptions or misconceptions about the environment can be countered by teachers who are sympathetic to and knowledgeable about local environmental issues, and who see themselves as vital components of the whole system that makes up the environment. These are the transformative leaders that we seek to send to the field of education. The DAET-R can provide initial data useful to teacher educators who work with preservice teachers in the area of environmental education. Through pre-assessment, faculty can facilitate preservice teachers' conscious examination of their existing mental models by bringing these to the surface for reflection. Raising doubt is the first step in creating shifts and changes in the existing mental models (Duffy 2003).

Mental models are never complete, but continue to expand and improve as new information is assimilated. Preservice teachers need a variety of opportunities to incorporate new learning into developing schema and expand their understandings as they experiment with new ideas and concepts. Therefore, preservice teacher education programs must infuse environmental education into a variety of specific learning experiences within the context of coursework in the program. Professional development models for preservice teachers should engage participants in reflective conversations about their own understandings about the environment. As preservice teachers evaluate their new ideas and understandings about the environment, they are also receiving feedback through interactive dialogue and practice. The professional development model for environmental education needs to include both connections to the individual understandings of the preservice teachers as well as how this newly constructed knowledge links to instruction with young children.

It is our belief that by challenging early childhood preservice teachers' existing concepts or images of the environment through engaging and systematic professional development, they will develop strong, appropriate mental models. It will be with accurate mental models that these transformative leaders in environmental education move into the field as professional educators of young children. It will be these teachers who, in turn, are now able to effectively work with young children as they construct their own mental models of the environment.

Appendix A: Draw-an-Environment Test

Date: _________ ID#____________________

In the space below draw a picture of what you think the environment is. Below that, please provide your definition of the environment (in words).

My drawing of the environment is:

My definition of the environment is:

(Adapted from Wee, B., Harbor, J., and Shepardson, D. (2004), November). Multiculturalism in Environmental Science: A snapshot of Singapore. Paper presented at North American Association for Environmental Education Conference, Biloxi, MS)

Appendix B: Draw-an-Environment Test – Rubric (DAET-R)

Date: _________ ID#:____________________

		Present	Interactions with other factors	System interactions made explicit	
Factor	0 Points	1 Point	2 Points	3 Points	Score
Human	Drawing does not contain pictures of humans	Human(s) drawn without any apparent interaction with other factors	Human(s) drawn interacting with other humans and/or another factor (e.g., human fishing or walking on a bridge), but without special emphasis placed on the influence of the interaction on the environment	Humans drawn with obvious deliberate emphasis placed on interaction with one or more factors and the influence of that interaction on the environment through the use of special indicators such as conceptual labels and/or arrows	
Living	Drawing does not contain pictures of living organisms	Living organisms (e.g., plants and animals) drawn without any apparent interaction with other factors	Living organisms drawn interacting with other living organisms and/ or another factor (e.g., animals grazing), but without special emphasis placed on the influence of the interaction on the environment	Living organisms drawn with obvious deliberate emphasis placed on interaction with one or more factors and the influence of that interaction on the environment through the use of special indicators such as conceptual labels and/or arrows	
Abiotic	Drawing does not contain pictures of abiotic factors	Abiotic items (e.g., mountains, rivers, Sun, or clouds) drawn without any apparent interaction with other factors	Abiotic items drawn interacting with other abiotic items and/or another factor (e.g., wind blowing a palm tree), but without special emphasis placed on the influence of the interaction on the environment	Abiotic items drawn with obvious deliberate emphasis placed on interaction with one or more factors and the influence of that interaction on the environment through the use of special indicators such as conceptual labels and/or arrows	

(continued)

(continued)

		Present	Interactions with other factors	System interactions made explicit	
Factor	0 Points	1 Point	2 Points	3 Points	Score
Human-built or designed	Drawing does not contain pictures of human-built factors	Human-built or designed items (e.g., buildings, automobiles, and bridges) drawn without any apparent interaction with other factors	Human-built items drawn interacting with other human built items and/or another factor (e.g., smokestack emitting smoke into the air), but without special emphasis placed on the influence of the interaction on the environment	Obvious deliberate emphasis placed on one human-built item interacting with one or more factors and the influence of that interaction on the environment through the use of special indicators such as conceptual labels and/or arrows	
				Total possible points: 12	
				Total points:	

Directions: Assign points for each Factor – Human, Living, Abiotic, Built – based on whether the factor is merely present in the drawing (1 point), interacting with other Factors in the drawing (2 points), or interacting with special additional emphasis placed on the influence of the interaction on the environment (3 points). Factors that are not drawn do not receive points (0 points). Factors must be drawn to be scored. Implied relationships do not receive a score. For example, if a subject draws a house but there are no drawn humans, it cannot be assumed the subject infers humans in the drawing. Diagrams without drawings of factors receive a score of "0."

Conceptual Label

A label that depicts interactions between one or more factors and an influence of that interaction on the environment is considered a conceptual label. For example, smog indicates interactions between abiotic, human, and built factors. A cloud labeled as water cycle instead of just cloud indicates interaction between abiotic and living factors. Trash and garbage indicates interaction between human and abiotic factors and the influence of that by-product on the environment.

Identification Label

Identification labels are different from conceptual labels in that they merely identify the object (tree, dog, house, etc.).

References

Alerby, E. (2000). A way of visualizing children's and young people's thoughts about the environment: A study of drawings. *Environmental Education Research, 6*(3), 205–222.
Arnheim, R. (1969). *Visual thinking*. Berkeley, CA: University of California Press.
Barman, C. (1996). How do students really view science and scientists? *Science and Children, 34*(1), 30–33.
Calderhead, J., & Robson, M. (1991). Images of teaching: Student teachers' early conceptions of classroom practice. *Teaching and Teacher Education, 7*, 1–8.
Chambers, D. (1983). Stereotypic images of the scientist: The draw-a-scientist test. *Science Education, 67*(2), 255–265.
Duffy, F. M. (2003). I think, therefore I am resistant to change. *National Staff Development Council, 24*(1), 30–36.
Eisner, E. (1997). Cognition and representation: A way to pursue the American dream? *Phi Delta Kappan, 78*(5), 349–353.
Farland, D. (2006). The effect of historical, nonfiction trade books on elementary students' perceptions of scientists. *Journal of Elementary Science Education, 18*(2), 33–49.
Finson, K. (2002). Drawing a scientist: What we do and do not know after fifty years of drawings. *School Science and Mathematics, 102*, 335–346.
Finson, K., Beaver, J., & Cramond, B. (1995). Development and field test of a checklist for the Draw-A-Scientist Test. *School Science and Mathematics, 95*, 195–205.
Flick, L. (1990). Scientist in residence program improving children's image of science and scientists. *School Science and Mathematics, 90*, 204–214.
Garet, M. S., Porter, A. C., Desimone, L., Birman, B. F., & Yoon, K. S. (2001). What makes professional development effective? Results from a national sample of teachers. *American Educational Research Journal, 38*(4), 915–945.
Global Learning and Observations to Benefit the Environment (GLOBE). (2008). GLOBE Vision. Retrieved March 17, 2008 from www.globe.gov
Goodenough, F. (1926). *Measurement of intelligence by drawings*. New York: Harcourt Brace.
Johnson-Laird, P. N., & Byrne, R. (2003). *Mental models website*. Retrieved August 24, 2008, from http://www.tcd.ie/Psychology/Ruth_Byrne/mental_models/
Kahle, J. (1988). Images of science: The physicist and the cowboy. In B. J. Fraser, & G. J. Giddings (Eds.), *Gender issues in science education* (Monograph in the Faculty of Education Research Seminar and Workshop Series). Australia: Curtin University of Technology.
Knight, M. & Cunningham, C. (2004, July). Draw an engineer test (DAET): Development of a tool to investigate students' ideas about engineers and engineering. *Proceedings of the 2004 American Society for Engineering Education Annual Conference and Exposition*. Salt Lake City, UT.
Loughland, T., Reid, A., & Petocz, P. (2002). Young people's conceptions of environment: A phenomenographic analysis. *Environmental Education Research, 8*, 187–197.
Lusebrink, V. (2004). Art therapy and the brain: An attempt to understand the underlying processes of art expression in therapy. *Art Therapy, 21*, 125–135.
Mead, M., & Metraux, R. (1957). Image of the scientist among high-school pupils: A pilot study. *Science, 126*, 384–390.
Moseley, C., & Norris, D. (1999). How preservice teachers perceive scientists. *Science and Children, 37*, 50–56.

North American Association for Environmental Education. (2004). *Guidelines for the preparation and professional development of environmental educators*. Washington, DC: Author.
Pajares, M. F. (1992). Teachers' beliefs and educational research: Cleaning up a messy construct. *Review of Educational Research, 62*(3), 307–332.
Perrotta, B., Moseley, C., & Cantu, L. (2008). Preservice teacher's perceptions of the environment: Does ethnicity and historical residential experience matter? *Journal of Environmental Education, 39*(2), 21–32.
Preskill, H., & Torres, R. T. (1999). *Evaluative inquiry for learning in organizations*. Thousand Oaks, CA: Sage Publications.
Project WILD. (2004). *Project WILD K-12 curriculum and activity guide*. Houston, TX: Council for Environmental Education.
Rickinson, M. (2001). Learners and learning in environmental education: A critical review of the evidence. *Environmental Education Research, 7*, 207–320.
Rogers, D. B., & Dunn, M. (1997). And never the twain shall meet: One student's practical theory encounters constructivist teacher education practices. *Journal of Early Childhood Teacher Education, 18*(3), 10–25.
Schibeci, R., & Sorensen, I. (1993). Elementary school children's perceptions of scientists. *School Science and Mathematics, 83*, 14–19.
Senge, P. M. (1990). *The fifth discipline: The art and practice of the learning organization*. New York: Doubleday.
Shepardson, D. (2005). Student ideas: What is an environment? *Journal of Environmental Education, 36*(4), 49–59.
Shepardson, D., Wee, B., Priddy, M., & Harbor, J. (2007). Students' mental models of the environment. *Journal of Research in Science Teaching, 44*, 327–348.
Thomas, J., & Hairston, R. (2003). Adolescent students' images of an environmental scientist: An opportunity for constructivist teaching. *Electronic Journal of Science Education, 7*(4), 1–20.
Thomas, J. A., & Pederson, J. E. (2003). Reforming elementary science teacher preparation: What about extant teaching beliefs? *School Science and Mathematics, 103*(7), 319–330.
Thomas, J., Pederson, J., & Finson, K. (2001). Validating the Draw-A-Science-Teacher-Test Checklist (DASTT-C): Exploring mental models and teacher beliefs. *Journal of Science Teacher Education, 12*(4), 295–310.
Van Driel, J. H., Beijaard, D., & Verloop, N. (2001). Professional development and reform in science education: The role of teachers' practical knowledge. *Journal of Research in Science Teaching, 38*, 137–158.
Van Manen, M. (1990). *Researching lived experience: Human science for an actions sensitive pedagogy*. London: State University of New York Press.
Vygotsky, L. (1971). *The psychology of art*. Cambridge: The MIT Press.

Pedagogy, Environmental Education, and Context: Promoting Knowledge Through Concept Mapping

Barbara Austin and Nina Schmidt

Using Concept Maps with Environmental Questions

We chose to develop and implement this project in our science methods course for several reasons. Because of limited course time to spend on this project, we needed students to learn both pedagogy and content simultaneously. We hoped that an experience in which authentic content learning took place would promote the learning of pedagogy and thus increase the propensity to use effective science teaching methods once they graduated. The first goal, our content goal, was for students to learn the integrated nature of science. To accomplish this, we chose to situate our pedagogical instruction within the context of a regional environmental question. Science is the study of the natural world, but the natural world does not fall into the neat disciplinary boundaries of chemistry, physics, botany, geology, microbiology, etc. By engaging in a project about the environment, the integrated, rather than disciplinary, nature of science becomes obvious to even the most unengaged student. Additionally, through this project, we wanted to support science and environmental literacy as the primary goal of secondary science instruction. The environment provides a perfect vehicle for doing so. Our second goal, a pedagogy goal, was to demonstrate how to make science relevant to secondary students by using a regional environmental question as a context for teaching content. Our part of Arizona, like many parts of the county, has a very fragile and delicate ecosystem. Environmental issues are commonly reported news items and are part of the local culture. Thus, the environment is a familiar topic in which to situate learning of science content. Our third goal was to teach preservice teachers how to integrate concepts into the curriculum that are central to the enterprise of science, but are

B. Austin (✉)
Center for Science Teaching and Learning, Northern Arizona University, PO Box 5697, Flagstaff, AZ 86011, USA
e-mail: Baa49@nau.edu

N. Schmidt
Environmental Health Specialist, 1665 N. Turquoise Dr., Flagstaff, AZ 86001, USA
e-mail: nlschmidt@inbox.com

A.M. Bodzin et al. (eds.), *The Inclusion of Environmental Education in Science Teacher Education*, DOI 10.1007/978-90-481-9222-9_15,

often on the periphery of secondary school science. These concepts are found in many states' content standards and the National Science Teacher Association (NSTA) Standards for Science Teacher Preparation (National Science Teacher Association 2003). They include personal and technological applications of science (NSTA standard 1c), analysis of problems including risks, costs, and benefits (NSTA standard 4b), and relating science to community resources and resolution of issues (NSTA standard 7b). Our final goal was to demonstrate a learning technology, Cmap, that can be used to teach the integrated nature of science (goal one) and the benefits of collaborative learning. Cmap is a free concept mapping program developed by the Institute for Human and Machine Cognition (see http://cmap.ihmc.us/conceptmap.html). Cmap can be used individually or collaboratively to facilitate visualization of a complex system, such as the environment. This visualization can promote improved understanding of a system and assist with informed decision-making. The complex system we used was the environment and the visualization system was Cmap. Our preservice teachers used Cmap to increase their understanding of environmental issues through the process of constructing concept maps of specific and local environmental questions. By involving our preservice teachers in a real-world problem that requires integrated science content knowledge, that is, environmental questions, we were modeling a context for teaching science that they could implement directly in a secondary science classroom.

Promoting Science Literacy, Environmental Literacy, and Pedagogical Content Knowledge

A common thread in each one of these goals was to increase the science literacy, environmental literacy, and pedagogical content knowledge of our preservice teachers, thus diminishing the gap between science education research and practice. Science literacy is a knowledge base essential for the perpetuation and growth of our society. Environmental literacy is a knowledge base essential for the continued existence of life on Earth. Pedagogical content knowledge is a knowledge base essential for successful teaching of science, including science and environmental literacy.

Science literacy can be defined as the knowledge and scientific habits of mind that facilitate the ability to make informed decisions at the doctor's office, voting box, and marketplace (e.g., American Association for the Advancement of Science 1993; Maienschein 1998). For example, at the doctor's office, knowledge of science and specifically how medical knowledge is developed assists parents in assessing the risk versus benefit of vaccinating their children or older patients in determining which treatment for heart disease is likely to have the highest rate of improvement for their age, gender, and lifestyle. At the voting box, science literacy helps citizens evaluate claims made by politicians, such as the negative impact that signing the Kyoto protocol would have had on the economy versus the benefits to the environment. Science literacy assists consumers in making decisions at the marketplace.

For example, what does irradiation do to fruits and vegetables and would I want to or object to purchasing them? Are the mercury switches that cause kids sports shoes to light up when they walk healthy for kids to wear?

Environmental literacy is similar to science literacy in that it involves both content knowledge and habits of mind in making personal decisions and contributing to group decision-making by being active participants in the civic process (Environmental Literacy Council 2002). Similar to science literacy, environmental literacy allows individuals to make informed decisions at the marketplace and voting box (North American Association for Environmental Education [NAAEE] 2007). For example, environmental literacy informs decisions about whether cloth or disposal diapers (or leather or recycled shoe soles) have less impact on the environment. This knowledge base includes knowledge of living systems, the relationships between living and nonliving systems, and awareness of the economic, aesthetic, and social factors that influence personal and public decision-making. In promoting environmental literacy, teachers must understand the importance of community-based learning, interdisciplinary teaching, and the role technology plays in the collection, analysis, and communication of information about the environment (NAAEE 2007). Promoting environmental literacy has taken on a new urgency. The global society will survive if most people do not understand the scientific evidence and claims developed from experiments at the Large Hadron Collider (science literacy). However, humans might not survive ignorance about the claims made about global climate change and act accordingly (environmental literacy).

Promoting pedagogical content knowledge (PCK) is the thread that is required in all of the activities we incorporate in our science methods class. In past semesters, we used concept mapping to demonstrate how concept mapping technology, specifically Cmap, facilitates construction of knowledge, development of expert knowledge, and high-quality implementation of collaborative learning. In contemplating how to meet our other learning goals of demonstrating the integrated nature of science and increasing environmental literacy, we realized that modifying our concept mapping project to use the environment as the topic would result in a win-win situation for us and our students by enhancing the learning tasks without adding any additional work.

Developing Expert Knowledge Through Concept Mapping

Generating universal explanations for observable events and then projecting outcomes from various decisions is a characteristic that distinguishes experts from novices. In a synthesis of literature on learning, Bransford et al. (2000) identify several characteristics of experts, three of which are: (1) notice and abstract meaningful patterns; (2) use an overarching theoretical framework to organize patterns; and, (3) use contextual clues to flexibly retrieve and apply knowledge. A central purpose of teaching is to advance learners along the trajectory from novice to expert knowledge, abilities, and dispositions. In other words, learning environments should promote the ability to recognize patterns (the basis of knowledge), organize and situate those patterns within a conceptual framework, and decipher contextual cues in order to retrieve and apply knowledge appropriately.

To promote learning environments that accelerate learners along the novice--expert continuum, teachers must be aware of and use tools that develop these three expert characteristics. One such tool is concept mapping, as conceived by Novak (1990). Jonassen (2000) suggests that concept mapping is a powerful tool for stimulating learning. There are many versions of concept mapping, but the original method as developed by Novak (1990, 2005) appears to have the most promise for developing the three characteristics of expert knowledge. According to Novak, concept maps are built through constructing propositions, which contain three parts: a concept followed by linking words and a second concept such that all three parts form a sentence. Without the linking words, the learner is not fully articulating and representing knowledge of relationships between concepts. Second, propositions are arranged hierarchically with the most general concepts on top and the most specific at the bottom. The hierarchical arrangement visualizes the conceptual framework. Third, the map should include crosslinks between knowledge units on different branches, further revealing interrelationships between concepts that demonstrate abstractions of patterns (concepts). Concept mapping promotes learning that is consistent with the definition of expert knowledge (Bransford et al. 2000) in that it requires learners to recognize meaningful patterns (identify concepts). As learners create and link propositions, they reveal and develop their conceptual framework.

As noted earlier, we used Cmap, a free concept mapping program. Cmap has many features that make it a valuable technology tool for secondary teaching. For example, the program allows the user to add many types of resources to concepts and linking words including annotations, notes, pictures, Web links, word documents, and audio and visual files. Several features of Cmap enhance the use of collaborative learning in the classroom. Students can collaborate synchronously and asynchronously from different computers on the same map. There is a chat feature that permits students to chat about the map they are creating synchronously from different computers. Also, the program has a recorder feature that allows an instructor to see the creation history of the maps. The instructor can use the recorder to identify how much each student has contributed to a particular map to monitor individual student participation for accountability purposes.

Concept Mapping the Environment in Preservice Science Teacher Education

This project was conducted in a 400-level secondary science methods course comprising 16 undergraduate and graduate preservice science teachers from the four major science endorsements areas offered at our university: biology, chemistry,

Table 1 Chronology of project

Session	Task	Group
One	Develop a concept map[a] for a regional environmental question	Interdisciplinary groups – four students per group
Two (2 class meetings later)	Review map. Generate a list of additional content knowledge needed to generate a better map	Interdisciplinary groups
Three (1 class meeting later)	Students given a partially completed map about one of the regional environmental questions and asked to complete it	Discipline-specific groups
Four (5 class meetings later)	Review map from session three and identify lab activities for teaching concepts from the map	Discipline-specific groups

[a]Students were given a list of ten words which they could use in the construction of their maps

earth science, and physics. The class met for two-and-a-half hours once a week for 14 weeks.

As noted earlier, approximately 80% of our course time is spent on lab-based teaching. Consistent with the definition given in America's Lab Report (Singer et al. 2006, pp. 31–32), we defined labs as hands-on labs and investigations, field work, simulations, interaction with real-world data (not student-collected), working with large databases, and remote access to science instruments and observations. Student assignments include designing two lab-based lessons, designing a field-trip that includes a lab experience as the central learning event, and compiling an annotated list of ten resources for teaching science through laboratories. Students cannot use more than two lab activities from one type of source; examples of source types are professional journals, government-funded projects, the internet, textbooks, other science books, or parks and museums. These resources become a part of students professional teaching portfolios.

The project was carried out over four sessions; a brief description of each session is provided in Table 1.

Promoting a "Need to Know"

Following a conceptual change model of constructivist learning theory (Posner et al. 1982), we wanted to put our students in a situation of knowledge discomfort in which they would unequivocally experience holes in their science content knowledge. Therefore, we devised two regional environmental questions. The purpose of using the environment was to reveal the integrated nature of science through studying the environment as discussed previously. We wanted the questions to be regional to demonstrate that science is relevant to everyday life and decisions made by voters and consumers. The two questions we devised for Flagstaff were:

What influences the Flagstaff water supply (quality & quantity)?
Since Flagstaff is in the desert, water supply is an issue that everyone understands as being important and vital to well-being and growth of the community and the regional ecosystem.

What influences Ponderosa pine forest health?
Pine forest health influences the incidence and severity of forest fires, a well-known topic in this area of Arizona. Additionally, many students are familiar with a local pest, the bark beetle, and its devastation of local forests.

Every region of the country has similar questions that most secondary students will be familiar with. We wanted our students to realize that they lacked certain science knowledge for making relevant, everyday science-based decisions.

Session One. In session one, two of the groups developed maps for the water question, two for the forest question. Even though three of the four groups had a difficult time constructing the maps, each topical map was surprisingly similar. Figure 1 presents one of the maps generated in response to the Flagstaff water supply question.

Students seemed surprised that mapping these questions was difficult. In the mapping debrief discussion, most of them expressed comfort with their level of science content knowledge. They also understood that the questions being asked were important for the general public to understand and were science-based. Students arrived at the desired conclusion that secondary science should be taught to enable the average citizen to use science for understanding and making decisions about important regional environmental concerns, such as those presented in the mapping exercise. Our students recognized deficiencies in the way science had been taught to them at both the secondary and college levels. We felt that this discussion was a good start towards conceptual change in recognizing the integrated rather than disciplinary nature of science.

Session Two. Students generated a list of concepts that they would need to know in order to better map these questions. On the water question, many of the concepts students identified were science-based including variation in infiltration rate as a function of soil type, how quickly water moves over various types of groundcover, variation in infiltration rate as a function of water quality, and the effect of infiltration on water quality. Other concepts they thought would help map the question were socioeconomic in nature including how the city sets the cost of water, typical water usage patterns for homes and businesses, variations in water usage patterns of different communities, and city water use. Students who worked on the forest health question came up with an analogous list of science and socioeconomic concepts.

As with session one, we were happy with the progress we were making toward helping our students understand that their conception of science and science instruction developed over many years of personal experience needed major modification. In our debrief discussion, we focused on the disconnection between

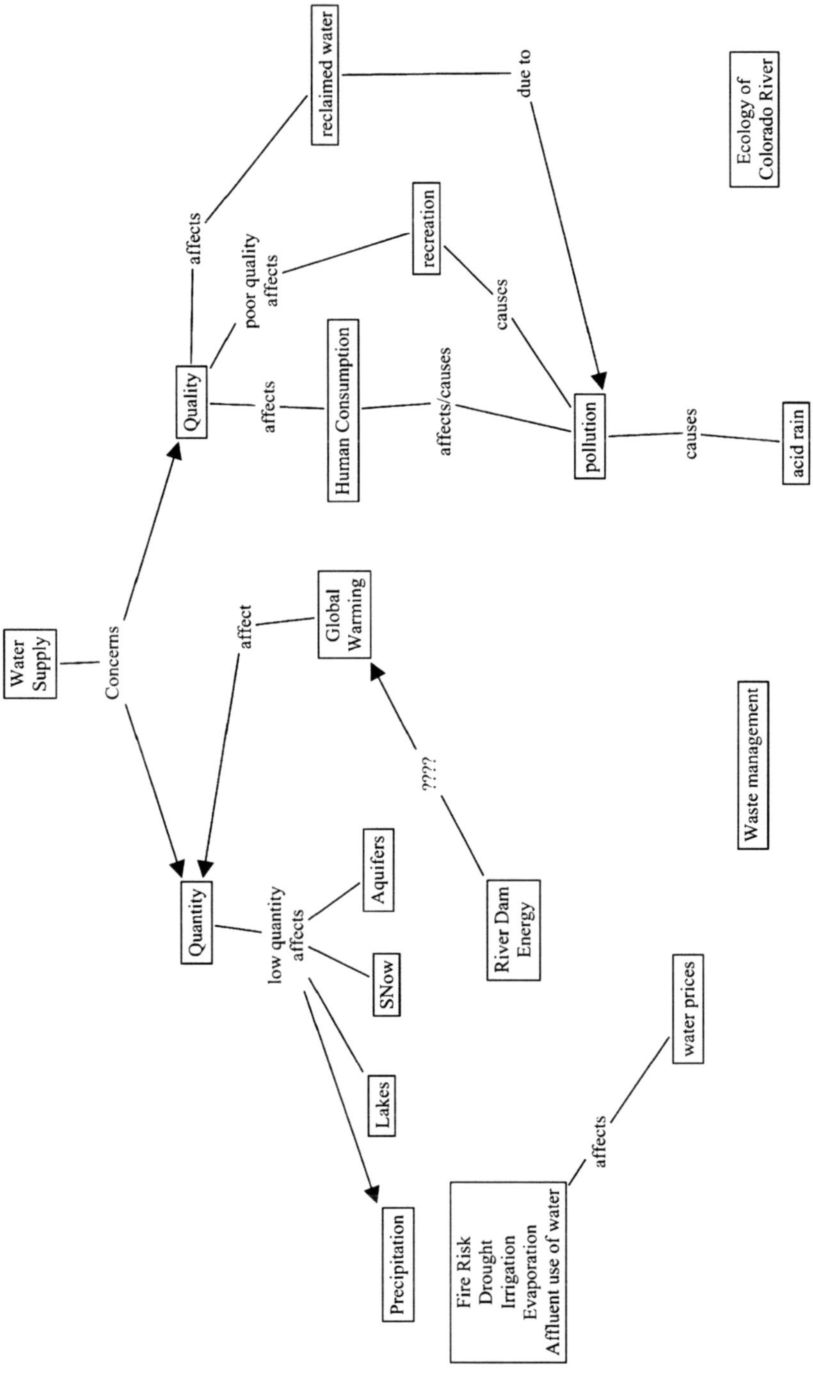

Fig. 1 Student-generated concept map about what influences the Flagstaff water supply

authentic science and environmental literacy and traditional science instruction. We could tell from the conversation that they were developing a vision for science teaching but two barriers came up in conversation that were a matter of concern for many students. The first barrier was the disciplinary nature of secondary science departments and curriculum. The second barrier was the disciplinary nature of the Arizona science content standards. Several students with ecology backgrounds brought up the idea of using the environment as a context for teaching all of the content areas. We concluded by saying that we would look more deeply at that idea next time.

Scaffolding Construction of Knowledge About Curriculum Design

Session Three. Because of the limited time available for this project, we decided to jump-start the process of using concept mapping in the design of curriculum to address the barriers mentioned in the session two debrief. We developed a two-level map for planning curriculum that not only used the regional issue as the organizing idea but also included concepts from the Arizona science content standards (Fig. 2). With the regional issue as the central focus of the course, our students would see how content from other disciplines aligned to their environmental question and ideally would also include socioeconomic content. In levels two and beyond, they could demonstrate to administrators and parents that they were addressing science content standards.

Students divided into disciplinary groups and began to insert concepts they would teach in their content area. An example of work from the biology group is provided in Fig. 3.

Session Four. Between sessions three and four, students continued to work on course assignments including the field trip lesson plan and finding resources for teaching labs. The goal for session four was for students to identify ways to teach the curriculum they mapped in session three. Using the features of Cmap, they added links, files, annotations, and notes about teaching to their map. Many of the items they added came from their lab-based lesson plans, field-trip, and lab resources assignments.

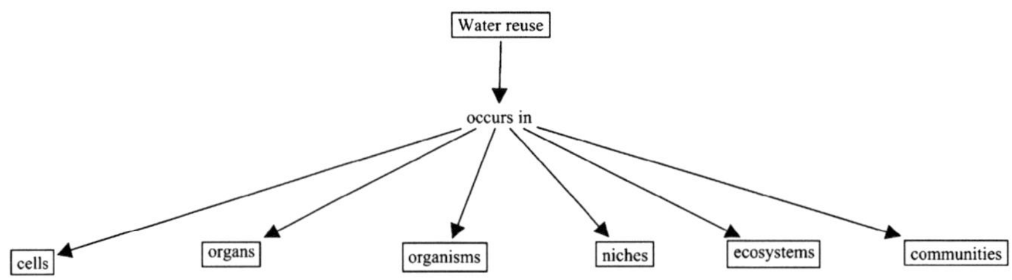

Fig. 2 Two-level curriculum organizer

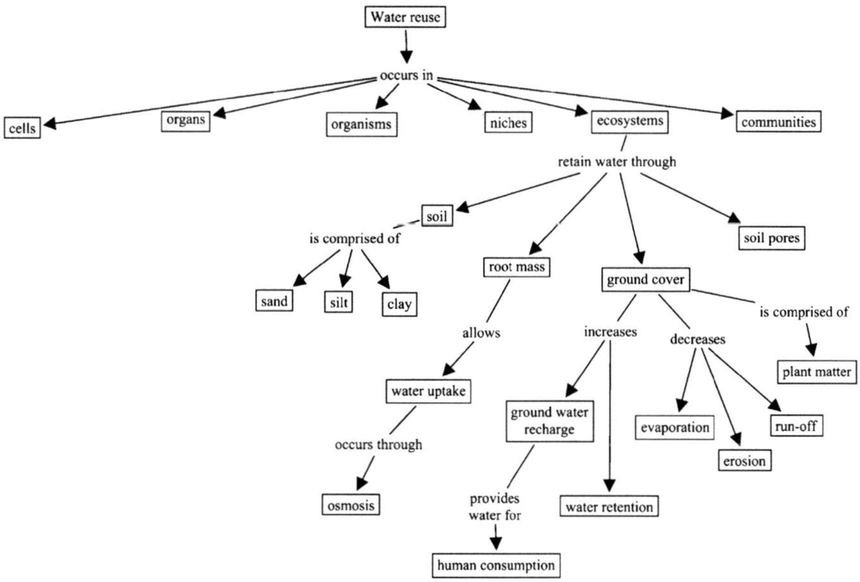

Fig. 3 Biology student group map of biology curriculum

Outcomes

As noted earlier, our goals for this project were numerous and complex in that we wanted students to develop both content and pedagogical content knowledge. We feel that the complexity of the project appropriately supported the multiplicity of learning that we wanted our students to gain from this project.

Promoting Science Literacy

Our primary learning goals for science literacy were for students to understand the integrated nature of science and to experience the relevancy of science in making everyday decisions. Both the topic (the environment) and the pedagogy (concept mapping) promoted understanding of the integrated nature of science. Mapping a regional environmental question requires understanding of living systems (biology and chemistry), nonliving systems (chemistry and earth science), and energy and matter flow (all disciplines). Mapping aids in understanding the relationships between the concepts and therefore the disciplines. In the project debrief discussions, students stated that most real-world science problems require knowledge from multiple disciplines. One talked about how the field of cellular biology has advanced in the last 20 years due to the application of chemistry techniques. A geology student also noted how ideas from microbiology have influenced hypotheses of how certain minerals form. The map also helped students document the relationship between science teaching contexts that are relevant to students and future science and environmental literacy and the content for which the teachers would be responsible for teaching (the standards).

Discussing the Arizona science content standards gave us the opportunity to foreshadow the major assignment that they would have to complete during student teaching and that is based on the NSTA NCATE standards that were referred to earlier. We strongly encouraged our students to consider how using the environment as an organizer for designing curriculum would enable them to demonstrate personal and technological applications of science, analysis of problems, and incorporation of community resources. These NSTA standards are consistent with the goal and purpose of increasing both science and environmental literacy.

Promoting Environmental Literacy

The concept mapping project occurred concurrently with our other course assignments. Compared to other semesters, more of the lab-based lesson plans, field trips, and lab resources included environmental topics and were specifically designed to promote environmental literacy. We found that students were utilizing and referencing environmentally themed articles and issues from practitioner journals including *The Science Teacher*, *Science Scope*, and the *American Biology Teacher*. Many of the lab-based lessons were based on articles from practitioner journals. This project clearly increased student awareness of these valuable professional resources. Additionally, in previous semesters, the only online environmental education resources cited by students were Project WILD (http://www.projectwild.org/) and Project WET (http://www.projectwet.org/). In this semester, students included many sites with environmental education resources for teachers including NAAEE (http://www.naaee.org/), the Environmental Literacy Council (http://www.enviroliteracy.org/), and the United States Environmental Protection Agency (http://www.epa.gov/).

This project also influenced the field trip topics. Nearly all the field trip lesson plans were based on environmental questions. These lessons included water quality in the various surface waters, species diversity after forest fires, species diversity in their schoolyard, and studies of microclimates. Students also incorporated community resources at higher rates this semester; new locations unseen in previous semesters included a visit to a herbarium, a biodiesel-production facility (local hobbyist), and the USGS.

In previous semesters, there was no obvious focus on the lesson plans, the field trips, and the lab resources assignment. This project gave students a framework for selecting topics for their assignments. We hope that the curriculum concept map gave them a framework for how, why, and where to use these materials so that they will actually use them.

Promoting Pedagogical Content Knowledge

Our PCK goals for the project promoted the use of educational technology and collaborative learning. In the session one debrief discussion, students noted several experiences with collaborative concept mapping:

1. Because of the interdisciplinary nature of the task, no single person could take over the task. All contributed.
2. Because every person in each group studied a different content area, everyone was able to contribute to generating the list of concepts and creating propositions.
3. Because the instructions were to make an interdisciplinary map, no single person had enough knowledge to generate every proposition on the map.

The instructor reviewed the list of experiences and gave a short presentation on the similarities between what the students had experienced and the research for supporting collaborative learning (Fuchs et al. 1997; Slavin 1991). This very personal experience with discomfort in mapping an unfamiliar topic mediated by peer support and peer contributions in the development of an authentic group product seemed to improve the opinion of some students about the benefits of collaborative learning. Mapping was seen as a viable avenue for introducing collaborative learning in the classroom.

Summary

The fulcrum of success of this project was getting students to reconsider the structure of scientific knowledge. Students realized that the reason they had difficulty with mapping regional environmental questions was because the science instruction they had received did not promote science or environmental literacy. By asking students to use their content knowledge in the design of a concept map (organize their knowledge), we reinforced the idea that the teaching they had experienced did not develop the type of expert knowledge required from ordinary citizens. Ordinary citizens must vote on science-based issues and make decisions at the marketplace about which choices have the least negative environmental impact; secondary science teaching must develop knowledge that informs this decision-making.

Putting students into disequilibrium about their science education enabled us to change the nature of conversation about how to teach science and what science to teach (disciplinary or integrated). Concept mapping was an excellent vehicle for demonstrating the complexity of the environment and the benefit of organized knowledge. Our students became vested in what we were trying to teach them: content is not a series of topics or chapters in a book but rather a valuable resource for making informed decisions as adults. They realized that lecture, lab, worksheet, lab report, followed by exam are endpoint learning activities that do not assist learners in transferring their knowledge to new situations and solving unfamiliar problems.

These activities have a place in learning environments, but are more effective if set within the context of an overarching problem. Teacher education, like teaching, is a complex process that requires expert enactment of knowledge from many content areas. We believe that mapping regional environmental questions will narrow the research-practice gap for many of our students and support their development as informed classroom decision-makers.

References

American Association for the Advancement of Science (1993) Benchmarks for science literacy. American Association for the Advancement of Science, Washington, DC

Bransford JD, Brown AL, Cocking RR (2000) How people learn: Brain, mind, experience, and school, Expth edn. National Academies Press, Washington, DC

Environmental Literacy Council. (2002). What is environmental literacy? Retrieved February 10, 2009, from http://www.enviroliteracy.org/subcategory.php?id=1

Fuchs D, Fuchs LS, Mathes PG, Simmons DC (1997) Peer-assisted learning strategies: Making classrooms more responsive to diversity. American Educational Research Journal 34(1):174–206

Jonassen DH (2000) Computers as mindtools for schools: Engaging critical thinking, 2nd edn. Merrill, Columbus, OH

Maienschein J (1998) Scientific literacy. Science 281:917

National Science Teacher Association. (2003). Standards for Science Teacher Preparation. Retrieved February 3, 2009, from www.nsta.org/pdfs/NSTAstandards2003.pdf

North American Association for Environmental Education. (2007). Standards for the initial preparation of environmental educators. Retrieved February 10, 2009, from http://www.naaee.org/

Novak JD (1990) Concept maps and vee diagrams: Two metacognitive tools for science and mathematics education. Instructional Science 19:29–52

Novak JD (2005) Results and implications of a 12-year longitudinal study of science concept learning. Research in Science Education 35(1):23

Penick JE, Yager RE (1988) Science teacher education: A program with a theoretical and pragmatic rationale. Journal of Teacher Education 39:59–64

Posner GJ, Strike KA, Hewson PW, Gertzog WA (1982) Accommodation of a scientific conception: Toward a theory of conceptual change. Science Education 66:211–227

Singer SR, Hilton ML, Schweinberger HA (2006) America's lab report. National Academies Press, Washington, DC

Slavin RE (1991) Synthesis of research of cooperative learning. Educational Leadership 48(5):71–82

Unraveling the Scientific, Social, Political, and Economic Dimensions of Environmental Issues Through Role-Playing Simulations

James G. MaKinster

Introduction and Background

Most college-level science courses fail to provide students with a true appreciation of the nature of science (Nelson et al. 1998). Often very structured and fairly intense, they tend to cover a lot of "material" that instructors consider essential. Such courses typically include laboratory sessions that tend to rely on a series of isolated activities. Consequently, students often come to view science as a body of information or "knowledge" to be memorized, while failing to discover and experience other dimensions of science and scientific inquiry (reviewed in Abd-El-Khalick and Lederman 2000; McComas 1996). A number of national initiatives are focusing on changing the nature of college-level *science* teaching (Shulman 2000). These include interdisciplinary science and environmental education courses for future teachers, which can serve as effective contexts for expanding students' conceptions of science.

One powerful means to create a more authentic environment for students to learn science and expand their conceptions of the nature of science is within simulations, in which students take on specific roles (Aubusson et al. 1997; Pennock and Bardwell 1994). *Role-play simulations* typically occur over an extended time period. They expand more traditional concepts of role-playing and debates by creating a context that is more dynamic and authentic (Webb 2002). Such simulations involve multiple groups of individuals, each with particular focus, which reflect specific backgrounds, histories, and competing agendas. Students take on specific roles within these groups that require significant personal investment as they try to understand the nature, thinking, beliefs, and perspectives of their character (McKeachie 1994; Pennock and Bardwell 1994). These types of role-play simulations can result in dynamic environments that oblige

J.G. MaKinster (✉)
Hobart & William Smith Colleges, Geneva, NY, 14456, USA
e-mail: makinster@hws.edu

A.M. Bodzin et al. (eds.), *The Inclusion of Environmental Education in Science Teacher Education*, DOI 10.1007/978-90-481-9222-9_16,

participants to critically examine their own perspectives and understandings, as well as those of their peers. Consequently, participants develop more robust understandings of abstract concepts (Webb 2002), such as the political and social nature of science.

This chapter focuses on a role-play simulation within an optional college-level interdisciplinary science course designed for preservice teachers and liberal arts students. The role-play simulation utilized a United States Senate Subcommittee hearing on the use of Bt (*Bacillus thuringiensis*) genes in corn. Bt genes in corn enable the plant to produce an insecticide within its vegetative structures. The curricular design of this simulation was based on the work of Harwood, MaKinster, Cruz, and Gabel (2002), who used a Senate Subcommittee hearing to explore global climate change. However, unlike the Harwood et al. project and some of the more focused simulations within the Project Wild (Council for Environmental Education 2003) and Project Wet curricula (Project WET 2003), the time frame for this simulation extended over a period of 3.5 weeks. This allowed students to explore the material in greater depth, and incorporate numerous teaching strategies and topics of current interest in science education (simulations, role-playing, driving questions, oral presentations, technology integration, portfolios, reflection, and concept mapping).

The goal of this chapter is to present and analyze an experience that enables preservice students not only to expand their conceptions of science and environmental inquiry, but also to understand better how science is applied to real-world environmental issues. This unit can be replicated using other environmental topics and in other types of courses. Senate Subcommittee hearings are appropriate for college or university science courses, secondary science courses, science methods courses, and interdisciplinary science courses. At the end of this chapter, other simulation curricula are discussed that exemplify the potential of this pedagogical strategy in the context of emerging technologies. In addition, there is a discussion that describes how the use of such simulations can broaden our definition of what it means to inquire into an environmental issue in productive and meaningful ways.

Unit Description

The role-play simulation unit utilized a United States Senate Subcommittee hearing on the use of Bt (*Bacillus thuringiensis*) genes in corn. This unit consisted of eight, 85-min class periods and involved several instructional strategies (Table 1). *Bacillus thuringiensis* is spore-forming bacteria that poison many types of insects, primarily butterflies and moths. It became a popular sprayed pesticide because it poisoned fewer non-target species than chemical insecticides; however, like chemical pesticides, precipitation could wash it off from the plants. A major solution to this problem was the development of genetically engineered corn that expressed the Bt gene within the plant tissue itself (Bessin 1999). This recombinant technology

Table 1 Schedule of topics and assignments for Senate Subcommittee hearing

Day	*Topic*	*Assignment*
1	Introduction/Concept map/Overview	Comparative reaction paper
2	General discussion and library research session (*Classroom and library*)	Summary of group position
3	General discussion and group work (*Classroom and library*)	Annotated bibliography
4	Portfolio discussion and PowerPoint presentations (*Computer lab*)	Outline of presentation
5	Feedback on presentation and group work (*Computer lab*)	First draft of PowerPoint presentation
6	Senate hearing – Day 1 – Initial arguments	PowerPoint presentations
7	Senate hearing – Day 2 – Follow-up questions, debate, and discussion	Legislative decision (senators only)
8	Senate Subcommittee Statement and debrief	Portfolios and individual reflections

enabled the plants' tissues to express the Bt toxins that are poisonous to insects. Interestingly, Bt toxins are not poisonous to humans due to differences in the pH levels in our digestive tracts.

Initial Concept Map

Prior to the Senate Subcommittee hearing simulation introduction, students were divided into groups of three and asked to make a concept map with different scientific disciplines as the central concepts. The concept maps were then combined to make a class concept map of "Science" (Fig. 1). This map almost exclusively described each science discipline relative to the sub-disciplines within that area. Several students stated that they focused primarily on how textbooks are organized or on the classes that they had taken in the past. This experience initiated a brief discussion about the nature of science and scientific knowledge. The students' ideas were validated, but the primary issue raised was that the concept map might be limited in scope when one considers the nature of science and how science is applied in the "real world." This conversation eventually led to a brief introduction of the science behind Bt genes in corn and a reading assignment for the next class.

Introducing the Unit

To begin exploring and discussing some of the issues surrounding Bt genes in corn, students read two papers, *Transgenic Pollen Harms Monarch Larvae* (Losey et al. 1999) and *False Reports And The Ears Of Men* (Shelton and Roush 1999). The first paper stated a number of concerns regarding the use of Bt Corn and its potential

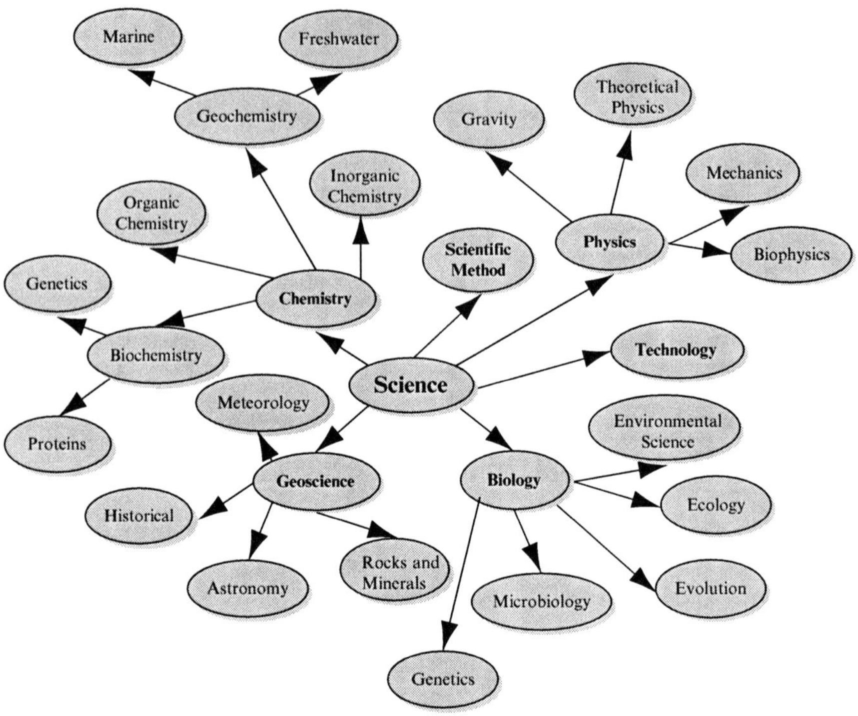

Fig. 1 Initial concept map constructed around the topic of "Science"

impact on monarch butterflies eating vegetation along the edges of Bt Corn fields. The authors argued that a significant number of monarchs were being killed by Bt toxins that were carried to vegetation at the edges of the field by airborne corn spores. The second paper attempted to refute these claims. The students were engaged in a lively discussion about the two papers, but no clear consensus was reached. In fact, most of the students said they were unsure about whether or not Bt Corn was of significant concern.

The course instructor used this opportunity to introduce the Senate Subcommittee hearing simulation by describing the instructional tasks to be undertaken over the next several weeks. He described what the Senate hearing would entail, and presented the driving questions for the hearing. The following questions guided both the student group investigations and the Senate Subcommittee hearing:

1. Are there or are there not legitimate concerns regarding the use of Bt genes in corn?
2. What are the options to manage the use of Bt Corn?
3. What additional research or legislation might be useful in addressing the first two questions?

Three students were assigned the role of senators and the rest ($n = 16$) were divided into six special interest groups:

1. NY State Farm Bureau
2. Food and Agriculture Organization of the UN
3. Monsanto Company
4. Genetic Research Scientists
5. Greenpeace
6. Environmental Defense Organization

The special interest groups included local, international, corporate, scientific, environmental, and legal groups that have an actual interest in the issues surrounding genetically modified food. For the next class, each student group wrote a description of their group, the issues of concern for their group, and what it is that they do. The senators had to adopt a political party and affiliate themselves with a particular state. This provided the senators with a particular mindset in terms of a specific political party and set of constituents whose interests they were to represent during the hearing.

Preparing for the Hearing

In preparation for the Senate Subcommittee hearing, the special interest groups had four primary responsibilities. First, they had to solidify their positions on Bt Corn; second, groups conducted a thorough literature search for relevant information and research; third, they developed a PowerPoint presentation to use during the hearing to present their case; and finally, they compiled a portfolio of their work to serve as both a resource during the hearing and as a summative assessment for the experience. Several class periods and significant out-of-class time were used to accomplish these goals. Students received a brief introduction to PowerPoint, but most of the software instruction was provided during three brief (5–10 min) lectures. Students were given 10–12 min to present their case, including not more than eight to ten slides. They could also include five to six extra slides to answer potential questions posed by senators following their presentations or during the discussion on the second day.

Senate Subcommittee Hearing

On the first hearing day, the instructor reminded the senators of their responsibility to run the hearing. The instructor's role was simply to facilitate the technology and make sure that each person remained in their role during the hearing. Senators were told to "run as formal a hearing as possible."

Each special interest group used PowerPoint to present their position and discuss their principal points. The extent to which students were comfortable presenting to their peers varied. Several groups presented compelling arguments, while other presentations led to a barrage of clarifying questions by the senators. In general, the senators followed each presentation with two or three clarifying questions. Introductions, presentations by each of the six groups, and transition from one group to the next took almost one entire class session (90 min).

After class, the senators met the instructor to discuss their responsibilities. The senators were charged with reviewing what each special interest group had said and developing follow-up questions to ask each group. The senators were able to juxtapose groups against one another by facilitating a general discussion around a particular topic (such as ideas for legislation). For example, senators might ask the same questions to two groups (e.g., the Monsanto Corporation and the Environmental Defense Organization), allow them to respond, and then open the topic up for discussion among all the groups.

The second day of the hearing was much more dynamic and lively than the first. While the students formally presented their ideas on the first day, the second day was intended to generate a debate atmosphere and to challenge students to articulate and defend their positions. The senators recalled each group to ask clarifying questions about the previously presented arguments. If conflicts arose between the groups, then the senators would encourage each group to justify their positions. Many students relied on both the extra PowerPoint slides they created or referred to the articles and book chapters they had copied. In addition, their portfolios served as a valuable resource, which enabled them to find and organize the papers and supporting information they were using with relative ease. Finally, the senators had several more open discussions to present issues to the groups and to see where the deliberations led. When the class period was over, the senators thanked the groups and told them that they would present their final decision during the next class period. The instructor met again with the senators to make sure they understood their responsibilities. The senators wrote a two-page response to the three driving questions and decided which special interest group(s) "would have a hand in writing the future legislation" regarding the Bt genes in corn.

Senate Subcommittee Statement and Debrief

In the next class, the senators presented their findings and decisions regarding each of the driving questions. The class then discussed the outcomes from the perspectives of their special interest groups. Finally, the students were asked to drop their roles and engage in a reflective discussion about the entire experience. To prepare for this discussion, each of the students had written an individual evaluative reflection that addressed the following questions:

1. Overall, what did you think of the experience?
2. Currently, how well-informed do you feel about the topic?
3. How did your views regarding Bt Corn change during our experience?

4. What are your current opinions regarding Bt Corn specifically, and genetically modified foods in general?
5. How has this experience changed your view on the application of scientific knowledge to policy, political, and social decisions?
6. Are there any other comments or suggestions you would like to make?

These questions served to guide the discussion and enabled students to reflect on their experience prior to the in-class discussion.

In an effort to represent the experience in its entirety, the class developed a group concept map of the Senate hearing. First, the students worked in their special interest groups, or as senators, to develop a concept map that represented the issues that were of primary concern to their group. These maps were drawn on 2 ft by 3 ft pieces of newsprint. For example, the Environmental Defense Organization identified ecological concerns, economic concerns, human safety concerns, EPA/FDA, recommendations, resistance, benefits of Bt Corn, and the goals of Bt Corn as the areas of greatest concern to their group during the experience. Special interest groups then provided more detail in each of those areas on their maps by linking other concepts or topics to their initial issues. Once each of the special interest groups had completed their maps, the class was instructed to create a single large concept map by placing their individual maps relative to one another on the classroom wall (Fig. 2). This combined concept map was revised over several iterations and conversations. Once finished, it was used as the centerpiece for a discussion about the nature of science, primarily how scientific issues are shaped by economic, political and social forces.

Unpacking the Experience

Senate Hearing

The Senate Subcommittee hearing structure forced students to explore, appreciate, and articulate the perspectives of their special interest group, as well as develop a solid understanding of the science behind Bt Corn. Initially, students had to explore their special interest group and develop an understanding of what their position might be on Bt genes in corn. This research was done primarily through online resources. Some special interest groups had fairly specific information on their website that was applicable to Bt Corn and the hearing, whereas other students needed to infer what their particular group's position might be, based on the perspectives or mission of the group. Once students identified their group's position on the issues at hand, the students began to search the literature for relevant information and research to build their case.

On the first day of the hearing, the special interest group presentations served to create a foundation for a more lively debate on the second day of the hearing (Table 2). The presentations were somewhat formal in nature and seemed to reflect

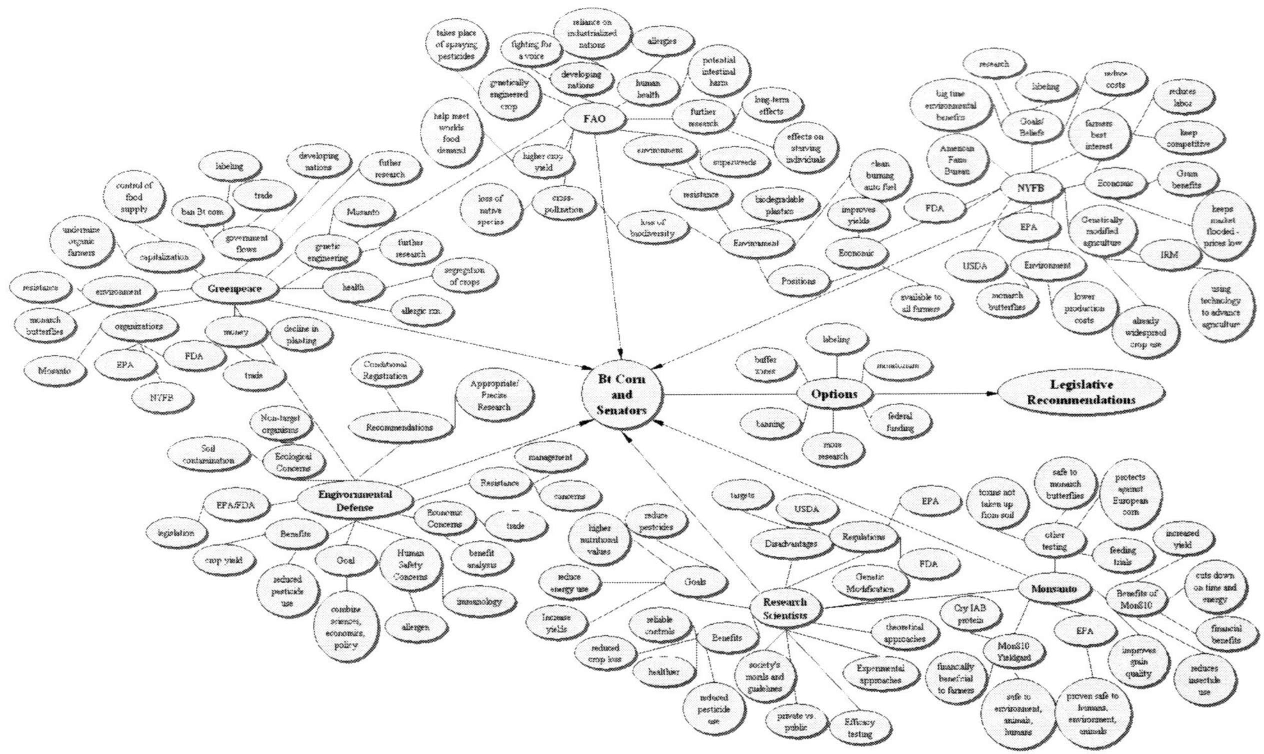

Fig. 2 Simplified electronic version of the class concept map of Senate Subcommittee hearing on Bt genes in corn

Table 2 Primary arguments made by special interest groups during the first day of the hearings

	Special interest group	Primary arguments
Pro	*Monsanto Company*	The Monsanto strain of Bt Corn was tested for 20 years before EPA approval and poses little risk
		Our Bt Corn product, YieldGard, results in a greater yield per acre
		Bt Corn reduces amount of insecticide used
		The use of Bt Corn improves grain quality
	NY State Farm Bureau	Bt Corn is beneficial for farmers
		A moratorium on Bt Corn would be damaging economically
		The impact on butterflies is negligible
		There are a number of environmental and economic benefits
	Genetic research scientists	Bt Corn is a reliable control of target pests
		Bt Corn has reduced crop loss and pesticide use
		Bt Corn has increased yields
		Bt Corn has provided farmers with the ability to use marginal land
	UN Food and Agriculture Organization	Bt Corn has devastating effects on native species
		Use of Bt Corn results in a loss of biodiversity
		Much of the impact of Bt Corn is within impoverished countries
	Environmental Defense Organization	The originally claimed benefits in the reduction of pesticides proved to be misleading
		Research on the impact of Bt Corn on butterflies was poorly designed and implemented
		Bt research on impacts was conducted on aquatic animals instead of animals actually affected by corn pollen
	Greenpeace	Bt kills Monarch butterflies
		There are concerns about transfer from plant chromosomes to soil bacteria
		Use of Bt Corn increases cost for farmers
		Use of Bt Corn decreases trade domestically and internationally
		Control of food supplies will be in the hands of a few large companies
Con		An increase in large-scale farming has led to rural unemployment

a certain amount of competition among the groups. All of the groups made good to excellent presentations. They were clear, straight-forward, and organized in a logical manner. The senators took notes throughout the hearing and tended to ask two or three clarifying questions at the end of each presentation.

The following day, the senators asked the follow-up questions. This required students to have a solid understanding of the material they were presenting and the primary literature they were using to support their case. The basis for students' presentations came in a number of different forms. Much of the information had been distilled into the PowerPoint presentations themselves. More detail was available within the copies of manuscripts that the students included in their portfolios. These research papers and the figures within were often at the heart of particular arguments. For example, the New York State Farm Bureau used a figure that documented a national decline in herbicide use due to the introduction of Bt hybrids in 1997. The senators asked a number of clarifying questions about this figure, in addition to asking about its source. This "cross-examination" was a great example of how the students were forced to explain their assertions and how the senators attempted to tease apart the arguments of each special interest group. For example, at one point during the second day, the senators asked Greenpeace to restate their position about the negative effects of Bt Corn in light of what Monsanto representatives had argued about the safety of their product.

> Senator: We would like to hear from both Greenpeace and Monsanto again. On Tuesday, you (Greenpeace) talked about the impacts that Bt Corn has had on butterflies and so forth. We'd like you to respond to what you've heard from Monsanto thus far. Have your views changed?
>
> Greenpeace: We feel that the science behind Monsanto's statements is questionable. We don't know how butterflies or plants in the surrounding fields and such will be affected.
>
> Monsanto: I'd like to respond.
>
> Senators: Go ahead.
>
> Monsanto: We based our position around the strain MON 810. Much of the research done on Bt Corn was done with other strains. These are very different and less safe that MON 810. There've been 20 years of testing on this strain by the EPA with no environmental concerns. Some of the other strains may pose a risk and, as we said, they'll likely be withdrawn from the market.

This interaction gave Monsanto a chance to reiterate its strongest point that their argument was based on MON 810, a strain that has, so far, been shown to be safe. When the argument started going back and forth, the Monsanto representatives stated repeatedly that they were referring to MON 810 and not some of the other strains.

Senate Subcommittee Statement and Findings

The three senators used what they learned over the course of the 2-day hearing to craft their statement, which was guided by the hearing's three driving questions. First, the subcommittee found that there were "several legitimate concerns regarding the use of Bt-genes in corn." These included the possibility of Bt genes acting as a food allergen, adverse effects on malnourished children, inadequate labeling, impacts on non-target organisms such as the Monarch butterfly, and the loss of native crops due to cross pollination. Second, they recommended "labeling of genetically-modified products (at least up to primary goods), provisions for a buffer zone surrounding Bt Corn fields (so that cross-pollination and resistance may be slowed or even prevented), and a required registration process." Finally, the senators decided that more research was needed, which should be especially focused on the:

> consequences of using Bt Corn, whether it is beneficial or detrimental to the environment, surrounding ecosystems, and human consumption.
>
> In order to meet the interests of all groups involved, including environmental organizations, company suppliers, farmers and individual states, we have found that the best solution would be to provide funding for further research, and regulate fields so that potential concerns may be avoided as much as possible.

The Senators did an excellent job of synthesizing a tremendous amount of information and making recommendations that recognized the complexity of this issue.

Concept Maps

The Senate Committee Hearing created a real-world context that enabled these future teachers to develop more robust conceptions of the nature of science and the complexities of controversial environmental issues applied within political and social arenas. The students' first concept map, drawn prior to the Senate Subcommittee hearing, represented their initial conceptions of science. It focused almost exclusively on the science disciplines and specific areas within each discipline (Fig. 1). After the senate Subcommittee hearing, a similar collaborative activity led to the creation of a "new concept map" (Fig. 2). This concept map focused on the nature of their experience throughout the hearings. By creating this map, students were able to capture, articulate, and discuss the epistemological complexities embedded within this single environmental issue and discuss how the Senate hearing changed their conceptions of science more generally. Science was now seen as cutting across social, economic, political, and often, very personal boundaries.

> Upon completion of the hearing and evaluating each issue, my views regarding the application of scientific knowledge to policy, political, and social decisions have changed. As a scientist in training, I realize the impact fundamental research can have on such policies... in addition, politicians need to evaluate this work thoroughly so that they may make the best overall legislative decisions... Scott – Greenpeace

This shift in thinking made it very challenging for the senators to arrive at answers to the driving questions for the class. Consequently, the senators invested a significant amount of time in preparation for the hearing's second day, and they had a real "need to know" when asking the different groups follow-up or "clarifying" questions. The driving questions served to both inform and organize the questions asked by the senators.

Student Experience

At times, the nature and structure of the role-play experience caused the students to truly struggle in a variety of ways. A primary goal of inquiry teaching is to create a learning environment in which students are challenged, yet have the supports they need to be successful (Crawford 2000). Some students struggled with the fact that their position within the role-play experience did not necessarily match their personal position on the issue. This experience provided an opportunity for them to consider views other than their own, while re-examining their own views on this issue at the same time. More importantly, the lack of lectures on the detailed science content behind Bt Corn forced students to develop their own understandings. The depth and accuracy of these understandings manifested themselves in their presentations and throughout the hearing. The students often had to explain concepts such as transmission or cross-pollination in considerable detail. The understandings that they demonstrated were usually far beyond the basic scientific information covered during the class.

To achieve success, students had to construct and articulate logical and rational arguments. Not surprisingly, some of the students were better able to communicate their ideas than others, but all of the PowerPoint presentations reflected a clear line of thought or argument that was grounded in relevant evidence. The nature and origin of the evidence varied from group to group and became of concern to the senators at various points during the hearing. For example, some of the evidence used by Greenpeace came from newspaper articles, whereas the New York State Farm Bureau used EPA sources, Monsanto, and other research studies published in peer-reviewed journals. By asking each group pointed questions about their sources, the senators were able to better understand the types of sources that were grounding their arguments.

The students developed robust interdisciplinary perspectives about the issues surrounding Bt Corn. There were several advantages in terms of a group responding to the senators in the context of "competing" with the other special interest groups. This sense of competition required each group to not only argue their own position, but also to understand, appreciate, and evaluate the position of the other groups, so that they could frame their arguments in a manner that was compelling to the senators. As a result, both the senators and special interest groups came to see the issues surrounding Bt genes in corn as embedded within cultural, political, economic, and social agendas.

> The benefits and disadvantages of Bt Corn certainly became a large part of how I perceived the issues. As a Senator, I had to not only keep in mind the interests of my state, but also and more importantly, keep an open mind to how it affects people of different groups, whether they are national, cultural, or even part of a minority. Partaking in this role thus allowed me to make an informed decision about my personal position on the issue. I experienced a lot of fluctuations before making a conclusion, finding it difficult to decide whom to take seriously with so many conflicting findings. Kaitlin – New York State Senator

The role of this student as a senator is clearly reflected in her response. How Bt Corn affects the lives of real people is of central concern to her. Her response also reflects the fact that different groups of people can be affected very differently by a particular policy or piece of legislation. Most students came to appreciate the impact that scientific issues can have on our lives.

> The Senate Hearing on Bt Corn integrated science and technology while at the same time addressing ethical and political issues. Science can be applied to every aspect of society and is used to make decisions that affect all our lives. As a political science major, I recognize the importance of a scientific background in our nation's leaders. Technology and science go hand in hand, but science is diverse and can affect people's lives in different ways. Erica – Food and Agriculture Organization of the UN

Student Understandings

The role-play experience and the follow-up discussion helped students to better understand the nature of science and scientific knowledge. Students saw science as much more than simply a body of knowledge. The shift in their views on science reflected their broadened outlook and understanding of the impact of science on the world.

> Science is much more a part of this world than simply the disciplinary area. It is an interdisciplinary perspective that intertwines all aspects of life. Julie – New York State Farm Bureau Representative

The concept map that the students made at the end of the semester best represented the way the students came to view science and the application of scientific knowledge to environmental issues. The development of the group and whole-class concept maps enabled the students to see the entangled and interconnected nature of the issues at hand. The question of genetically modified food, and Bt Corn specifically, served as useful examples of how technological innovation can have far reaching social and political ramifications.

> The concept map did a nice job capturing my views regarding the application of scientific knowledge to policy, political, and social decisions have changed. As a scientist in training, I realize the impact fundamental research can have on such policies. However, I had never looked at it through the eyes of a politician, until now. So much research is done in this world on an infinite number of elements, that often times similar research is going on somewhere else that completely contradicts one's findings. This does not mean that one's work is insignificant. What it does mean however, is that politicians need to evaluate this work so that they may make the best overall legislative decision. Kaitlin – Senator

Kaitlin did a nice job highlighting the fact that there are often competing research agendas around controversial topics.

The students developed a greater appreciation about why some political groups may never agree on a particular issue. However, decisions still need to be made and policy makers may or may not have the time to hear every side of a particular story.

> This whole process has changed my opinions on scientific knowledge and its entanglements with public policy and environmental decisions. I think scientific research is very important, but that it must look at some of the social ramifications it can have. Not everyone is going to agree and support all research. Mike – Monsanto Representative

The role-play simulation helped students to see that politics is much more than how laws are made in Washington, D.C. Politics and political agendas refer to groups of real people who share common concerns and agendas. At times, such stakeholders might organize themselves into special interest groups in order to exert significant influence over legislation and funding. Having a diverse set of special interest groups assisted most students in developing an appreciation for the diversity of legitimate perspectives on a particular issue.

Finally, the students came to see how issues such as Bt Corn touch their own lives and the lives of people in their community. As one group represented the New York State Farm Bureau, students increased their awareness of how genetically modified foods affect local farmers. We were fortunate that one of the students in this group grew up on a New York state farm. She was able to talk to her father about the issues presented. This provided an additional level of authenticity to her experience.

Implications

The Senate Subcommittee hearing on Bt genes in corn facilitated these students in expanding their conceptions of science and helped them to develop a greater appreciation for how science is applied to environmental problems. Environmental role-play simulations are a powerful way to put students in a context that challenges them on many levels. The context forces them to explore the science behind an issue, understand the social and environmental ramifications of a particular technology, and develop not only an appreciation for the topic, but an ability to make a convincing, logical, and coherent argument to their peers.

This simulation role-play falls within the realm of what Zeidler, Sadler, Simmons and Howes (2004) and others refer to as the *socioscientific issues domain* (SSI) of science education. This movement refers to scientific topics that lead to the "consideration of ethical issues and construction of moral judgments about scientific topics via social interaction and discourse." Issues that cut across these constructs can powerfully motivate students. More importantly, bridging the scientific, ethical, moral, and often very personal boundaries within which students operate can lead to greater and more broadly defined scientific literacy.

Experiences such as a simulation role-play around a socioscientific issue are extremely valuable for future teachers. Having a broader view of science and the application of scientific knowledge will help them think about ways to help their future students develop an appreciation for the nature of science and the real-world applications of science. Additionally, the unit incorporated a wide variety of teaching strategies and topics that are currently of interest in science education: simulations, role-playing, driving questions, oral presentations, technology integration, portfolios, reflection, and concept mapping. The ultimate goal of this experience was to model the structure and nature of an inquiry-based unit, which would aid these future teachers as they design lessons and units for their own classrooms.

Exploring Other Contexts and Questions Using Simulations

The same approach, context, and structure can be used in many contexts. I have engaged students in Congressional Subcommittee hearings on the Endangered Species Act, Genetic Testing, and the establishment of hog farms. The core aspects of this approach are the driving questions, the nature and focus of the student groups, and the level of responsibility given to the students. First, the driving questions presented earlier only need to be modified in terms of the topic. The final question of "what additional research or legislation might be useful" is the primary framework for student inquiry and their presentations. Everything they do, including addressing the other two questions, is in pursuit of this question. Second, the nature and structure of the student groups are very important. Students should be provided with adequate resources to understand the focus of their groups and to conduct their research. Furthermore, the stakeholder groups chosen for the hearing should accurately represent the diversity of perspectives on the issue to be discussed. Finally, the students should have as much autonomy and responsibility as possible. This will vary widely depending on the class, subject, and context, but the goal should be to enable students to develop presentations that reflect their own work and perspectives.

Several other curricula provide students with opportunities to role-play within a simulation. The *Environmental Issues* section of the EnviroSci Inquiry project at Lehigh University has several units that engage students in extended role-play simulation projects (EnviroSci Inquiry 2008). For example, the Abandoned Mine Drainage module involves students examining the nature and history of abandoned mine drainage issues in Pennsylvania. Students build on prior experiences and use a range of technologies to explore data relevant to their investigation. This experience culminates in a student debate focused on how best to clean up streams and rivers across Pennsylvania that are affected by abandoned mine drainage.

Hog Wild! and the Potential of New Technologies

Structures for science and environmental role-play simulations have even greater potential in the light of new information and communication technologies. One example is a recently developed curricular module entitled HogWild! (Wilson and MaKinster 2008). HogWild! is an interdisciplinary simulation that engages students in a town council hearing on the proposed establishment of a hog farm in a local watershed. After brainstorming the potential effects of a new hog farm, students use Google Earth to visit each of the stakeholders and then use ArcGIS software to create maps that support their positions. The HogWild! curriculum has been used in science courses for preservice teachers, high-school environmental science, and in a middle school by a social studies and English teacher. Each teacher had different conceptual goals and emphases; however, they were guided by the same set of goals and curricular expectations.

- What are the costs and benefits of establishing a hog farm in your watershed?
- Who are the stakeholders?
- What is your group's position on establishing a hog farm?
- How can you use maps and related data to support your position?
- How can you most effectively communicate your position?
- Should Hatfield Pork establish a new hog farm in your watershed? (hearing)
- What have we learned about this issue and what are some solutions to the issue?

Other new technological and assessment tools, such as blogs, wikis, and podcasts create additional pedagogical opportunities. Blogs can be used as a means to communicate with experts in the field or as a means for student groups to capture their research or findings. Wikis enable individual students or student groups to collaboratively contribute to a document that serves to reify their position on a particular issue. For example, in the HogWild! curriculum, each student group (Hog Farmers, Farm Bureau, EPA, Tourism Bureau, etc.) could develop a Wiki that serves to articulate their position and incorporated images, videos, graphs, and other information in support of their arguments. Finally, audio or video podcasts are yet another means by which students can share their ideas. The technology necessary is becoming increasingly accessible, both in schools and at home. Each of these approaches has its own set of challenges; however, there are free online services for blogs (e.g., Blogspot 2009) and Wikis (e.g., PB-Wiki 2009) that are being used by an increasing number of educators.

Environmental Inquiry

Ultimately, this use of role-play simulations informs the ways in which environmental and science educators ask students to explore environmental questions and, in particular, what it means to engage in *environmental inquiry*. Too often, science investigations are cast or discussed under the heading of environmental education, simply because they have relevance to environmental issues.

Environmental inquiry must enable students to examine the scientific, social, political, economic, and ethical dimensions of a particular issue. Such a lesson or unit should enable students to see and discuss each of these dimensions simultaneously. Ideally, students should encounter situations where they have to wrestle with diverse perspectives and apply them to a given situation; however, such an approach can be applied anywhere along the teaching continuum from lecture to open inquiry. When environmental issues are considered from only one or two of these perspectives, it is difficult, if not impossible to identify environmentally sound solutions that address or acknowledge the needs and concerns of everyone involved. As environmental educators, we must take the time to step back and reveal the hidden assumptions and dimensions within an issue in ways that help our students become not only better informed, but also better able to contribute to environmental solutions in their own schools, communities, and the world.

References

Abd-El-Khalick, F., & Lederman, N. G. (2000). Improving science teachers' conceptions of nature of science: A critical review of the literature. *International Journal of Science Education, 22*(7), 665–701.

Aubusson, P., Fogwill, S., Barr, R., & Perkovic, L. (1997). What happens when students do simulation role-play in science? *Research in Science Education, 27*(4), 565–579.

Bessin, R. (1999). *Bt-Corn: What is it and how it works*. Retrieved October 31, 2002, from http://www.uky.edu/Agriculture/Entomology/entfacts/fldcrops/efl.30.htm

Blogspot. (2009). *Blogspot website*. Retrieved March 22, 2009, from https://www.blogger.com/start

Council for Environmental Education. (2003). *Project WILD: K-12 curriculum and activity guide*. Houston, TX: Council for Environmental Education.

Crawford, B. A. (2000). Embracing the essence of inquiry: New roles for science teachers. *Journal of Research in Science Teaching, 37*(9), 916–937.

EnviroSci Inquiry. (2008). *Abandoned mine drainage module*. Retrieved October 5, 2008, from http://www.leo.lehigh.edu/envirosci/watershed/riverexp/mine/index.html

Harwood, W. S., MaKinster, J. G., Cruz, L., & Gabel, D. (2002). Acting out science: Using Senate Hearings to debate global climate change. *Journal of College Science Teaching, 21*(7), 442–447.

Losey, J. E., Rayor, L. S., & Carter, M. E. (1999). Transgenic pollen harms Monarch larvae. *Nature, 399*(6735), 214.

McComas, W. F. (1996). Ten myths of science: Reexamining what we think we know about the nature of science. *School Science and Mathematics, 96*(1), 10–16.

McKeachie, W. J. (1994). *Teaching tips: Strategies, research and theory for college and university teachers*. Lexington, KY: D.C. Heath.

Nelson, C. E., Nickels, M. K., & Beard, J. (1998). The nature of science as a foundation for teaching science: Evolution as a case study. In W. F. McComas (Ed.), *The nature of science in science education*. Dordrecht, Netherlands: Kluwer.

PBwiki. (2009). *PBwiki website*. Retrieved March 22, 2009, from http://pbwiki.com

Pennock, M. T., & Bardwell, L. V. (1994). *Approaching environmental issues in the classroom* (pp. 1–23). Ann Arbor, MI: University of Michigan.

Project WET. (2003). *Project Wet: K-12 Curriculum and activity guide*. Boseman, MT: The Watercourse.

Shelton, A. M., & Roush, R. T. (1999). False reports and the ears of men. *Nature Biotechnology, 17*(9), 832.

Shulman, L. S. (2000). From Minsk to Pinsk: Why a scholarship of teaching and learning? *The Journal of Scholarship of Teaching and Learning, 1*(1), 48–52.

Webb, D. (2002). *Role-playing: A teaching strategy*. Retrieved July 11, 2003, from http://condor.admin.ccny.cuny.edu/~cb6879/spring2002/davidpaper.htm

Wilson, C., & MaKinster, J. (2008). *HogWild!: An interdisciplinary unit exploring the environmental issues surrounding the establishment of a hog farm in the Finger Lakes Region*. Retrieved from http://fli.hws.edu/gitahead/hogwild/teacher.html

Zeidler, D. L., Sadler, T. D., Simmons, M. L., & Howes, E. V. (2004). *Beyond STS: A researched-based framework for socioscientific issues education*. Paper presented at the National Association for Research in Science Teaching, Vancouver, British Columbia.

Exploring Environmental Education Through Ecofeminism: Narratives of Embodiment of Science

M.E. Spencer and Sherry E. Nichols

Partially hidden in a thicket of cattails at the edge of the ramp, a small pond slider lay silently dying. Quickly the slider was seized in a dip net. "Oh no! Something is wrong! Look at her shell and her mouth! What happened?" the teacher asked in a worried tone. The others had joined us in response to the sound of concern in her voice and to see what the problem was. An angry looking rusted fishing hook penetrated both of the slider's upper and lower jaws. Her algae covered carapace revealed a long jagged Y-shaped crack. Fragile, delicate tissues and organs could easily be seen. We placed some of the cattail bedding and a little lake water into her plastic container. With my help, the teacher gently placed the injured slider onto the soft mass of cattails. Compassionate, concerned looks could be seen on the faces of the other teachers in the class. "Help it. Take it. Can you glue it back and get the hook out?" their caring voices said to me.

The previous narrative is an excerpt from my doctoral dissertation in which I recognized problems associated with the use of a problems-based model within teaching and learning about [the][1] environment. Problems-based Learning (PBL) (Barrows 2000; Savin-Baden 2003) was adopted as the central pedagogical strategy of the environmental education course I was preparing to teach, since at the time it was commensurate with my explicit views of learning. Environmental science was selected as a means of framing the context and content for learning within the course.

Aspy, Aspy, and Quimby (1993) called for further investigation into the application of PBL to other academic subjects at all levels of learning; however, little research, which critically evaluates the underlying tenets of PBL, has been done. Learning and learners can be empowered or impeded, especially when PBL is used in relation to environmental issues (Spencer 2005). Environment, when viewed as

[1]Brackets denote our view of environment as embodied not something separate from us.

M.E. Spencer (✉)
University of Montevallo, 2970 Pelham Parkway, Pelham, AL 35115, USA
e-mail: spencerme@montevallo.edu

S.E. Nichols
University of Alabama, 230-E Graves Hall, Tuscaloosa, AL 35487-0232, USA
e-mail: snichols@bamaed.ua.edu

A.M. Bodzin et al. (eds.), *The Inclusion of Environmental Education in Science Teacher Education*, DOI 10.1007/978-90-481-9222-9_17,

problematic through an anthropocentric view, is positioned as something to be controlled and used by man for his needs, purposes, and whims, much in the way that women were and sometimes still are thought of in parts of our world (Marchant 1980; Warren 1990). Children and women are often marginalized and oppressed through hierarchical structures. To forge a new perspective of PBL and environmental education, we examined PBL through a theoretical lens of ecofeminism.

This study emerged out of planning and teaching a graduate research in elementary science teaching and environmental education course at Greentree University, which is a public liberal arts institution in the South with an annual average enrollment of 3,000 students. The university campus was used as an ecosystem model, which was based on the Campus Ecosystem Model (CEM) developed at Florida Gulf Coast University (Tolley et al. 2002) where students experienced environmental issues through the teaching and learning processes of PBL. Greentree University's ecosystem appeared ideal owing to the presence of a large lake, golf course, hundreds of trees, and a strong diverse population of plant and animal species. Additionally, there was building and expansion taking place on the campus, which provided environmental issues and experiences for inquiry through PBL. A fundamental assumption of this work regards learning as a lived or more accurately a living experience and as such is situated in a past, present, and future context and must be acknowledged as so.

This study explored the tensions associated with elementary teachers' embodiment of science, specifically environment, and science teaching and learning through a PBL approach. Embodiment can be best described as continuous on-going action involving physical, emotive, and psychological consciousness mediated through sociocultural relations and practices. During conversations with teachers about notions of embodying science through environmental education, I slowly began to recognize practices of domination in the teaching profession as well as throughout the study of environmental education. This knowledge prompted my research into the possibilities of using ecofeminism as a framework for environmental pedagogy.

Theoretical Framework and Ideas

According to Warren (1989) and Warren and Cheney (1991), an ecofeminist is any person who actively works to deconstruct the "wrongful and inter-connected dominations of women and nature [and] bring about a worldview that is not based on socioeconomic and conceptual structures of domination" yet rather focuses on a reconnection between humans and nonhuman nature. Ecofeminism provides a way for teachers to embody environmental science through holistic perspectives that include much more than just the mind, but rather promotes awakening of the mind, body, emotions, perceptions, and consciousness. Ecofeminism and embodiment recognize and embrace the interconnections among all living beings and nonliving entities. Warren (1993) acknowledged that "relationships are not something extrinsic to who we are...they play an essential role in shaping what it is to be human" (as cited in Kronlid 2003, p. 127).

In Alabama where great emphasis is placed on reading and writing, very little attention is paid to the teaching of environmental education. When environmental education is taught, often there is a promotion of an environmental ethic that fails to recognize the interconnectedness of the domination of nature and oppression of women. Using ecofeminism as a lens in this study influenced our methodology and allowed us to explore the interconnectedness between human and nonhuman nature and issues of power and oppression.

Seventy percent of Alabama is covered with timber, of which over 78% is privately owned and managed. Groups of landowners employ powerful lobbyists to ensure a "positive" portrayal of the forestry industry including having strong input into environmental education courses of study and materials. For example, the study of forestry management is defined by many as the application of business and technical practices to ensure the maximum financial benefit for the land owner. From a purely business perspective, this idea may seem rather benign. When students are taught about forestry, rarely are they prompted to ask whether a forest needs "managing" and who would have the power to do so and why. Ecofeminists and List (2000) would argue that understanding of interconnected ideas of domination and oppression of women and nature need to be understood if "we are to live in harmony and practice a different kind of forestry" (p. 20).

Study Methodology

The use of narrative methodology in this study allowed for multiple ways of framing and understanding learning. Clandinin and Connelly's (2000) notion of narrative as a three-dimensional landscape provided a backdrop for conceptualizing teacher learning in a larger study (Spencer 2005)– one which could account for tensions of time, space, and place associated with teachers' embodiment of science and learning to teach science. One case narrative about the participant, Sally, provides a focal point for this chapter as we explore the following discussion points: (1) In what ways do teachers embody science in their personal and professional lives? (2) In what ways might ecofeminism explicate our assumptions of PBL as a pedagogical strategy for environmental education? (3) In what ways might teachers embody environment through environmental education? Ultimately, our insights gained from the study lend support for the redesign of science content and methods courses for teachers where former disciplinary-based content is reconnected through environmental education. Accordingly, courses must be taken to enable learners to experience science in the truest sense of the word. Learners must be situated as humans *being* part of the natural environment as the following narrative illustrates.

Roly-polys

> *There they were...those chubby swollen ankles flopping over the sides of the new white Keds. Attached to the ankles were long fabric-enrobed legs with a brief hint of expanding hips and a belly. She lay under the shrubs excitedly calling to her classmates, "They are here!"*
>
> *Earlier, Sally with her large round life-filled belly gazed at the dry soil from her statuesque height and emphatically stated, "There are no living things here; not in this dirt!" "Get a little closer," I suggested as I plopped myself down by a bush. Lowering her body awkwardly to the ground she appeared uncertain, nervous, and concerned about "what was lurking in the bushes." "Let's look together," so we rolled onto our stomachs and wiggled our feet, knees, and hips toward the row of shrubs. Propped on our elbows, our faces near the soft ground, we could smell the earthiness and feel of leaves and branches as they tickled and ensnared our hair. "Are you sure there are no snakes?" she questioned in a whisper. "No," I replied as I grabbed her hand to comfort her. Be still. Listen. Watch. Breathe. As light filtered down through the canopy, we began to see slight movements around us. Ants on branches flew rapidly through the air as she flicked them away. A cricket hopped on her hand and she squealed. She audibly noted that the baby in her womb jumped and kicked.*
>
> *I saw them. Here they came...the gray armored tanks on bunches of little legs. "Look! Roly-polys! I played with these as a child. We'd get them and they'd roll up in balls on your hand. Take one now," she said to me. "Now, be real still and it'll crawl on you. Feel it tickle your skin with its tiny little feet! Smell it!" I lifted the small bug to my nose and inhaled deeply. "Stinks, don't it?" She laughed as I agreed and made a wrinkled face.*
>
> *We watched for quite a while and others began to join our adventure under the bushes. I wiggled out; others rolled in becoming much like roly-polys under the bushes. Sally made them all capture other little roly-poly beasts, wait, and smell them. Nestled closely together like peas in a pod, they all eventually performed to her direction. One could hear excited squeals often accompanied by laughter. As she finally emerged from the bushes, with lots of assistance and leaves and twigs in her hair, she laughed. "There is life everywhere, even in dirt, under bushes," she exclaimed. Our clothes, hair, faces, conversations, and recollections of long past experiences with bugs were exhibitions of the adventure we had had. As we made our way back, some casually brushed the dirt and plant debris from their clothing. Others lightly wiggled their fingers in the air or on a friend's neck, imitating the tiny legs they had experienced. More laughter and squeals filled the fall air.*

I wondered why Sally was apparently enjoying crawling under the bushes when she had been so concerned about being on the ground in the first place. How was her intuitive and experiential knowledge of science awakened and valued rather than merely being questioned in light of text-based book knowledge? Why and how do elementary teachers of science respond to a perceived "content knowledge expert"? As humans such as Sally experience the nature they are part of, why is there a perceived need to take care of nature as if it requires human caretaking?

Sally's narrative illustrates a reconnection with nature as if she had been separate from it at some point. This perceived separation from nature is reflected through personal and socio/culture aspects of embodiment. When first exploring the life under bushes, she was apprehensive and doubtful, yet as she began to relax and observe, she found wonderful forms of life.

The body is our primary means of accessing a world at all, yet Sally was reticent to physically experience the ground and its inhabitants – *Lowering herself awkwardly to the ground she appeared uncertain, nervous, and concerned about "what was lurking in the bushes."* As Sally sat on the dirt, felt the cool, damp soil on her skin, smelled the earthiness of the ground, and expressed an awareness of the shrub's limbs as they touched her hair she was embodying the experience in a physical, sensory manner. This turning point where corporeal embodiment was happening led to her further aspects of intercorporeal embodiment as she not only relaxed as she felt the roly-polys March across her skin, but also as she recruited and encouraged others to do the same– *Sally stayed under the bushes to talk and share with the others, became comfortable in her surroundings, and eagerly began to share her prior knowledge and experiences.* While the roly-polys traversing her hands and lower arms initially alarmed her, she eventually described this sensation as "tickling." As living beings, our sensory awareness and capacity for corporeally experiencing play a large role in our embodiment of the world.

Understanding of lived experiences, felt meanings including but not limited to the physical, and the construction of meaning occur through embodied action. Merleau-Ponty emphasized the importance of experience as the foundation of knowledge (Lyon 1999). When we think of embodiment, we are at first reminded of the physical dimensions and sensory experiences of being in the world. Physical feelings revolve around sensory modalities of sight, touch, taste, smell, and hearing.

We agree with Langer (1989) that the body is not merely a sensory vessel, but our "point of view upon the world" (p. 25). Some might rationalize the physical activities described in the narrative to be simply active learning experiences. While we would not deny the corporeal, cognitive, and physical processing of information that was taking place here, attention needs to be given to the other embodied aspects of meaning making such as emotions, interconnections, memories, as well as the sociocultural context of learning science. Thought and emotion are embodied and mediated through physical and sociocultural experiences. Aspects of embodiment (e.g. emotion, perceptions, memories) became explicit and were revealed in the following shared memories, voice intonations, and imaginings of Sally as she interacted with her environment and the other human and nonhuman beings in it. Listening to Sally, one can sense her excitement and enthusiasm as she explored anew the tiny arthropods. Two examples of embodiment are seen in the following statements she made: "*Feel it tickle your skin with its tiny little feet!*" and "*There is life everywhere, even under the bushes*"! Her appreciation of the roly-polys was palpable as she shared remembrances of playing with them as a child.

Sally never overtly expressed value of the creatures as living beings having a place in nature under the bushes, but she did not toss them away as she had previously done with the ants. Her fear and anxiety, which reflected her perceptions of nature under bushes, was not only verbally expressed, "*Are you sure there are no snakes*?" but also caused me to comfort her with a light touch of reassurance. Her manner of being-to-the-world strongly reflected physical and emotive experiences and sociocultural activities in the context of meaning making, which reflects what Merleau-Ponty referred to as "embodied action."

Embodiment is also mediated through what Lemke (2001) is described as sociocultural experience:

> The unit of analysis in research on science learning should not be limited to studies of the Cartesian mind of individual students, rather science education needs to be examined as symbolic activity enacted in a material context and a socioculturally specific community. (p. 307)

Sociocultural aspects of this experience were evident when Sally stayed under the bushes to talk and share with the other teachers as they too came and went. As Sally became comfortable in her surroundings, she eagerly began to share her knowledge and prior experiences of the behavior and characteristics of the roly-polys with others as they crawled under the bushes. Talking excitedly about the bugs documented her developing a sense of agency related to her own and others' learning. Strathern (1996) and Blackman (2001) offered embodiment as the "here and now presence of people to one another" (p. 2). Since individual beings do not exist alone, their shared memories and feelings illustrate embodiment of the present and past moments as learning.

Sally was somewhat reticent about experiencing bugs and getting on the ground to do it. Yet, she did not appear concerned when she realized we were all dirty after the adventure. Using the roly-polys and her memories of childhood, she unknowingly taught others about these bugs through her own physical, sensory embodiment, and through shared embodied memories. "Teaching is embedded in cultural meanings, sedimented in history, and reinforced by ideology and emotion" (Grumet 2003, p. 249). She smelled the "stink" and encouraged others to do the same. Roly-polys were welcomed to roam over her skin. Her anxiety of "being in nature" lessened as can be seen in her actions, comments, and laughter. *Sally made them all capture other little roly-poly beasts, wait, and smell them. Nestled closely together like peas in a pod, they all eventually performed to her direction. One could hear excited squeals often accompanied by laughter.*

Seeking to understand how PBL instruction intersected with the teachers' learning, both authors drew upon theoretical notions of embodiment and ecofeminism. Science is often posited as a body of knowledge, consisting of unconnected disciplines, to be imparted to learners. Elementary teachers tend to enjoy the biological sciences best. Often what is taught in biological "life science" tends to focus on a micro world that is quite literally out of humans' touch. When larger living beings are explored, they are typically dissected and dead, which leaves one with the impression that "life science" is actually a study of dead things– dead things that are separate from us and the rest of nature. All in all, biology is the study of dead things removed from their lived context and all of the interconnections– an apt metaphor that frequently parallels the learning of science.

Tolley (2003) asserted, "The field of science once perceived as a source of enjoyment and opportunity for women, increasingly appeared inappropriately unfeminine" (p. 173). It appears as if very little, if any, time during the school day is spent exploring and reconnecting to nature. Teachers and their students spend most of their time in classrooms that are sterile and empty of most forms of nonhuman nature. If there are plants and animals in the classroom, they are usually store bought, exotic species. Many chemical measures are taken to rid the school of insects and other "pests." Designating field mice, garter snakes, and spiders among other nonhuman beings as "pests" allows for the perpetuation of oppression through our very language (Berman 1994).

Conservation Biology, Bioregionalism, Deep Ecology, and Ecofeminism

Similar to the history of science education, the history of environmental education has been subject to changes in driving philosophies. Corcoran and Sievers (1994) identified four philosophical perspectives related to the teaching and learning of environmental education: conservation biology, bioregionalism, deep ecology, and ecofeminism. The Society for Conservation Biology formed during the 1980s still seeks to "arrest destructive activity, preserve biological diversity, and develop a responsible and effective approach to biotic conservation" (Corcoran and Sievers 1994, p. 4). Conservation biology's focus on the interrelatedness of everything in the natural world and its situation in various socioeconomic contexts is a view shared by some in environmental education. Those interested in conservation biology focus on sustainability and seek to save species from extinction, prevent loss of important habitats, and provide a view of environment "not as desiccated scientific facts, but as components of the vibrant rush of the real earth story" (Corcoran and Sievers 1994, p. 5). Project WILD (Western Regional Environmental Council, 1991), WOW the Wonders of Wetlands (2003), and Project Learning Tree (American Forest Foundation 2003) are examples of environmental education curricula that are built on the philosophies of sustainability and conservation (Western Regional Environmental Council 1983). These curricula are widely used throughout the United States to teach environmental education to students of all ages. Chapter 3 provides additional information about Project WILD and Project Learning Tree.

When one thinks of a bioregion, the images of flora, fauna, soil, climate, and geology come to mind; however, another essential part is the human culture that has evolved within and is connected to that specific place. "This essential human element is what distinguishes the concept of a bioregion from similar ecological entities which traditionally treat humans and their cultures as interlopers rather than (Slattery, Kesselheim, Higgins, & Schilling, 2003) as integral components of a natural community" (Traina and Darley-Hill 1995, p. 1). Bioregionalism draws attention to the situatedness of humanity in specific places. Attention is given to living within a specific biome and "in harmony within the natural systems upon which we depend" (Corcoran and Sievers 1994, p. 5). Bowers (1987) claimed that the understanding and concern for our sense of place is at the heart of bioregionalism.

Ecology is an area of scientific study and specifically a "branch of biology dealing with living organisms' habitats, modes of life, and relations to their surroundings"

(Berkowitz 1993; Sandilands 1994, p. 167). Deep ecology, a philosophy that developed during the environmental crisis of the 1970s, values all species of life on earth equally. "The human race is merely one piece of the puzzle and not the center of divine existence" (Hogan and Priest 1996). Gaia theory provided a base for deep ecology as Lovelock (1988) encouraged humans to begin to "see the world as a living organism of which we are a part, not the owner, not the tenant, nor even a passenger" (p. 12).

Just as there are many forms of feminism, so too are there many philosophies of ecofeminism, yet all tend to focus on naturism as a central feminist issue (Henderson 1997). According to Warren (1997), naturism is "the unjustified domination of nature," which reflects the ways that women have been and continue to be dominated (p. 4). Ecofeminism resides at the junction of feminism, science and technology, and native perspectives. Through acknowledgement of all parts and processes of the Earth having equal value, ecofeminism embraces the interconnectedness and value of every living being and nonliving entity (Marchant 1980).

When nonhuman nature is viewed as a means of sustenance, a source of beauty, or a thing to be controlled through what is erroneously defined as caring, then an anthropocentric worldview is all that is possible. Plumwood (1997) claimed that "the critique of anthropocentrism, however, unlike the other critiques of centrism, continues to be denied legitimacy in many quarters" (p. 328). Anthropocentric worldviews, even if veiled in notions of caring for environment or nature, perpetuate ideas of domination and quite possibly oppression. Nonhuman nature, from an ecofeminist perspective, does not exist in a structure that allows for domination; rather nonhuman nature is deeply and inextricably interwoven with human nature. Environmental education when taught or experienced from an anthropocentric stance rejects embodiment of environment through ways other than what would be deemed acceptable scientific practices and beliefs including the hierarchy of nature.

We concur with Berman (1994) in that our actions as humans against or for the nonhuman beings in our world depend on our values and beliefs. We often fail to value what we do not understand: things such as garter snakes. Many teachers tend to exhibit an anthropocentric worldview; in other words, "human-centered" rather than life-centered (Plumwood 1997; Warren 1990). During one of the final interviews, Sally shared how her thoughts had started to change: "For the first time I am beginning to see the big picture connection between science and myself." As we ponder the interconnectedness between humans and all nonhuman nature and nonliving things, we find ourselves in agreement with Kaufman et al. (2003), "The dominant culture does not value embodied knowing, and hence we believe that this ability is not ordinarily explored or expanded as we grow" (p. 138). Science educators and others concerned with science teaching and learning must begin to recognize multiple ways of knowing environment other than conventional ways, which often problematize nature and silence the voices of those who are aware of alternative notions of embodied environment.

Science educators and others concerned with science teaching and learning must begin to recognize multiple ways of knowing environment other than conventional ways, which often problematize nature and silence the voices of those who are aware of alternative notions of embodied environment. Many activities in the *Project* curricula (such as Project WILD and Project Learning Tree) espouse mere conservation of the environment or focus only on local bioregional aspects of environment and may fall short of the goal of educating an environmentally literate citizenry. From an ecofeminist perspective, environmental education curricula and programs including science taught in elementary schools in the United States today fail to emphasize "the importance of the continuance of the Earth's delicate balances and the interconnectedness and sacredness of all life" (Giuliano 1998, p. 14). Elementary science teachers who teach about environment must "help all students to embrace the concept of the web of life and the interdependence of all things" (Giuliano 1998, p. 15) through pedagogical strategies and lived experiences that promote embodiment of environment.

Discussion

From embracing ecofeminism as a powerful tool and lens with which to critique pedagogical practices and to "extricate patriarchy from nature" (Warren 1990), we now turn to the "problem" of using PBL in environmental education. PBL has been extensively used as a pedagogical strategy in medical, nursing, law, engineering, and education colleges and K–12 schools. However, research related to the use of PBL with elementary environmental education remains nonexistent to our knowledge. Part of the appeal of using PBL with environmental education was its underlying philosophical stance of understanding emerging from interactions with content in context. Additionally, PBL's focus on students' active construction of knowledge through collaborative learning about "real-world" problems and the synthesis of knowledge across disciplinary boundaries appeared to contain many of the attributes of what is commonly referred to as "best practices" in science teaching and learning (e.g. active engagement with content in an "authentic" context, student ownership, and responsibility for one's own learning).

Through the process of engaging in PBL activities, the participants had not only learned the steps in the pedagogical strategy, but also that nature was a problem to be solved by humans. In this study, Greentree University's campus had been used as an ecosystem model, where participants had been taught to identify "problems" in and with nature that could be solved, without paying little or any attention to the source of the problem in this case – human actions. For example, one problem the participants engaged in solving was the erosion around a new building site on campus. Another PBL activity was the runoff of pesticides, herbicides, and fertilizer from a nearby golf course into the lake where the slider was found. Never was the question raised regarding the impact of overpopulation of humans on the environment; it was just seen as "progress and growth."

As I modeled and used PBL with my participants, 1 day I became acutely and painfully aware of the lessons about feminism and nature that I had unknowingly taught. Believing that my participants' perceptions were being respected and that learning and senses of efficacy related to science teaching and learning were increasing, I erroneously thought feminism must be at the heart of PBL. The turtle narrative, presented at the beginning of this chapter, provides examples of the promotion of anthropocentric ideals and the reification of the domination of nature. Participants' separation from nature, even though not in the physical sense, contributed to their understanding of nature as something we should or could control. Utilization of patriarchal structures, such as a strong focus and control of time on task and adherence to preset goals that participants had no control over or contribution to, had further inhibited corporeal and intercorporeal embodiment of science teaching and learning and environment. PBL provided as a teaching and learning strategy positioned science and nature as a set of problems to be solved and further reinforced the perceived divide between humans and nonhuman nature.

Making a turn from what has been a theoretical discussion, we would now like to propose practical consideration for environmental education premised on ecofeminist philosophy. While problems-based pedagogy is appropriate for many areas of science education, we argue that it is antithetical to humans' meaningful understanding of the environment. Accordingly, for environmental and science educators and science teacher educators, we would offer ecofeminism as a supporting philosophy to guide many of the existing models presently in use. To provide an example, we offer a learning cycle/5E instructional model (Bybee 1997) based on ecofeminism. The following provides guiding questions, which might be considered in planning components of a 5E learning experience.

Explore:	Will the experience provide direct physical contact with nonhuman things?Do the experiences promote emotional responses?
Engage:	Do the experiences avoid hierarchy and dominance of humans over nonhumans?Will the experience encourage recognition of connectedness rather than separation?
Explain:	Do the experiences promote shared moments of insight and understanding?Are insights fluid and generative rather than a pursuit to a final understanding?
Elaborate:	Do the experiences promote opportunities for revisiting, revising, and extending ideas?Do the experiences generate more questions?
Evaluate:	Is there recognition of learning as cyclical and continual?

At this point, we were tempted to *re*-illustrate or create a graphic *re*-presenting an ecofeminist learning cycle model. In doing so, however, this would only reinstate a patriarchal and reified vision, which is antithetical to our premises of ecofeminist environmental education. Accordingly, we provide a narrative that reflects the sort of dialogue that we referenced earlier in this chapter. As we revisit the statement made earlier …dialogue and action must be taken to enable learners to experience science…and situate humans *being* part of the natural environment, we now conclude by offering the following insights into a notion of an ecofeminist learning cycle.

Teachers in the class gave the slider to me expecting me to "fix it." Looking back on that particular experience, I now see that they appeared willing to give up their power of knowledge to me, the "expert." From a feminist stance, this was not what was expected nor desired. After the slider died and had been de-fleshed, I placed the skeleton on my desk. Some of the teachers expressed sadness, anger, curiosity, and indifference. Angry teachers were quite vocal as they expressed their feelings of disgust, disillusionment, and madness directly at me for not "fixing" slider. Now looking back at that interaction, I believe they had made a turn– they did not relinquish power over their own experience and learning, instead they espoused their feelings and expected to be heard. From a feminist view, there was no longer a barrier between me, the teacher, and them, the learners. In essence, what took place within our community of learners served as a parallel to what and how we could envision learning in and about environment.

References

American Forest Foundation. (2003). *Project Learning Tree: Environmental education preK-8 activity guide*. Washington, DC: American Forest Foundation.

Aspy, D. N., Aspy, C. B., & Quimby, P. M. (1993, April). What doctors can teach teachers-based learning. *Educational Leadership, 50*(7), 22–24.

Barrows, H. S. (2000). Foreword. In D. J. Evenson & C. E. Hmelo (Eds.), *PBL: A research perspective on learning interactions* (pp. vii–ix). Mahwah, NJ: Erlbaum.

Berkowitz, A. R. (1993). New opportunities for ecology education in the United States. In M. Hale (Ed.), *Ecology in education* (pp. 45–59). Cambridge, UK: Cambridge University Press.

Berman, T. (1994). The rape of mother nature? Women in the language of environmental discourse. *Trumpeter, 11*(4), 173–178.

Blackman, L. (2001). *Hearing voices: Embodiment and experience*. New York: Free Association Books.

Bowers, C. A. (1987). *Elements of a post liberal education*. New York: Teachers College Press.

Bybee, R. W. (1997). Achieving scientific literacy: From purposes to practices. Portsmouth, NH: Heinemann.

Clandinin, D. J., & Connelley, F. M. (2000). *Narrative inquiry: Experience and story in qualitative research*. San Francisco: Jossey-Bass.

Corcoran, P. B., & Sievers, E. (1994). Reconceptualizing environmental education: Five possibilities. *Journal of Environmental Education, 25*(4), 4–8.

Giuliano, J. A. (1998). *Teaching as if your life depends on it: Environmental studies as a vehicle for societal and educational transformation*. Unpublished dissertation, University of The Union Institute, Cincinnati, OH.

Grumet, M. R. (2003). Afterward: My teacher's body. In D. P. Freeman & M. S. Holmes (Eds.), *The teacher's body: Embodiment, authority, and identity in the academy* (pp. 249–258). New York: State University of New York.

Henderson, K. A. (1997). Ecofeminism and experiential education. *The Journal of Experiential Education, 20*(3), 130–133.

Hogan, J., & Priest, S. (1996). Deep ecology: Toward eco-equalism. *Practical and Informative Adventure Education, 1*(1), 26–27.

Kaufman, J. S., Ewing, M. S., Montgomery, D. M., Hyle, A. E., & Self, P. A. (2003). *From girls in their elements to women in science: Rethinking socialization through memory-work*. New York: Peter Lang.

Kronlid, D. (2003). Ecofeminism and environmental ethics: An analysis of ecofeminist ethical theory (Doctoral dissertation, Uppsala University, Sweden, 2003). *Dissertation Abstracts International, 64*, 03.

Langer, M. M. (1989). *Merleau-Ponty's phenomenology of perception: A guide and commentary*. Tallahassee, FL: Florida State University Press.

Lemke, J. (2001). Articulating communities: Sociocultural perspectives on science education. *Journal of Research in Science Teaching, 38*(3), 296–316.

List, P. C. (2000). *Environmental ethics and forestry: A reader*. Philadelphia: Temple University Press.

Lovelock, J. E. (1988). *The ages of Gaia: A biography of our living earth*. New York: Norton.

Lyon, M. L. (1999). Emotion and embodiment: The respiratory mediation of somatic and social processes. In A. L. Hinton (Ed.), *Biocultural approaches to the emotions* (pp. 182–214). Cambridge, UK: Cambridge University Press.

Marchant, C. (1980). *The death of nature: Women, ecology, and the scientific revolution*. San Francisco: Harper and Row.

Plumwood, V. (1997). Androcentrism and anthropocentrism: Parallels and politics. In K. J. Warren (Ed.), *Ecofeminism: Women, culture, nature* (pp. 327–353). Bloomington, IN: Indiana University Press.

Sandilands, C. (1994). Political animals: The paradox of ecofeminist politics. *Trumpeter, 11*(4), 167–172.

Savin-Baden, M. (2003). *Facilitating problem-based learning illuminating perspectives*. Philadelphia: The Society for Research into Higher Education and Open University Press.

Slattery, B. E., Kesselheim, A. S., Higgins, S. H., & Schilling, M. R. (2003). *Wow! The wonders of wetlands*. Bozeman, MT: Environmental Concern/The Watercourse.

Spencer, M. E. (2005). *Exploring elementary science teaching and through an eco-feminist perspective: Narratives of embodiment of science*. Unpublished dissertation, University of Alabama at Tuscaloosa, Tuscaloosa, AL.

Strathern, A. J. (1996). *Body thoughts*. Ann Arbor, MI: The University of Michigan Press.

Tolley, K. (2003). *The science education of American girls: A historical perspective*. New York: Routledge Falmer.

Tolley, S. G., Everham, E. M., McDonald, M. R., & Savarese, M. (2002, March/April). The campus ecosystem model: Teaching students environmental stewardship. *Journal of College Science Teaching, 31*(6), 364–369.

Traina, F., & Darley-Hill, S. (1995). *Perspectives in bioregional education*. Troy, OH: North American Association for Environmental Education.

Warren, K. J. (1989). Rewriting the future: The feminist challenge to the malestream curriculum. *Feminist Teacher, 4*(2), 46–52.

Warren, K. J. (1990). The power and promise of ecological feminism. *Environmental Ethics, 12*(2), 132–146.

Warren, K. J., & Cheney, J. (1991). Ecological feminism and ecosystem ecology. *Hypatia, 6*, p. 1.

Warren, K. J. (1993). A feminist philosophical perspective. In C. J. Adams (Ed.), *Ecofeminism and the sacred* (pp. 119–132). New York: The Continuum.

Warren, K. J. (Ed.). (1997). *Ecofeminism: Women, culture, nature*. Bloomington, IN: Indiana University Press.

Western Regional Environmental Council. (1983). *Project WILD Secondary Grades 5–12*. Boulder, CO: Western Regional Environmental Education Council.

Western Regional Environmental Council. (1991). *Project WILD*. Boulder, CO: Western Regional Environmental Education Council.

The Value of Nonformal Environmental Education-Based Professional Development in Preservice Science Teacher Preparation

Tamara E. Peffer and Alec M. Bodzin

Chris, a newly certified middle school science teacher, accepts her first teaching position as a 7th grade biology and 8th grade earth science teacher at a small secondary school (grades 7–12) in Pennsylvania. Soon after her hiring, Chris meets with the science department chair and learns that the district has decided to overhaul the school curriculum to better address the newly enacted Pennsylvania Standards for Science and Technology, and Environment and Ecology (E&E). The school recently adopted the latest edition of a basal textbook program. The department chair is quite excited about the adoption since ancillary materials specific to global environmental topics are included and the program is now aligned to both national and PA standards. Chris is informed that she needs to ensure that her course-planning outline includes E&E materials that align to the state standards.

After reviewing the standards document, Chris notes that the E&E standards require that she develop lessons to address agriculture, ecology, economics, environmental health, endangered species, integrated pest management, human interactions with the environment, and environmental laws and regulatory bodies. The E&E standards strongly encourage the exploration of topics across various sciences and into social science disciplines that were not emphasized during her science teacher preparation program. The standards also emphasize the development of problem-solving skills. Chris is now acutely aware that integrating environmental education into her classroom will involve integrating themes such as personal choice, cultural influence, value judgments, economics and social consequence into her instruction.

A quick review of the textbook reveals it was clearly designed for a general audience even though it stated alignments to PA standards. While the topics of tropical rainforests and desertification are of great concern on a global scale, the text materials do not provide adequate content or support on how to investigate local ecosystems or regional environmental issues in the students' watershed. Chris realizes these curricular materials alone will not help her students develop respect for the art of compromise or understand the community-based decision-making processes necessary to solve environmental issues.

T.E. Peffer (✉)
Lehigh University, College of Education, 120 Center Street, Reading, PA 19606, USA
e-mail: tep205@lehigh.edu

A.M. Bodzin
Lehigh University, College of Education, A113 Iacocca Hall, 111 Research Drive, Bethlehem, PA 18015, USA
e-mail: amb4@lehigh.edu

A.M. Bodzin et al. (eds.), *The Inclusion of Environmental Education in Science Teacher Education*, DOI 10.1007/978-90-481-9222-9_18,

Chris looks for additional E&E instructional materials to integrate into her curriculum. Chris goes to the National Science Digital Library website, *a resource suggested by her science methods course instructor. A search for environmental health resources for grades 6–8 returns 1,470 responses. However, Chris's preparation time is limited and she is overwhelmed by all the available information. She decides the best approach is to just stick with the school's adopted textbook program, even though the textbook lacks local environmental examples. Chris thinks to herself, "My newly designed curriculum "meets" the content portion of the academic standards, but it still feels disconnected to the students' daily lives. How do I help my students identify personal connections to the curriculum when I haven't yet discovered this myself?"*

The above vignette illustrates the importance of integrating environmental education (EE) into science teacher preparation. One way of encouraging integration of EE into science teacher professional development is to collaborate with nonformal environmental educators. Nonformal EE (also known as informal EE), refers to EE efforts that are based outside formal science classrooms (NAAEE 2004a). Organizations that provide nonformal EE services include but are not limited to national, state, and local parks, government organizations, conservation districts, nonprofit conservation groups, university extension programs, marine and aquatic research facilities, and even corporate outreach programs. These organizations provide programming that includes professional development and environmental education curricular materials for teachers, and experiential programs for school students and the general public. Some of these organizations have home facilities located in urban, rural, and suburban areas. Others may work off-site in local schools or at various community locations to provide both professional development and community programming opportunities. In addition to print and video materials, nonformal EE programming is also provided through distance education in the form of live webcasts and web-based materials. Nonformal environmental education practitioners (henceforth EE educators) come from a variety of professional backgrounds. They are often intimately familiar with local resources and can recommend instructional materials designed to address local environmental topics and issues whose context can be applied globally.

Preservice teacher preparation programs can be enhanced to better promote both science and environmental literacy by integrating nonformal EE into the preparation experiences. It has been our experience that preservice science education courses that are modified to include nonformal EE provide novice teachers with unique interdisciplinary content and novel instructional strategies. Exposing preservice teachers to nonformal EE gives them additional skills, abilities, and resources necessary to integrate EE concepts and processes more seamlessly into science curricular contexts.

This chapter discusses the potential role of nonformal EE in science teacher preparation programs. As we characterize nonformal environmental educators and their available resources, we demonstrate how facilitative relationships can support science teacher education. We also illustrate how EE methodologies can be integrated into science teacher preparation programs. Finally, the benefits of collaborative relationships between nonformal EE educators and science teacher educators are described. This discussion includes strategies that can be used to overcome institutional obstacles that science teacher educators are likely to experience when attempting to integrate EE into preservice programs.

Who Are Nonformal EE Educators and Why Should Science Teacher Educators Work with Them?

According to a recent survey of over 400 nonformal EE educators, almost half have taught in formal K-12 school settings (Peffer et al. 2008). One-third of those surveyed have teaching certificates, and over half of these certified teachers are qualified to teach a science discipline. Other backgrounds of nonformal EE educators included advertising, business, management, engineering, environmental areas, biology, health services, military, zoo keeping, interior design, land use planning, anthropology, mining, geology, and research writing.

Nonformal EE educators work with learners of all ages and are not limited to working with just school-aged children. It is not uncommon to find nonformal EE educators facilitating awareness lessons with preschoolers one day and then conducting a workshop for a senior citizen watershed monitoring team on the following day. Many educational programs developed by nonformal EE educators are intended for K-12 learners. According to the aforementioned survey, over 87% of nonformal EE educators work with elementary, middle-level, and high-school students.

Nonformal EE educators regularly attend and participate in professional development networks to keep current in their knowledge of environmental issues and local natural resources. For example, many state-level EE organizations provide annual professional development opportunities for their educative staff. These events allow nonformal EE educators to share experiences and discuss issues in their profession. They attend sessions on diverse topics across a variety of disciplines including history, geology, biology, entomology, hydrology, and dendrology. In addition, they learn instructional strategies that are appropriate for educating audiences of various ages and backgrounds.

Well-practiced nonformal EE educators are also adept at addressing personal choice issues (for example, deer management, land use, population issues, global climate change, and wildlife conservation) objectively without imposing their own bias. For many years, EE was criticized for attempting to sway students' opinions and promoting action by using emotional tactics (Gigliotti 1999). Professional nonformal EE educators recognize the need to maintain neutrality and avoid such criticisms to ensure that students have opportunities to investigate their own beliefs and understandings. By including nonformal EE in preservice preparation, novice teachers are immersed in the understanding of these bias-related issues. They become sensitive to their own biases and learn through modeling the practices of how one negotiates conflicting viewpoints in an instructional context.

Nonformal EE integration also provides new teachers with many opportunities to utilize inquiry learning across multiples disciplines, thus increasing the potential for environmental and science literacy of both teachers and their future students (Christenson 2004; Ramsey 1993).

The theoretical framework that supports nonformal EE interdisciplinary facilitative practices has deep roots in the works of Swan, Stapp, and Hungerford (Hungerford et al. 1980; Stapp et al. 1969; Swan and Stapp 1974). Using a wide assortment of teaching methodologies in diverse learning environments, EE practitioners seek to educate the public about issues within the biophysical environment, improve awareness of how to solve its associated problems, and work toward solutions to these difficulties (Stapp et al. 1969). This philosophy is also reflected in important science education reform documents including *Benchmarks for Science Literacy* (American Association for Advancement of Science 1993) and the *National Science Education Standards* (National Research Council 1996). These reform initiatives encourage diverse, actual exploration of the natural world while emphasizing the need to investigate science through multi- or interdisciplinary approaches to develop essential skills required from literate citizens. Environmental educators who embrace the historic philosophies of EE utilize an interdisciplinary approach through all phases of the EE continuum: awareness, knowledge acquisition, problem-solving skill development, issues awareness, and action implementation (Stapp and Cox 1974). These are each important pedagogical facets for teachers to implement in a curricular context to promote scientific and environmental literacy.

State of EE in Teacher Preparation

While the inclusion of nonformal EE-based professional development in teacher preparation programs is highly recommended by associated professional organizations (NAAEE 2004b) and recently adopted by the National Council Accreditation of Teacher Education (NAAEE 2007), it is not a standard practice. Unfortunately, most preservice teacher education programs do not incorporate EE into their programs of study (McKeown-Ice 2000). Very few institutions highly integrate EE into their preservice teacher preparation curricula (Heimlich et al. 2004; Powers 2004). According to a 2004 study, less than one third of the responding institutions offered courses that provided coursework in environmental issues for preservice teachers (Powers 2004). This limited exposure to EE in preservice preparation may be an important contributing reason that teachers may be reluctant to incorporate EE into their classroom instruction.

The discomfort with inclusion of EE in teacher preparation programs appears to be a systemic problem in schools of higher education. A 2004 study reported that only 14.8% of the 499 higher education institutions surveyed included a specific EE methods course in their teacher preparation program (Heimlich, et al. 2004; McKeown-Ice 2000). When EE methods were incorporated into discipline-specific methods courses (such as environmental science, biology, and social studies), the inclusion of EE-based principles and methods, resources for teachers, EE trends and issues, and environmental content coverage was low. Many responding teacher educators were not familiar with the majority of the nationally recognized EE curricular programs and resources listed on the survey.

Low levels of inclusion of nonformal EE in preservice teacher preparation may result in low confidence and comfort levels of novice teachers to integrate EE pedagogical practices into their classroom instruction. EE practitioners encourage open-ended, cross-discipline, inquiry-based professional development embedded with strong environmental content. Preservice teachers who have not experienced this learning environment often limit their classroom instruction to traditional methods within the disciplinary boundaries of their certification area (Yilmaz-Tuzun 2008). By exposing preservice teachers to nonformal EE instructional methodologies, we can assist new science educators to support their classroom students' needs for effective learning, development of basic concepts, essential learning skills, and an appreciation for the natural environment, thus fostering environmental and science literacy (Hammerman and Hammerman 1985).

Resource and Facility Diversity

As we saw briefly in Chris's story, the volume of resources listed from a national database on just one environmental topic can be quite intimidating. Perhaps, if Chris had been introduced to the resources and services provided by nonformal EE educators during her preservice training, she may already have an existing set of knowledge to know that she could tap into local EE educators for support. Chris would have become aware that EE educators are willing to collaborate and support classroom teachers. In addition, she would know that their services are usually free of charge since they are often supported by national- and state-level government funds.

Often, many people picture EE programming as occurring only in a rural or wooded setting. This is not the case; environmental education can be taught anywhere. Nonformal EE facilities include many sites located in urban and suburban locales including national parks, environmental centers, river centers, county parks, and museums. In addition, EE programs may also be offered in classroom settings. EE organizations provide a variety of learning resources that include distance education, web-based materials, video, and printed materials. They develop and implement a variety of teacher professional development experiences ranging from short, topic-focused workshops to sustained and supportive long-term professional development initiatives. Providers of nonformal EE range from international organizations to local grassroots efforts. Some organizations may specialize in a particular EE area such as watersheds and water quality, land use practices, global climate change, energy use, wildlife and habitat conservation, and integrated pest management. Nonformal EE educators are trained to provide educational resources to their constituency. Often, they act as a clearinghouse for specialized curricular materials and as a lending source for instructional equipment that is not readily available in school settings.

At the national government level in the USA, the Environmental Protection Agency (EPA), National Oceanic and Atmospheric Administration (NOAA), National Aeronautic and Space Administration (NASA), and the Department of Energy (DOE) have developed EE curricular materials on a variety of topics including but not limited to watersheds, groundwater, pollution, energy use, recycling, and climate change. These organizations contain education divisions that provide professional development opportunities in both formal school settings and informal settings.

Many national, state-level, and regional nonprofit groups and academic institutions develop and make available EE curricular resources and materials. A few well-known EE curricular projects include: the Leopold Education Project: Lessons in a Land Ethic (Leopold Education Project 1996), 4-H Environmental Education Programs (National 4-H Curriculum 2008), Project Learning Tree (WREEC 2004), Project WET (Nelson et al. 1995), Project WILD (Council for Environmental Education 2003), and Project Food, Land, and People (Food, Land, and People 2004). State-level government EE programs support these national initiatives often through the parks system, a regulatory department, or an education office. For example, in Pennsylvania, several government departments and commissions (i.e. Department of Conservation and Natural Resources, Bureau of State Parks, Department of Education, the PA Fish and Boat Commission, and the PA Game Commission) provide professional development and access to the aforementioned national curricular projects as well as more localized EE resources. Some state-level government organizations have developed regionally specific curricular modules to support these national projects (for example, the *Project* curricula [Marshall and Censky 2001; Steinhart and Vathis 1997]), PA State Parks Watershed Education (PA DCNR 1998), PA Land Choices (Wisser 2005), PA Song Birds (Mowery et al. 1998), Activities for Environmental Learning (Hopkins 1984). Most of these curricular materials are developed by teams of nonformal EE educators and formal science educators and include state standards alignment and ancillary curricular materials to support learning across various grade levels.

Many of the programs mentioned above require participation in a professional development workshop to obtain guidance in how to use the materials. Most of these workshops are free or involve a nominal fee to educators to cover some material costs. Curricular materials supported with grant funding through foundations often disseminate their materials through collaborative partnerships with state and local nonformal EE providers. As a result, many nonformal EE educators at state and local levels are trained as workshop facilitators and teacher professional developers for many national and regional programs and become part of that program's network of profession development facilitators.

EE Teaching and Learning Methodologies

EE curricular program materials often use diverse teaching and learning methods that may be nature-based. Such programs frequently involve hands-on exploration in classroom and in the field including schoolyard field trips or visits to various natural settings in the community. EE educators examine issues from diverse perspectives by employing game-like simulation activities, or use role-playing debate simulations. Issue awareness, acquisition, and synthesis of content knowledge are encouraged through integration of creative writing, art expression, and song analysis into curricular contexts. These teaching methodologies are quite different from traditional didactic instructional methods that commonly occur in traditional science classrooms (Shymansky et al. 1982). Most EE curricular programs are designed to flow along the EE continuum from awareness and knowledge, to the practice of problem-solving skills, discussions of issues and personal value systems, and action. They also provide learning activity structures that are cognitively appropriate according to the age-level and abilities of the students. EE learning activities recognize the need to construct meaningful understanding at the personal level by providing appropriate scaffolds for students to relate their prior knowledge and personal experiences to new scientific knowledge and myriad experiences (Bretz 2001; Varelas et al. 2008).

Many nonformal EE programs initially develop awareness through exploratory investigations or outdoor "game-like" simulations. While the students play and investigate, they develop conceptual understandings of the environment based on their experiences. The educator acts as a coach, providing scaffolds and questioning to promote learning. By maintaining a positive learning experience and demonstrating ease with the natural environment, nonformal EE educators can also act as a conduit to reduce negative emotions about outdoor experiences. Reducing emotional and physical stress in the natural learning environment enables preservice teachers to reach higher levels of cognitive processing (Liu et al. 2004). As a result, students are given opportunities to internalize and interpret their own experiences, thus encouraging a personal connection thought to be a prerequisite for the development of environmental literacy (Rockcastle 1989).

Preservice teacher preparation programs can take advantage of "game-like" simulations to introduce EE pedagogical practices that bridge multiple disciplines and encourage creative applications of content knowledge. Below, we present four simulations: *O-Deer* (Council for Environmental Education 2003), *Migration Headache* (Council for Environmental Education 2003), *An Eventful Journey* (Kane et al. 2003), and *Forest Consequences* (WREEC 2004) that we have successfully used with preservice teachers to promote environmental awareness, knowledge, problem-solving, and discussions of issues and personal value systems. In the context described below, a nonformal EE facilitator provided the lead instruction with preservice teachers, modeling specific management techniques for instructional implementation.

Oh-Deer and *Migration Headache* involve role-playing scenarios embedded into a running game. Participants adopt a role of a predator, a prey animal, or a habitat component. During the simulation, students experience the effects of various environmental limiting factors including food, water and shelter availability, disease, habitat loss and degradation, weather conditions, accidents, pollution, and hunting. In *Oh-Deer*, the limiting factors are predominantly natural in origin. The students designated as habitat components determine the resource availability at the beginning of each simulated year, and contribute to deer population fluctuations in response to the availability of the limiting factors.

Migration Headache depicts the trials and tribulations of migrating shore birds. In this activity, learners experience habitat scenarios that serve to limit the success rates of shore birds. In this game, changes in habitat, human influences, and natural disasters alter survival rates. In both activities, students must move from one side of the playing field to another, trying to find food or shelter while avoiding predators. Students use graphs and tally charts to monitor the rise and fall of population levels for predators and prey as limiting factors are introduced and carrying capacities are surpassed. Following each activity, students are asked to interpret the simulation results, noting changes in populations based on carrying capacity, and offer suggestions for personal behavior changes that could positively reverse or alter the habitat degradation experienced in the game. These activities present the complex concepts of carrying capacity and limiting factors in a manner that personalizes abundance, shortage, and overconsumption. They also address emotionally charged topics including wildlife management issues and land use choices at a level that students can understand. In each simulation, students develop important environmental concepts through their experiences in the game.

An Eventful Journey is a simulation designed to enhance learners' understanding of environmental issues and impacts that Neotropical birds experience during their semiannual migration. This activity requires no prerequisite content knowledge beyond the game rules. In the activity, students move through a series of stations laid out like a giant game board in a classroom or other area. The students read a message card and are instructed to perform an action based on the type of issue they run into. Some birds get caught in storms, others experience oil spills or encounter researchers who delay their migration by tagging them with a monitoring device. Depending on the obstacle, students are instructed to move ahead, back, or act out a consequence. For instance, a bird that benefits from a feeding station will move ahead several spaces while a bird that eats caterpillars that have been sprayed with pesticides will be asked to rub their tummy and moan multiple times before they can move forward. The simulation enables students to become intimately familiar with how human activities affect migratory birds. A variety of topics including habitat improvement and loss, pesticide use, overconsumption, building hazards, scientific monitoring, and hunting are addressed during the simulation. To conclude the activity, students are presented with questions to determine which events could be preventable. Students are also prompted to suggest personal actions that could reduce human impacts or possibly aid animals that are victims of a natural disaster. This may result in follow-up discussions to inspire student action.

Forest Consequences is a simulation activity that addresses forest management and land use concepts. This activity not only promotes awareness and content knowledge development but also engages learners with issues discussion and values judgments, important components of environmental literacy. Students take on the role of town council members in a mid-sized, middle-class town who must make land use decisions on a 250-acre property called Morris Woods. Students must consider the natural and historic heritage of the property as they decide whether or not a mall and housing development should be built on the property. Students adopt community stakeholder roles and discuss different proposed uses of the site. In this process, they evaluate a series of land use choice facts and opinions, discuss advantages and disadvantages of development proposals, and analyze costs and benefits of each proposed idea. In groups, students determine which proposals to accept, reject, or modify. If they choose not to accept any proposal, they must offer an alternative proposal that must gain each group's approval. We have found this role-playing activity to be quite effective with preservice teachers with regard to having them adopt a different perspective on an issue that might be different from their own. It provides them an opportunity to alter their personal behavior and see alternative viewpoints toward an issue.

Benefits of Collaborative Relationships

The inclusion of nonformal EE into preservice teacher preparation presents many pedagogical and professional benefits (Ballantyne 1995). Preservice teachers exposed to diverse EE instructional practices are provided access to new learning resources and strategies for integration into classroom instruction. They are also introduced to many science and environmental professionals to help extend their educational network. Many of these professionals have had prior classroom experience and understand the obstacles teachers face as they try to integrate EE into the curriculum. Nonformal EE educators also assist science teacher educators with learning about many nonformal EE resources. As a result, science teacher educators are better able to mentor their preservice teachers to efficiently access and use multitudes of EE resources. In addition, collaborative inclusion of nonformal EE is very likely to increase preservice teachers' environmental literacy and enhance their aptitude to transform newly acquired EE pedagogical content knowledge into actual classroom practice.

Curricular co-planning with nonformal EE educators can assist science teacher educators to enhance their instructional goals to develop teachers who will be both scientifically and environmentally literate. Co-planning is a noteworthy strategy since EE educators offer a wide range of services and provide unique facilitation perspectives that are not typically found in preservice teacher education programs of study. Moreover, incorporating collaborative activities demonstrates to preservice teachers that EE integration is an effective pedagogical practice to promote interdisciplinary science learning.

Many nonformal EE educators make commitments to science teacher educators and their preservice teachers that last for many years. We contend that preservice teacher exposure to nonformal EE providers in their program of study is quite important. This inclusion promotes the development of professional collaborations with supportive environmental community members and organizations. Awareness of and initial introduction to nonformal EE services opens new doors to preservice teachers as they begin to develop their teaching style and start identifying resources they view as being important to have as a future classroom teacher. As a result, novice teachers might be more likely to integrate interdisciplinary contexts into their future science instruction and become less likely to be "tied down" to their prescribed curriculum.

Preservice teachers exposed to nonformal EE activities during teacher preparation personally experience the effectiveness of EE teaching methodologies. This exposure allows them to develop new interdisciplinary environmental content knowledge and confront their own knowledge deficits and misconceptions. In addition, they become more aware of their future students' potential misconceptions (Halim and Meerah 2002). We contend that novice teachers gain a deeper appreciation for diverse teaching methods and interdisciplinary approaches to learning. In addition, they develop critical thinking skills about environmental topics and issues.

Overcoming Obstacles to Integrate Nonformal EE in Preservice Teacher Education

There are many, but not insurmountable, obstacles that prevent the integration of EE professional development into science educator preparation programs. Unfortunately, many science teacher educators lack knowledge of EE epistemologies and available resources. Some also have misunderstandings about EE pedagogical practices and venues that are used for nonformal EE instruction. A lack of a common disciplinary-based language between nonformal EE educators and science teacher educators exists and often contributes to hindering the inclusion of nonformal EE into teacher professional development. University politics, personal viewpoints about the environment, time to incorporate EE into an already very full teacher preparation curriculum, cultural attitudes, and instructors' preference to teach science-specific disciplinary content are additional barriers to EE inclusion (Powers 2004). Equipment availability at some institutions and lack of knowledge about conducting environmental field studies also serve as obstacles. Consequently, some science teacher educators may underestimate the potential benefits of EE inclusion due to a preconception that it does not have a place in a preservice preparation program. These misconceptions and knowledge deficits may impede the establishment of collaborative relationships between nonformal EE educators and science teacher educators.

Overcoming these barriers for the benefit of future science educators is critical. Interdisciplinary co-teaching among faculty members of different disciplines is an ideal way to incorporate EE into a preservice teacher education program. Many EE concepts embed both science and social studies disciplines. For example, teaching and learning about complex environmental issues links together the content and pedagogy of the natural and social sciences. Such pedagogical instruction demonstrates an understanding of the interdisciplinary nature that is inherent to the EE discipline. Nonformal EE educators can serve as a facilitator to work with university faculty in such collaborations to assist with the inclusion of EE into a teacher preparation program.

Unfortunately, established and perceived "course ownerships" among disciplinary-based university faculty may prohibit integrating EE across one's program from the start. In such cases, EE with assistance from a nonformal EE educator can be easily integrated into one's own methods course in multiple ways, either through a single class visit or through multiple visits throughout a semester. Most nonformal EE educators are quite willing to provide both short-term and long-term workshop formats as part of a methods course. In addition, they are very open to co-planning these sessions to ensure that institutional teaching certification requirements are met.

Concluding Thoughts

Environmental education is continuously evolving and its efforts are quite diverse. In 2003, EE was recognized as being the "most widely nondisciplinary field spread now in all the countries of the world, and one of the first ones to be considered as a general educative need for all the inhabitants of the world" (IV Congresso Iberoamericano de Educacion Ambiental 2003). Recently, the North American Environmental Education Association (NAAEE) and the National Council for Accreditation of Teacher Education (NCATE) have recently approved *Standards for the Initial Preparation of Environmental Educators* to further promote the incorporation of EE into teacher certification areas (Simmons 2008). We are hopeful that this initiative will ultimately result in enhanced exposure of preservice teachers to EE content, pedagogy, resources, and professional networks. We are optimistic that EE will become an essential component in many science teacher preparation programs.

References

American Association for Advancement of Science. (1993). *Benchmarks for Science Literacy: Project 2061*. New York: Oxford University Press.

Ballantyne, R. R. (1995). Environmental teacher education: Constraints, approaches and course design. *International Journal of Environmental Education and Information, 14*(2), 115–128.

Bretz, S. (2001). Novak's Theory of education: Human construction and meaningful learning. *Journal of Chemical Education, 78*(8), 1107–1117.

Christenson, M. A. (2004). Teaching multiple perspectives on environmental issues in elementary classrooms: A story of teacher inquiry. *The Journal of Environmental Education, 35*(4), 3–16.

Council for Environmental Education. (2003). *Project WILD: K-12 curriculum and activity guide*. Houston, TX: Council for Environmental Education.

Food, Land, and People (Ed.). (2004). *Project food, land and people: Resources for learning* (2nd ed.). Chandler, AZ: Food, Land, and People.

Gigliotti, L. M. (1999). Environmental education: What went wrong? What can be done? *Journal of Environmental Education, 22*(1), 9–12.

Halim, L., & Meerah, S. M. (2002). Science trainee teachers' pedagogical content knowledge and its influence on physics teaching. *Research in Science & Technological Education, 20*(2), 215–225.

Hammerman, E., & Hammerman, D. (1985). What basic needs are met through outdoor education? *Outdoor Communicator, 16*(2), 28–31.

Heimlich, J. E., Braus, J., Olivolo, B., McKeown-Ice, R., & Barringer-Smith, L. (2004). Environmental education and preservice teacher preparation: A national study. *The Journal of Environmental Education, 35*(2), 17–21.

Hopkins, S. (Ed.). (1984). *Activities for environmental learning*. Harrisburg, PA: Pennsylvania State Department of Environmental Resources, Bureau of State Parks.

Hungerford, H. R., Peyton, B., & Wilke, R. J. (1980). Goals of curriculum development in environmental education. *Journal of Environmental Education, 11*(3), 42–47.

IV Congresso Iberoamericano de Educacion Ambiental. (2003, June 2–6). *Delegates' resolution of the IV Iberoamerican Congress on environmental education*. Paper presented at the IV Congresso Iberoamericano de Educacion Ambiental, Havana, Cuba.

Kane, P., Rosselett, D., Schierloh, J., & Anderson, K. (2003). *Bridges to the natural world* (2nd ed.). Bernardsville, NJ: New Jersey Audubon Society.

Leopold Education Project (Ed.). (1996). *The Leopold Education Project: Lessons in a land ethic*. St. Paul, MN: Pheasants Forever.

Liu, M., Bera, S., Corliss, S. B., Svinicki, M. D., & Beth, A. D. (2004). Understanding the connection between cognitive tool use and cognitive processes as used by 6th graders in a problem-based hypermedia learning environment. *Journal of Educational Computing Research, 31*(3), 309–344.

Marshall, C., & Censky, E. (Eds.). (2001). *Pennsylvania amphibians & reptiles: A curriculum guide*. Harrisburg, PA: PA Fish and Boat Commission.

McKeown-Ice, R. (2000). Environmental education in the United States: A survey of preservice teacher education programs. *Journal of Environmental Education, 32*(1), 4–11.

Mowery, M., Alberici, T., Bainey, L., & Kromel, T. (1998). *Pennsylvania Songbirds: A K-12 Teacher's guide for activities in the classroom*. Harrisburg, PA: Pennsylvania Game Commission.

National 4-H Curriculum. (2008). *About the National 4-H curriculum*. Retrieved March 29, 2009, from http://www.4-hcurriculum.org/default.aspx

National Research Council. (1996). *The National Science Education Standards*. Washington, DC: National Academy Press.

Nelson, D., Higgins, S., Kasselheim, A., Robinson, G., & Project WET. (1995). *Project WET curriculum & activity guide*. Bozeman, MT: Project WET International.

North American Association for Environmental Education. (2004a). *Nonformal environmental education programs: Guidelines for excellence*. Washington, DC: North American Association for Environmental Education.

North American Association for Environmental Education. (2004b). *Guidelines for the preparation and professional development of environmental educators*. Washington, DC: Author.

North American Association for Environmental Education. (2007). *NCATE standards for the initial preparation of environmental educators*. Washington, DC: Author.

Peffer, T., Duffield-Smith, J., & Bodzin, A. M. (2008). *Technology integration in nonformal environmental education*. Unpublished research paper, Lehigh University, Bethlehem, PA.

Pennsylvania Department of Conservation and Natural Resources (PA DCNR). (Ed.). (1998). *Pennsylvania State parks watershed education*. Harrisburg, PA: Pennsylvania Department of Conservation and Natural Resources (PA DCNR).

Powers, A. L. (2004). Teacher preparation for environmental education: Faculty perspectives on the infusion of environmental education into preservice methods courses. *Journal of Environmental Education, 35*(3), 3–11.

Ramsey, J. M. (1993). The effects of issue investigation and action training on eighth-grade students' envrionmental behavior. *Journal of Environmental Education, 24*(3), 31–36.

Rockcastle, V. (1989). What our students should know and be able to do. *Nature Study, 43*(1/2), 8–9, 22.
Shymansky, J. A., et al. (1982). How effective were the hands-on science programs of yesterday? *Science and Children, 20*(3), 14–15.
Simmons, B. (2008). Our next big step: Preparing teachers to be environmental educators. *NAAEE Communicator, 38*(1), 1, 17.
Stapp, W. B., & Cox, D. A. (Eds.). (1974). *Concerning spaceship earth* (Vol. 1). Dexter, MI: Thomson-Shore.
Stapp, W. B., et al. (1969). The concept of environmental education. *Journal of Environmental Education, 1*(1), 30–31.
Steinhart, D. R., & Vathis, P. (1997). *Pennsylvania Department of Education's "Office of Environment and Ecology" Agriculture K-12 Curriculum Supplement - Act 26*. Harrisburg, PA: Pennsylvania Department of Education.
Swan, J. A., & Stapp, W. B. (1974). *Environmental education: Strategies toward a more liveable future*. New York: Wiley.
Varelas, M., Plotnick, R., Wink, D., Fan, Q., & Harris, Y. (2008). Inquiry and connections in integrated science content courses for elementary education majors. *Journal of College Science Teaching, 37*(5), 40–48.
Western Regional Environmental Education Council (WREEC). (Ed.). (2004). *Project learning tree: Environmental education (PreK- 8) activity guide*. Washington, DC: American Forest Foundation.
Wisser, N. (Ed.). (2005). *Pennsylvania land choices: Activites for grades 6–12 and adults - building connections to natural resources, communities and the planning process*. Harrisburg, PA: Pennsylvania Department of Conservation and Natural Resources.
Yilmaz-Tuzun, O. (2008). Preservice elementary teachers' beliefs about science teaching. *Journal of Science Teacher Education, 19*(2), 183–204.

Using Environmental Education *Project* Curricula with Elementary Preservice Teachers

Adele C. Schepige, Patricia D. Morrell, Cindi Smith-Walters, Kim Cleary Sadler, Miriam Munck, and Donna Rainboth

With the addition of the North American Association for Environmental Education (NAAEE) environmental education standards to the National Council for Accreditation of Teacher Education (NCATE) teacher preparation accreditation standards and the frequent occurrence of environmental issues in the news, science educators who prepare elementary teachers are facing an ever growing challenge as they design the scope and sequence of topics in their courses. Researchers suggest that teacher preparation has a powerful influence on whether teachers implement

A.C. Schepige (✉)
Division of Teacher Education, Western Oregon University, 345 N Monmouth Ave, Monmouth, OR 97361, USA
e-mail: schepia@wou.edu

P.D. Morrell
University of Portland/School of Education, 5000 N. Willamette Blvd, Portland, OR 97203, USA
e-mail: morrell@up.edu

C. Smith-Walters
Biology/MTSU Center for Environmental Education, Middle Tennessee State University, PO Box 60, Murfreesboro, TN 37132, USA
e-mail: csmithwa@mtsu.edu

K.C. Sadler
Biology / MTSU Center for the Study of Cedar Glades, Middle Tennessee State University, PO Box 60, Murfreesboro, TN 37132, USA
e-mail: ksadler@mtsu.edu

M. Munck
Eastern Oregon University, PO Box 100, Pendleton, OR 97801, USA
e-mail: mmunck@eou.edu

D. Rainboth
College of Education, Eastern Oregon University, One University Blvd, La Grande, OR 97850, USA
e-mail: drainbot@eou.edu

A.M. Bodzin et al. (eds.), *The Inclusion of Environmental Education in Science Teacher Education*, DOI 10.1007/978-90-481-9222-9_19,

Environmental Education in their own instruction (Tilbury, 1992; Cutter, 1998 as cited in Miles & Cutter-Mackenzie, 2006). However, Miles & Cutter-Mackenzie, (2006) found that "despite national and international policy rhetoric about the importance of preservice teacher preparation in environmental education, preservice teachers' preparedness for teaching environmental education is overwhelmingly low" (p. 140). They further speculate that EE in preservice teacher education has remained unchanged for the past 2 decades. We must overtly address environmental education in teacher preparation programs if we are to meet "the priority of priorities" for environmental education (UNESCO-UNEP) and expect EE to find its way into K-12 classrooms (Volk & Hungerford, 1990; McKeown-Ice, 2000). Three problems to overcome are (a) teacher self-efficacy in addressing EE topics; (b) lack of preservice and inservice teacher training in EE; and (c) availability of classroom resources (Oguz et al., 2004; Stepath, 2004; Miles & Cutter-Mackenzie, 2006).

In overcoming these problems, one must consider that elementary preservice teachers typically take a single elementary science methods course, if they take one at all. Adding a focus on EE, along with science pedagogy, national standards, and other varying state-specific requirements, arguably would likely be a daunting task.

Stoner (1986) suggests that one approach is to integrate content and science pedagogy along with EE within the framework of a traditional methods course. She recommends methods courses be designed to incorporate *Project WILD* (WILD) (Council for Environmental Education, 2004) and *Project Learning Tree* (PLT) (American Forest Foundation, 2006) *Guides* curricula. In a national study about EE and teacher preparation, Heimlich, Braus, Olivolo, McKeown-Ice, and Barringer-Smith (2004) reported that *Project WET* (WET) (The Watercourse and Council for Environmental Education, 1995), WILD, Aquatic WILD (Council for Environmental Education, 2003), and PLT along with *Ranger Rick Nature Scope* were the most commonly recognized EE resources used in teacher preparation. Their study also indicated that content methods courses were perceived as the best place to house EE. Powers (2004) reported on interviews with 18 elementary science and social studies methods instructors about their practices and barriers to including EE. Impediments to EE in methods classes included time (also noted in a study by Heimlich, et al., 2004), a pressure to keep disciplines separated, opportunities to work with real elementary children, finding a mentor teacher as a role model, and the general hesitancy of elementary teachers to want to teach science.

Few studies have focused on the use of the *Project Guides* in preservice settings. Most information about the implementation and impact of *Project Guides* comes from the evaluations and reports found on the respective websites. Most of that is directed toward inservice teachers. This chapter showcases four case studies of integrating EE through the use of the *Project Guides* into preservice teacher coursework at four different universities. Emphasis is placed on those techniques that draw on EE and the *Project Guides* to strengthen science content knowledge, develop science process and inquiry skills, integrate literacy, and introduce fieldwork. For some background information on the *Project Guides*, see the Chapter Summer Methods in Summer Camps: Teaching Projects WILD, WET, and Learning Tree at an Outdoor Environmental Education Center.

Case One: Life Science Content Course Designed with *Project Guides*

Life Science for Elementary Teachers (BIOL 3000) at Middle Tennessee State University is a life science content course combining lecture, laboratory, and outdoor field experiences. It is taught in 2-h blocks three times a week, or twice a week in 3-h blocks. The course is designed to teach life science content using interactive activities both inside and outside of the classroom. Students purchase an introductory biology text as a reading reference and receive at no cost the curricular *Project Guides* (PLT, WET, Aquatic WILD, and WILD) at strategic points in the semester. The *Project Guides* are an integral component of the course and are used by the instructors to support life science content delivery and learning. Moreover, the *Project Guides* concurrently provide modeling of appropriate pedagogical practices for novice teachers. These resources utilize a multidisciplinary approach to teaching and each guide uniquely addresses course EE topics related to life science in an interactive manner. The *Project Guides* are also used in other ways to augment content and prepare students for a career in teaching.

BIOL 3000 does not approach biology in the manner of a traditional semester-long college biology course. The scope and sequence of the course begins with the big picture (biodiversity, populations, and communities) and ends with the molecular basis for life. In contrast, most of the biology courses begin by examining the small (molecules, organelles, cell) and move toward the broad (ecosystems and the environment). Ecology, often the last few chapters in a college biology text, may be either completely ignored or receive limited attention by the instructor and consequently the students. In contrast, BIOL 3000 was designed specifically for preservice elementary teachers. The first two weeks are devoted to helping students develop scientific literacy by interactively modeling and reinforcing science process skills. Eight weeks, the majority of the course curriculum, are dedicated to biodiversity, ecological concepts, the environment, and organismal biology. Understanding the big picture in biology and the complexities of ecosystems and the environment provides a balanced foundation for the final four weeks that are dedicated to topics related to the molecular basis of life.

The *Project Guides* are used in a number of ways. They are a source for background reading; an example is the narrative titled *The Ecosystem Concept* found in the appendix of WILD. For one assignment, students select an organism and research its ecology and prepare a "foldable" graphic organizer detailing habitat, niche, and related information. The animal list from which they choose is extensive, containing organisms both familiar (raccoon, grasshopper, little brown bat, box turtle) and those not so recognizable (pearly mussel, cicada killer, vole, shrew). Each student selects a different organism allowing the class to meaningfully explore symbiotic relationships with guide activities, such as PLT's *Web of Life*. In this activity, students research and simulate a food web as they discover ways in which plants and animals are connected to each other. *Project Guides* activities are sometimes used to introduce a topic prior to lecture, thereby establishing a "need to know" within the minds of students. An example is *The How Wet Is Our Planet* from Aquatic WILD, where students calculate the amount and distribution of water on the Earth in oceans, rivers, lakes, groundwater, icecaps, and the atmosphere, then make inferences about the importance of responsible water use. The activity is additionally used to introduce the water cycle and other nutrient cycles upon which all life depends. Some activities are used as a way to evaluate student learning after a concept or topic is covered. In the WILD activity *Oh Deer!*, students take on the roles of deer or habitat components such as food, water, or shelter in an effort to see how resources and their availability impact wildlife populations. Additionally, *Oh Deer!* is used to introduce and reinforce a range of concepts, such as predator/prey relationships, carrying capacity, how humans and/or nature influence limiting factors, and population growth curves. Data collected during a portion of the activity related to deer population numbers may be given to students for analysis as a part of an exam.

The *Project Guides* have been integral parts of the course design for BIOL 3000 for more than 15 years. In each semester, new strategies for incorporating the *Project Guides* are developed based on the time of year, the location of the course, and the instructor. Table 1 shows a sample of activities from the *Project Guides,* which have been used to support lecture and life science content. (Activities mentioned throughout this paper are described in Appendix A.)

In addition to using the guides to support combined lecture, laboratory, and field experience, each preservice teacher creates their own life science resource trunk from the *Project Guides*. The trunks are a "ready to use" resource full of teacher-made, scavenged, and purchased materials that can be used to teach science within a classroom. Examples of teacher-made materials might be animal picture sets, a working model lung, or even a plant press. Scavenged materials might include twig puzzles, collections of stuffed or plastic animals, baby food jars, or buttons to sort. Purchased materials run the gamut from books and field guides to puzzles and games. Preservice teachers in the course choose two lesson activities from each *Project Guide* and package a copy of the activity along with a durable classroom set of materials needed to conduct the lesson (we call these *Activity Packs*). Student handouts or cards must be laminated and enough sets included for cooperative groups. A huge challenge is that many favorite activities do not require materials

Table 1 Project-guide activities and related environmental content

Content	Guide	Activity
Biomes	WILD	*Move Over Rover*
Energy flow within an ecosystem	WILD	*Owl Pellets*
Nutrient cycles	WET	*Incredible Journey*
	PLT	*Water Wonders*
Predator/Prey relationships	WILD	*Quick Frozen Critters*
Population genetics	WILD	*Bottleneck Genes*
Science process skills	WET	*H_2O Olympics*
Sexual reproduction	WILD	*Are You Me?*
Symbiotic relationships	PLT	*Dynamic Duos*

beyond what would be found in a classroom (chalkboard, markers, ruler, and tape). The syllabus bans using these types of lessons in the *Activity Packs*. The *Packs* are designed to use free or easy-to-obtain materials and packaged in such a way that they can be taken from the trunk and immediately used. Lastly, each of the *Activity Pack* lessons used must cross-reference to specific grade levels of the Tennessee State Science Curriculum Framework. Cross-referencing serves multiple purposes. It not only provides experience to the preservice teacher in using state mandated standards, but also allows them to develop an understanding of the importance of standards-based teaching and learning.

E-mails and letters are received every year from grateful students who were prepared for interactive lessons with their own classrooms through the development of the resource trunk and *Activity Packs*. Feedback has been overwhelmingly positive from former students with regard to using the materials in school settings.

Preservice teachers in BIOL 3000 have taken 8 h of science, which are the prerequisite science courses for the university general education requirement. Yet, when surveyed, a majority admit to a fear of teaching science and being unprepared to teach science content. With these things in mind, a simple survey instrument was developed to determine student perception of ability to teach specific topics covered in the course. The survey is administered on the first and last day of the class; several subtopics are listed and students are asked to circle a smiling face, a neutral face, or a sad face to indicate their perceived confidence in teaching each topic. Overall, pre-responses have consistently shown that only 30% of our students feel confident to teach the listed topics; post responses show that more than 80% of our students feel confident to teach the topics examined in BIOL 3000 and supported by use of the *Project Guides*. It is not the intent of the authors to suggest *Project Guides* use is the implicit reason for the huge positive gain in perception of students' ability to teach these topics, but to show that students leave BIOL 3000 with a greater confidence in their ability to teach the subject matter.

Case Two: Science Inquiry with the *Project Guides*

"If a single word had to be chosen to describe the goals of science educators during the 30 year period that began in the late 1950s, it would have to be inquiry" (DeBoer 1991, p. 206). The *National Science Education Standards* highlight learning science through investigative or problem-solving strategies as the central component of science instruction (NRC 1996). Certain traits characterize inquiry instruction from other science instruction: questioning and connecting personal understanding with scientific understandings, designing experiments, investing phenomena, and construction meaning from data and observations (NWREL 2005). The *Project Guides* are extremely useful in connecting preservice teachers to inquiry teaching strategies. Knowing that teachers tend to teach as they were taught (Lortie 2002; Cuban 1984), modeling inquiry teaching with exemplary environmental curricula to preservice teachers becomes even more advantageous.

Teaching with inquiry has been described as difficult because of the required teacher skill and understanding of inquiry pedagogy. In order to teach inquiry effectively, teachers need to have an understanding of the inquiry process, content knowledge, and how to facilitate students as they conduct investigations (Furtak 2006; Johnson 2004). In an effort to clarify inquiry and make the process easier for teachers and students to master, inquiry has been broken into a developmental continuum. At the teacher-directed end of the continuum is *structured inquiry*. Students are provided information and questions to investigate, in addition to the design of the investigation. In the middle of the continuum, *guided inquiry* invites the teacher and students to share the responsibilities for the inquiry process. The teacher may pose the question and provide a portion of the design of the investigation. The students finish the design, collect and analyze the data, and draw conclusions. At the student-centered, *open inquiry* end of the continuum, students ask questions of their choice and interest. They research their questions, and design and conduct experiments with the teacher acting only as a facilitator (NWREL, 2005). The inquiry continuum provides both teachers and students with stepping-stones from guided to structured inquiry to open-ended inquiry.

A study was conducted at Eastern Oregon University to determine the effect of using *Project Guides* lessons to teach science inquiry to preservice teachers and to determine to what extent preservice teachers could utilize the *Project Guides* to teach environmental content to children using inquiry. Specifically, the research questions were: To what extent are the *Project Guides* useful in teaching preservice teachers science inquiry strategies? To what extent will preservice teachers utilize *Project Guides* lessons and science inquiry to teach science content to children?

The study participants were 50 graduate and undergraduate teacher education students enrolled in science methods courses. During the course of instruction, preservice teachers learned to identify the characteristics of inquiry and the four primary domains of inquiry including forming a question, developing procedures, collecting data, and analyzing results. The students reviewed *Project Guides* lessons to determine if lessons contained inquiry features that could be modified to incorporate inquiry. The *Project Guides* lessons – *Cold Cash in the Ice Box* (WET), *A Grave Mistake* (WET), *Nature's Recyclers* (PLT), and *Are Vacant Lots Vacant?* (PLT) – were taught to preservice teachers to model how inquiry could be utilized when teaching environmental science. In this phase of the study, the preservice teachers participated in the *Project Guides* lessons as learners. After completing the lesson, through a discussion with their group members and professor, the preservice teachers reflected on inquiry elements in the lesson, where the lesson fit on the inquiry continuum, and how the professor modified the lesson to incorporate various essential features of inquiry, such as data collection and analysis, as shown in Table 2. The science concepts and content addressed by the lesson were also identified and discussed. In the second phase of the study, the preservice teachers identified the inquiry elements as well as the science concepts and content in *Is There Water on Zork* (WET) and *How Plants Grow* (PLT). They then redesigned the lessons to include more elements of open-ended inquiry.

Table 2 Inquiry elements from project-guide lessons

Lesson and guide	A Grave Mistake (WET)	Cold Cash in Ice Box (WET)	Water on Zork (WET)	Nature's Recyclers (PLT)	How Plants Grow(PLT)	Vacant Lots (PLT)
Inquiry elements	Analyze	Design Analyze Conclusions	Design Analyze Conclusion	Design Analyze Conclusion	Design Analyze Conclusion	Analyze Conclusion
Type of inquiry	Structured	Guided	Guided	Guided	Guided	Structured

Students' basic knowledge of inquiry teaching was assessed through an analysis of the students' and professor's discussion about the selected *Project Guide* lessons. Ninety-four percent of the students correctly explained the inquiry elements in the lessons. Based on students' knowledge of inquiry and their ability to adapt PLT and WET lessons to meet the requirements of inquiry, each preservice teacher selected a different PLT or WET lesson, identified the inquiry elements contained in the lesson, and explained how they would modify the lesson to become a more student-centered inquiry. Analysis of the assignment outcomes showed that 90% of the students demonstrated the ability to modify the strategies of existing lessons to be more inquiry-oriented.

The final portion of the study required preservice teachers to teach their modified inquiry-based lesson to children. After the teaching experience, additional perception data were collected about the usefulness of the *Project Guide* lessons in learning and teaching inquiry with children. A 1–10 scale was used, where 1 was not at all useful and 10 was extremely useful. In a brief interview, students were also asked to explain their numerical score choice. Analysis of students' response to the question, how useful were the *Project Guide* lessons in learning science inquiry strategies, resulted in a mean score of 9.1, indicating students believed the *Project Guides* to be very useful toward their understanding of inquiry. The analysis of the second question response, how useful were the *Project Guides* in teaching inquiry to children, resulted in a mean score of 8.8, indicating that students found the *Project Guides* very useful. Interview responses were positive for all students. Some students expressed their feelings about their understanding and inquiry teaching ability, while others recapped what they learned about inquiry. Table 3 presents selected student responses.

The study results indicated that the *Project Guides* provide excellent curricula for teaching inquiry-based instruction. Science methods instructors can utilize *Project Guides* lessons to effectively teach preservice teachers developmental inquiry methodology. The value of the *Project Guides* is even more apparent when preservice teachers are able to make use of *Project Guides* lessons and their knowledge of inquiry instruction, design instruction, and teach children using

Table 3 Narrative questions and selected responses

Question 1: How useful were the *Project Guide* lessons in learning science inquiry strategies?	Question 2: How useful were the *Project Guides* in teaching inquiry-based science to children?
Student: I feel I know a lot more about how to teach inquiry lessons	Student: My cooperating teacher was very impressed with what the children learned and how well they did inquiry.
Student: The (*Project Guides*) lessons provided me with the foundation for inquiry	Student: I enjoyed teaching with inquiry so much and my students did so well!
Student: I learned what to look for in lessons that could be changed for inquiry.	Student: I linked four (*Project Guides*) lessons to create an inquiry unit.
Student: Inquiry teaching was scary before I used the *Project Guides* to design inquiry lessons.	Student: I didn't think I could teach inquiry before I had the books (*Project Guides*).

inquiry-based strategies. The content, quality lessons, and hands-on strategies of the *Project Guides* provide an excellent base for learning and teaching scientific inquiry, as well as teaching standards-based EE content.

Case Three: Literacy and the Environmental Education *Project Guides*

Preservice elementary teachers at the University of Portland are introduced to the *Project Guides* in their science/mathematics methods class. The Projects are not an "add-on" to the curriculum to help students teach EE; rather, *Project Guides* lessons are used as the basis to illustrate concepts or demonstrate instruction strategies. The teacher licensing agency in Oregon mandates that preservice teachers incorporate literacy into their lessons; therefore, this case examined the use of the *Project Guides* to foster reading, writing, and listening skills. The purposes of the study were to determine: (a) if the lessons in the *Project Guides* could be successfully used as models to show the integration of EE and literacy for the preservice teachers and (b) how the preservice teachers viewed the usefulness of the *Project Guides*.

The participants involved in this study were 22 preservice elementary teachers enrolled in a 3-credit-h elementary science and mathematics methods course. The methods course is a required class taken during the Fall semester of the senior year. Prior to this course, as a part of their education program, the students have completed three science courses for non-science majors (Ideas in Physics, Human Biology, and Earth Science) and one course with a focus on literacy. The education program requires field experiences from the freshman through junior years. Concurrent with the methods course, these students spend 15 h per week in a part-time student teaching experience in an elementary school. All participants were on schedule to move to full-time student teaching the semester after the methods course, the last major requirement for degree completion and teaching licensure.

The curriculum for this class was redesigned to include introducing the preservice teachers to four *Project Guides*. Rather than being something "in addition" to the curriculum, activities from the *Project Guides* were used as appropriate within the framework of the course to model certain pedagogical strategies, with the added benefit of reviewing science content information with the students. An emphasis was placed on how literacy and science could be integrated with the *Project Guides* activities. For example, *The Incredible Journey* (WET) was used as the basis for discussing conceptual change. As a part of the lesson, students reviewed and discussed about 20 *easy reader books* to determine the value of using them with their students when discussing the water cycle. *Every Tree for Itself* (PLT) was used as the basis for modifying activities to be developmentally appropriate for a variety of age levels and make lessons personally relevant; we also reviewed pertinent *easy reader books* and discussed possible writing activities that could complement the lesson. WILD's *Color Crazy* was used when discussing interdisciplinary models.

One of the assignments for this course required students to choose an activity from any of the *Project Guides* and use it as the basis for a lesson plan, incorporating a literacy component into the lesson. For the purposes of this study, the lessons were examined noting the *Project Guides* activity and the type of literacy activity included in the plans.

Of the 22 plans submitted, 21 were analyzed; one student submitted a plan that integrated literacy into a science lesson, but did not base it on a *Project Guides* activity and was excluded from the analysis. Thirty-three percent of the students used an activity from PLT; 29% used WET; 24% used WILD; and 14% used *Aquatic* WILD. Of the literacy components integrated, 12 were of writing activities, four of reading, and five a combination of both reading and writing components. The majority of the literacy ideas incorporated into the lessons were either directly from the *Project Guides* or modifications of suggested activities/extensions.

At the end of the semester, the same students were asked during the class to respond anonymously in writing to questions concerning the science lesson plan assignment specifically and their views about the usefulness of the *Project Guides* in general. All 21 students present in class that day shared their views. Twenty of the 21 students felt that using the *Project Guides* activity as a base was a positive experience ("very helpful," "good"). These students felt that the assignment helped them to think about how literacy could be incorporated into a science lesson, with familiarizing themselves with the *Project Guides*, and modifying existing lessons. One student said, "I thought it was very helpful because it made students think about how literacy can be incorporated in a variety of lessons." The student who responded negatively said "it was too restrictive" as the *Project Guides* did not "fit" with his/her current student teaching assignment.

About 50% of the students stated that they used the *Project Guides* for assignments other than the lesson plan during the course. Six said they used activities from the *Project Guides* as a part of a science resource collection they needed to compile for the course, three used them as a source for a presentation for the course, and one used a lesson for an assignment for an assessment course. Additionally, five students (24%) used activities from the *Project Guides* in their field placement.

All respondents thought the *Project Guides* were useful and indicated they would use them in the future (In a follow-up at the end of the following semester, 50% of the students reported using the *Project Guides* in some way in their full-time student teaching placement). The students commented that the *Project Guides* were well organized, easy to use, adaptable, and had educationally fun activities:

> I like the organization of them and the set up of the lessons. They present many adaptable ideas.
>
> Because the lessons are so detailed and adaptable, it will save me lots of time.
>
> I especially am interested in teaching science and I love nature. These books will promote those goals. I (can also use them) to integrate multiple subject areas.

Incorporating activities from the *Project Guides* into an existing science methods class was an efficient way to provide preservice teachers with the content background, pedagogy, and resources needed to encourage them to teach EE to their students, while modeling how literacy and science could be interwoven in lesson planning. The students' lesson plans showed that they could easily adapt *Project Guides* lessons to suit their own needs and integrate literacy into those plans. All the preservice teachers recommended that the *Project Guides* continue to be used with future methods classes.

Case Four: Using the *Project Guides* for Field Work

Environmental education in the elementary science/math methods course at Western Oregon University has two major emphases: using field work to teach science and having preservice teachers plan and teach lessons from WET, WILD, Aquatic WILD, and PLT in a variety of settings. All elementary preservice teachers in the undergraduate program are required to take this course prior to full-time student teaching.

Before teaching lessons, the preservice teachers learn about scientific inquiry, observation, and field work. We begin by using an inquiry model that is very similar to the one described in the second case. Next, the preservice teachers learn what an observation is and is not, and how it is linked to field work. According to the National Research Council (1996) Science As Inquiry standard, students in the grades k-4 should be expected to use simple instruments such as magnifiers, thermometers, microscopes, watches, and rulers in making observations to gather data and learn what constitutes evidence. Their teachers need to be able to do the same. Depending on the location chosen, time of the year, and the weather, different *Project Guides* activities can be specifically used to teach about observation outdoors. These include WILD's *Wildlife is Everywhere*, *Urban Nature Search*, and *MicroTrek Scavenger Hunt,* PLT's *School Yard Safari* and *The Closer You Look*, and WET's *Stream Sense.* Three different sets of observations at the same location and based on the same investigative question from a *Project Guides* activity are made. For example, when using WET's *Stream Sense*, the class goes to a local park that has a small stream. In small groups, they begin their observations using the Sensory Observation Sheet. When PLT's School Yard Safari is used, a small area on the college campus is used as the setting to look for evidence of wildlife, instead of using an elementary school. Between each observation, students share their observations. They are also provided with supplemental information, such as a packet about Lewis and Clark's discovery journey that contains copies of observational information and data that Lewis and Clark recorded at various places. The preservice teachers are asked to find evidence of the kinds of observations that Lewis and Clark made, and how they made and recorded them. Additionally, they determine what kinds of skills and knowledge that Lewis and Clark needed to make the observations. Qualitative and quantitative strategies are introduced as well as simple equipment and instruments to assist them in improving their observational skills. Preservice teachers then transfer those skills to the observations needed as indicated in the *Project Guides* activities. Their observations then become richer, more accurate, and include both qualitative and quantitative data.

After learning about observations, the preservice teachers are introduced to using the *Project Guides* in relation to field work, and how field work is different from a field trip. Field work was selected because of its natural association with scientific inquiry, environmental issues, and place-based education, all of which link to the necessity of providing children with real-life learning opportunities (NSTA, 2003) outside of the four walls of the classroom. Field work is very much contextual and includes components of inquiry. Inquiry outdoors can take place in a variety of ways that align with the ways scientists do field investigations, from typical hypothesis testing to descriptive, comparative, and correlational studies (Windschitl et al., 2007). *Field work in the Geography Curriculum* (Rice & Bulman, 2001) is used as a framework for this. The preservice teachers receive formal instruction about conducting and assessing field work. They are also asked to read and respond to the article by Christianson (2004) about elementary teachers' perspectives on trying to teach EE in an already crowded curriculum. The preservice teachers are then asked to design, pilot, and report the results of their own mini-field work unit appropriate for elementary-age children, which is aligned to standards and incorporates *Project Guides* lessons supplemented with other resources.

Finally, the preservice teachers experience teaching a field work based lesson from the *Project Guides* to children. They have the option of teaching at a half-day mini-outdoor school for third through fifth graders for a local elementary school adjacent to the city park or finding their own (pre-approved) setting. At the mini-outdoor school, the preservice teachers divide into teams and conduct a site visit. They then develop four field work stations united by a common environmental theme, such as habitats or water, which also integrate field work skills such as observation or sampling. During the mini-outdoor school, student groups rotate every 45 min through each station. At the completion of each station, preservice teachers use a variety of assessments to check student learning. To gauge what the preservice teachers were learning, they report information about which activities from which *Project Guides* were used, where the lesson was taught, and who the students were. They are also asked to reflect both on their students' and their own learning.

The following data are from 66 preservice teachers. Their reports show that:

1. Forty-five planned the entire mini-outdoor school and taught *Project Guides* lessons.
2. Nine taught a single *Project Guides* lesson in a student teacher setting that was not field-work-related.
3. One used WET as the primary curriculum for their student teaching unit.
4. Nine taught *Project Guides* outdoor lessons to after-school programs.
5. One taught a PLT outdoor lesson at a Police Activities League.
6. One used WET to teach about the water cycle to inservice teachers at a summer institute.
7. Three combined ideas from two different *Project Guides* to make one field work lesson.

Themes emerged from their reflections about what they learned. All reported positive experiences using the *Project Guides* activities to teach field work lessons. Many stated that teaching lessons outdoors was different than in a classroom, in terms of preparation, site visits, safety concerns, student excitement, and keeping students from getting distracted. Another theme was that the preservice teachers' own environmental content knowledge was lacking. They recognized the need for observational skills as being important for field work. The reflections showed that they know that evidence of children's learning content and skills can come from written artifacts, such as lists, sketches, and answers to questions on handouts. They also realized that evidence of learning comes from small group discussions, children's telling of different examples of the concepts, and using movement to act out dramatizations of predator/prey relationships. When planning, some preservice teachers found that they needed to modify *Project Guides* lessons to their specific situation or site. Most indicated that the suggested time of the lesson listed in a *Project Guide* was not in alignment with the actual implemented lesson time.

Project Guides activities can easily be used to support teaching EE through field work. Providing opportunities for preservice teachers to plan and teach using activities from the *Project Guides* is a valuable part of the process. Using the mini-outdoor school is quite effective because actual teaching opportunities are based on field work.

Conclusions and Implications

Integrating the *Project Guides* addresses the hurdles noted in the literature (Oguz, et al., 2004; Stepath, 2004; Miles & Cutter-Mackenzie, 2006) in preparing preservice teachers to teach EE, namely, teacher training, classroom resources, and teacher's self-efficacy. The four cases explore various ways in which the *Project Guides* can be successfully used with preservice teachers to strengthen both content and pedagogical knowledge. The activities in the *Project Guides* proved to be good models for instruction and useful teaching tools. The depth and breadth of the content and teaching strategies included in the *Project Guides* allow for a great deal of flexibility in their use. The instructors were able to seamlessly include the *Project Guides* into their teaching, and were successful in promoting both EE content and a variety of teaching ideas and strategies within their respective courses. Similarly, the preservice teachers found the *Project Guides* to be valuable resources for use in classroom teaching.

While the university instructors will continue to employ and refine the use of the *Project Guides* in their courses, there is little data on how the preservice teachers use these *Project Guides* in their student teaching experiences, and later in their own classrooms. Longitudinal research should be done to examine how the use of the *Project Guides* during teacher preparation courses plays out in the actual classrooms of beginning teachers and how it impacts students' understanding of environmental content. Examining the implementation of the content and pedagogies in actual elementary classrooms can provide guidance to how university instructors might make better use of the *Project Guides* as they prepare teachers for the important work of educating students.

Appendix A

Project Guides activities and descriptions

Lesson activity	Guide	Description
Are Vacant Lots Vacant?	PLT	Students examine a study plot for signs of life.
Are You Me?	WILD	Playing cards with pictures of "young" and "adult" animals are matched up.
A Grave Mistake	WET	Well-water data is analyzed to determine the source of community water contamination.
Bottleneck Genes	WILD	A fictional population of animals is reduced in number and students examine what happens to the genetics within the population over time.
Cold Cash in the Ice Box	WET	Students examine refrigeration by designing mini insulators to keep ice from melting.
Color Crazy	WILD	Students design a colorful animal to examine the role of color and its importance to survival of wildlife.
Dynamic Duos	PLT	Symbiotic relationships are identified and students describe how partners in these relationships affect one another.
Every Tree for Itself	PLT	A game is used to introduce tree growth requirements of sunlight, water, and nutrients.
How Plants Grow	PLT	Experiments are designed to test factors affecting plant growth.
How Wet Is Our Planet?	Aquatic WILD	The usable percentage of fresh water on earth is calculated.
Incredible Journey	WET	A simulation of water movement through a more complex water cycle or journey.
MikroTrek Scavenger Hunt	WILD	Students go outside to find different kinds of evidence that wildlife exists.
Move Over Rover	WILD	Animals and their appropriate habitats are discussed.
Nature's Recyclers	PLT	Students observe pill bugs to help determine their role in an ecosystem.
Oh Deer!	WILD	Students become deer in a role play as they investigate habitat components, carrying capacity, and limiting factors.

(continued)

Appendix (continued)

Lesson activity	Guide	Description
Owl Pellets	WILD	Owl pellets are dissected in order to discover food preferences, study rodent anatomy, and reveal the importance of food chains and decomposers in recycling nutrients within ecosystems.
Quick Frozen Critters	WILD	Students simulate fox and rabbit populations in this very active game. Afterward is a discussion about the similarity of the skills and traits that both predators and prey must have to survive.
School Yard Safari	PLT	On school grounds, students observe and find evidence of animals and relate that to habitat requirements.
Stream Sense	WET	At a stream site, students make observations using their senses, take measurements, and test water quality.
The Closer You Look	PLT	Drawings of trees made from memory and direct observation are compared.
Urban Nature Search	WILD	Students go on a nature search designed to increase student-observation skills in nature.
Water Olympics	WET	The marvelous physical properties of water are examined and how these properties affect life on Earth.
Water Wonders	PLT	Students explain how the water cycle is important to living things and how plants affect the movement of water in a watershed.
Web of Life	PLT	After researching forest organisms, a simulated food web is built and the relationship between organisms is emphasized.
Wildlife is Everywhere	WILD	Looking for direct and indirect evidence of life in various settings.

References

American Forest Foundation. (2006). *Project Learning Tree: Pre-K environmental education activity guide*. Washington, DC: American Forest Foundation.

Christianson, M. A. (2004). Teaching multiple perspectives on environmental issues in elementary classrooms, a story of teacher inquiry. *Journal of Environmental Education, 35*(4), 3–16.

Council for Environmental Education. (2003). *Project WILD: K-12 curriculum and activity guide*. Houston, TX: Council for Environmental Education.

Council for Environmental Education. (2004). *Project WILD Aquatic K-12 curriculum & activity guide*. Houston, TX: Council for Environmental Education.

Cuban, L. (1984). *How teachers taught: Constancy and change in American classrooms, 1890–1980*. New York: Longman.

DeBoer, G. E. (1991). *A history of ideas in science education*. New York: Teachers College Press.

Furtak, E. M. (2006). The problem with answers: An exploration of guided scientific inquiry teaching. *Science Education, 90*(3), 453–467.

Heimlich, J., Braus, J., Olivolo, B., McKeown-Ice, R., & Barringer-Smith, L. (2004). Environmental education and preservice teacher preparation: A national study. *Journal of Environmental Education, 35*(2), 17–21.

Johnson, K. (2004). The role of field paleontology on teachers' attitudes toward inquiry science. *NOVAtions Journal*, (2f).

Lortie, D. (2002). *Schoolteacher: A sociological study* (2nd ed.). Chicago: University of Chicago.

McKeown-Ice, R. (2000). Environmental education in the United States: A survey of preservice teacher education programs. *Journal of Environmental Education, 32*(1), 4.

Miles, R., & Cutter-Mackenzie, A. (2006). Environmental education: Is it really a priority in teacher education? In S. Wooltorton & D. Marinova (Eds.), *Sharing wisdom for our future. Environmental education inaction: Proceedings of the 2006 Conference of the Australian Association of Environmental Education.*

National Research Council (NRC). (1996). *National science education standards*. Washington, DC: National Academy Press.

National Science Teachers Association. (2003). *Standards for science teacher preparation*. Retrieved March 15, 2006, from http://www.nsta.org/preservice?lid=tnavhp

Northwest Regional Education Laboratory (NWREL). (2005). Is there only one way to do science inquiry? Retrieved March 15, 2006, from http://www/nwrel.org/msec/science_inq/answers.html

Oguz, A., Fortner, R., Adadan, E., Gay, K., Kim, C., Yalcinoglu, P., Bektasli, B., Cook-Hoggarth, K.L., McDonald, C., Mishler, K., & Manzo, L. (2004). *A look at environmental education through science teachers' perspectives and textbooks' coverage*. Paper presented at the Annual Meeting of the School Science and Mathematics Association, Atlanta, GA.

Powers, A. (2004). Teacher preparation for environmental education: Faculty perspectives on the infusion of environmental education into preservice methods courses. *Journal of Environmental Education, 35*(3), 3–11.

Rice, G. A., & Bulman, T. L. (2001). *Field work in the geography curriculum: Filling the rhetoric-reality gap*. Indiana, PA: National Council for Geographic Education.

Stepath, C. (2004). *Awareness and monitoring in outdoor marine education*. Paper presented at the Tropical Environment Studies and Geography Conference, Cairns, Australia.

Stoner, D., & California State University. (1986). *Integrating environmental education into preservice teacher preparation* (ERIC Document Reproduction Service No. ED276600).

The Watercourse & Council for Environmental Education. (1995). *Project WET*. Bozeman, MT: The Watercourse & Council for Environmental Education.

Tilbury, D. (1992). Environmental education within pre-service teacher education: The priority of priorities. *Environmental Education & Information, 11*(4), 267.

Volk, T., & Hungerford, H. (1990). Changing learner behavior through environmental education. *Journal of Environmental Education, 21*(3), 8.

Windschitl, M., Dvornich, K., Ryken, A., Tudor, M., & Koehler, G. (2007). A comparative model of field investigations: Aligning school science inquiry with the practices of contemporary science. *School Science & Mathematics, 107*(1), 382–390.

Situated Learning in Environmental Education: Using Geospatial Technologies with Preservice Secondary Teachers

Rita A. Hagevik, Harriett S. Stubbs, and Diane C. Whitaker

Introduction

With today's increased emphasis on standardized testing, most teachers probably spend little to no time outside allowing direct encounters with nature. Two major reform documents produced at the end of the last century, *Science for All Americans* (American Association for the Advancement of Science: Project 2061, 1990) and the *National Science Education Standards* (National Research Council 1996) continue to be an effective focus of discussion for preparing a scientifically literate society. Yet, these documents are silent with regard to explicitly advancing the study of natural history, spending time outside with direct encounters with nature, or simply being observant about the natural surroundings. While these documents do not ignore the natural world, they also do not advocate that our science students make regular, systematic studies of nature.

In our experience, we find that although science teacher educators mostly love and appreciate the out-of-doors, they have rarely had any sort of situated nature study – the structured use of a local environment to learn about organisms and their interrelationships and interactions. In fact, the study of nature has become increasingly overlooked, in a similar manner that habitats have become increasingly marginalized across the globe. How many graduating preservice educators become excited

---

R.A. Hagevik (✉)
The University of Tennessee, A404 Bailey Education Complex, Knoxville, TN 37996, USA
e-mail: rhagevik@utk.edu

H.S. Stubbs
Department of Mathematics, Science, and Technology Education, College of Education, & c/o GIS Program, College of Natural Resources, North Carolina State University, Box 7106, Raleigh, NC 27695-7401, USA
e-mail: h_stubbs@ncsu.edu

D.C. Whitaker
CALS Center for Applied Aquatic Ecology Education Specialist, North Carolina State University, Box 7510, Raleigh, NC 27695, USA
e-mail: dcwhitak@ncsu.edu

A.M. Bodzin et al. (eds.), *The Inclusion of Environmental Education in Science Teacher Education*, DOI 10.1007/978-90-481-9222-9_20,

about a new insect or a bird seen in the out-of-doors? How many have a sense of wonder or awareness of an animal sound or a different type of cloud in the sky? How many can identify native species in their area? How useful will this knowledge be to them as they begin their teaching and when their students want to know "what is this?" It is not important that teachers answer the question directly; what is important is that they possess skills necessary to guide inquiry based on a question or observation in nature. It is difficult to think of a more valuable education than one that develops curiosity and includes a deep awareness and knowledge about the environment in which one lives.

We propose that science teacher preparation programs should prepare our preservice science teachers to regularly use situated studies in nature as a routine part of their teaching (White 2005). We believe that doing so will make modern science education reform recommendations more explicit with regard to the study of the environment. In this chapter, research relating to the use of nature study in the curricula is explored. Practical suggestions are offered to effectively utilize Geospatial Information Technologies (GIT) in the process of nature study with preservice teachers. Using GIT in science education provides learning opportunities to get students outside, to teach them to be more observant about their natural surroundings, to engage them in meaningful data collection activities, and to help them through the power of visualizations and maps to make connections in the natural world.

Nature Study in Science Education

Many years ago, what we refer to as *nature study* or *natural history* was commonplace in schools around the nation. Students routinely followed their teacher into the woods or schoolyard, not just for the occasional field trip, but for regular, planned, and purposeful studies of nature that contributed to their total understanding of science. Historically, the direct study of nature was the focus of science education in K-12 settings. For example, Anna Comstock's (1911) classic book, *Handbook of Nature Study*, was a common text for K-12 teachers. The table of contents reveals detailed directions for collections and systematic study in nature that was once required for teacher certification. In *The School and Society* (1915), Dewey advocated an experiential approach to student learning in the local environment, "Experience (outside the school) has its geographical aspect, its artistic and its literary, its scientific and its historical sides. All studies arise from aspects of one earth and one life lived upon it" (1915, p. 91). What is sometimes known as *place-based education* usually includes the conventional outdoor education methodologies as advocated by John Dewey. This helped both teachers and their students connect with their particular corner of the world (Woodhouse and Knapp 2000).

Current science and science methods textbooks bear no resemblance to these historical approaches. The systematic study of nature has been replaced by technological, molecular, and microscopic (rather than macroscopic) analysis of scientific phenomena, far removed from natural settings. In short, science education has moved away from its roots. In a nationwide survey of teachers (Survey Research Center 2000), only about 10% had taken a course on environmental teaching methods and only 26% had any environmental science/ecology coursework at all. A survey conducted in 2000 revealed that of the preservice teacher education institutions responding, only 13% required a course in environmental education for preservice elementary education students (McKowen-Ice 2000). Another survey disclosed that after taking college science courses, many preservice elementary teachers admitted that they like science even less, and as a result they tend to teach less of any kind of science (Bayer Facts of Education Survey V 1999).

Findings from the *Environmental Attitude and Knowledge Roper Starch Survey* revealed that preservice science teachers who had at least one environmental education course, compared to those who did not, scored significantly higher levels of environmental literacy. The negative impact of not having integrated environmental education in college science classes is evident by the fact that the preservice science teachers who had taken no environmental education courses scored lower than the national sample (Robinson and Crowther 2001). A possible reason for this difference is that preservice teachers are not commonly exposed to the outdoors in the few lab-centered natural science courses they are required to complete. Noticeably missing from the curriculum are field-based courses to provide opportunities for developing a curiosity, awareness, and knowledge of the out-of-doors and interactions with natural systems.

Authentic contexts, such as environmental experiences for studying nature, can promote the engagement of students in inquiry. These courses help to develop science process skills and, in general, can promote a higher level of scientific literacy (National Research Council 1996). Other research specifically suggests that linking education and the environment can result in many positive measurable outcomes (Volk and McBeth 2005). The conceptual focus that uses the school's surroundings and community as a framework for learning has been called *Environment-Based Learning* (EBL) or using the *Environment as an Integrating Context* for learning (EIC) (State Education and Environment Roundtable 2008). Liberman and Hoody (1998) suggest that place-based education can improve student motivation, provide skills for lifelong learning, prepare students for later careers, and foster an attitude of respect and responsibility.

In a Wisconsin study (Archie 1998), the majority of preservice teachers stated they would never have taken a course in environmental education if it were not required. But after completion of the class, nearly 80% said that it had contributed as much or more to their education than any other course they had taken. Environmental education fosters a broader view of the world and the way it works. It could be a part of a comprehensive process to help people understand the environment, their place in it, and its related issues (Archie and McCrea 1998). The essential goal of environmental education is for people of all ages to know enough about environmental science and related social issues to make sound and well-reasoned environmental decisions (Roth 1992). Also, of course, knowing for the sake of knowing is not without merit. Consider the ancient idea of "listening to the land" and living in harmony with the land and each other

(Woodhouse and Knapp 2000). As stated in David Orr's book (2004), "open our souls to love this glorious, luxuriant, animated planet."

Using Geospatial Information Technologies (GIT) to Study the Environment

The recent report, *Learning to Think Spatially* (National Research Council 2006) states that geospatial technologies can support problem-solving in real-world contexts in K-12 education. GIT, of which Geographic Information Systems (GIS) is a part, has the potential to facilitate learning across a range of subjects, supporting interdisciplinary and multidisciplinary learning (p. 218). Since it is a tool for both scientific research and problem-solving, it provides a link between science and policy. The visualizations produced from scientific investigations that are linked to specific geographic contexts are often used to predict environmental outcomes. This makes GIT an important tool in environmental science and natural resource management. In fact, when students use GIT to address community projects, they are using the same tools that professionals use in their work.

GIT permeates our everyday lives as an integral part of international, national, regional, state, and local planning in many diverse areas and sectors including emergency services, utilities, transportation, communications, and natural resource planning. GIT is used for many societal tasks such as coordinating census data, routing buses, redistricting schools, and analyzing crime as well as for fire management and landslide warning systems. The power of GIT is in collecting, storing, manipulating, and displaying data referenced by a spatial or geographic component. Information is stored in layers linked to a map by geography. The user can change these maps or visual representations almost instantaneously. The layers may contain widely varying kinds of information that can be queried, combined, analyzed, and displayed to create new representations. Visual displays of locations and descriptions of data have proven valuable in solving multifaceted real-world problems such as those found in the environment. Some examples include studies of the prevention and control of invasive species, habitat preservation for endangered species, and monitoring and assessment of air and water quality.

GIT includes mobile technologies such as Global Positioning Systems (GPS) and maps and data that are often combined on a single handheld device. The internet has greatly improved access to data for the public. Google Earth, for example, displays data in an accessible online environment. Google Earth contains satellite images in an easy-to-use interface that allows an individual to zoom in and out and view physical features, vegetation, buildings, roads, and other geographic information. Google Earth is being used by NASA to show carbon dioxide distributions in the USA and by the National Oceanic and Atmospheric Administration (NOAA) to display marine data using a new application called Google Ocean. You can import your own GPS data, pictures, and movies into Google Earth. A new layer on Google Earth by David Tryse (Tryse 2009) shows the rates of deforestation across the world. You can use Google Earth to dive beneath the surface of the world's oceans to see coral reefs, trenches, and other marine features. You can view layers showing locations of shipwrecks, routes for ocean expeditions, and movements of GPS-tracked sea animals (including videos and images) from sources including the National Geographic Society, the Cousteau Society, the International Union for Conservation of Nature, the Monterey Bay Aquarium, and NOAA. Web-based applications such as Google Earth can display many types of environmentally related data. In an after-school science club, fourth-grade students used web-based GIS maps and Google Earth visualizations as effective tools to show how a pond was connected to other bodies of water (Bodzin 2008). Students were able to trace the flow of water on a map to better understand how the pond was a part of a greater watershed. GIT is truly a real-world technology with multiple applications in environmental education (EE).

Research on the use of geospatial technologies (GIT) in schools has shown that teachers and students are able to engage in data visualization and analysis, spatial interpretation, and real-world problem-solving (Alibrandi 1998; Audet and Paris 1996; Baker 2002; Kerski 2000; McWilliams and Rooney 1997; Stubbs et al. 2002). *Learning to Think Spatially* (National Research Council 2006) states that geospatial technologies (including GIS) meet four educational goals: (1) support the inquiry process; (2) be useful in solving problems in a wide range of real-world contexts; (3) facilitate learning across a range of school subjects; and (4) provide a rich, generative, inviting, and challenging problem-solving environment (p. 176). Additional research has further documented other important benefits of using GIS such as increased motivation (McWilliams and Rooney 1997), self-efficacy and attitudes toward technology (Baker 2002), acquisition of spatial analysis skills (Audet and Paris 1996), increased mathematics ability (Coulter 2003; Coulter and Polman 2004), and geographic and scientific content knowledge (Abraham 1998; Alibrandi 1998; Kerski 2003). Using these technologies in environmental education and science education brings the "world" to the learners' finger tips and enables students to better interpret what they see happening in the environment as they are experiencing it. In this way, GIT can enhance the outdoor experience, supporting data analysis, and problem-solving (Hagevik 2008; Hagevik et al. 2007).

Using GIT and Situated Learning to Promote EE in a Schoolyard

Kozma (2003) argues that when talking about situated learning, we need to take into account two systems, material systems and social systems. These systems have limitations and strengths, which are called affordances and constraints, that either enable or constrain the learning activity (Engestrom 1993; Kozma 2003; Lave and Wenger 1991). For instance, the affordances and constraints of material systems, in this case representations of data using maps, involve characteristics that permit or inhibit cognition (Vygotsky 1978). In the case of GIT, this includes visual features and relationships within and across multiple forms of data that represent the content. Such features can either inhibit or facilitate the process of knowledge construction.

Social systems include the "conventions of social practice, such as patterns of turn-taking in conversation, appropriate ways to interact conversationally when working together on a task, and the kinds of products that are expected or warrants of claims that are required in order to decide that a kind of task has been successfully accomplished or satisfactory results obtained in the community of practice" (Kozma 2003). In successful social systems, learning is characterized by one's ability "to become attuned to constraints and affordances of activity that results from interactions among people, and between people and their material and representational resources, as they engage in inquiry" (Kozma 2003, p. 206). Situated learning translates to teaching practice (Kozma 2003) when teachers provide learning opportunities for students that involve multiple representations of content (material systems) and the use collaborative learning activities (social systems). Learners develop an understanding of the content through these collaborative learning experiences.

An example of a collaborative learning activity that uses GIT to conduct an environmental inquiry in a schoolyard is the Mapping Our School Site (MOSS) project. The MOSS project (Hagevik 1999) uses the *Problem-Study* framework (Hagevik 2008) to define the environmental problem, generate conclusions based on evidences, and study and apply these conclusions to new understandings within the community and through community partnerships. MOSS begins with an ill-structured problem, *how do the living and nonliving components of the environment relate to each other*? Students select a 10 m × 10 m study site that has a variety of vegetation and possibly a water source. The site is measured, surveyed, and marked with stakes and string. Students use a data collection grid and a compass to spatially orient themselves. All points on the grid and associated data table are located by x, y coordinates shown in Fig. 1.

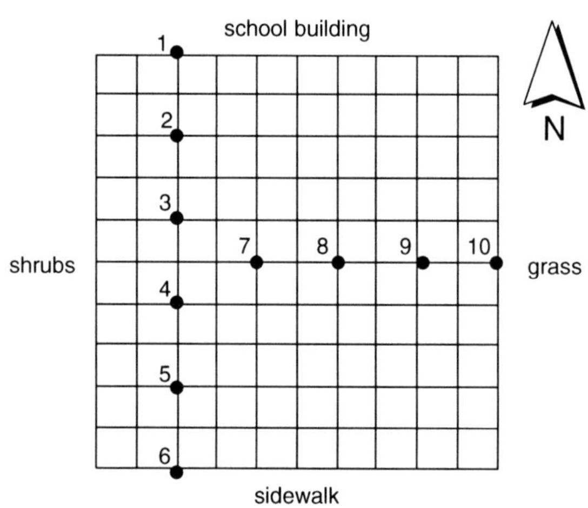

Fig. 1 10 m × 10 m grid. Points 1–10 represent pitfall traps

The grid is not georeferenced to real-world coordinates. However, it is possible to georeference the grid to an aerial photograph using a GPS unit. As a first step, students break up into smaller groups and collect data on five established environmental data collection protocols: vegetation/ground cover, animals, pitfall traps, trees, and abiotic. These protocols can be viewed in detail on the MOSS website (Hagevik 1999). Students use a Geographic Information Systems (GIS) software such as ArcGIS (ESRI 2005) desktop or MyWorld GIS to analyze spatial patterns and formulate a problem question as a group to investigate. Students use or modify additional scientific protocols, for example the GLOBE Program data collection protocols (UCAR/CSU 2004), to gather data to answer their problem question. After performing additional spatial analyses, students present their conclusions in a written report and oral presentation. Next, students further examine their data or other existing data to identify a participatory environmental activity. Examples include planting wildlife trees, making and placing bird boxes, performing bird counts, and reporting water quality data to the city (Hagevik 2007; Harte 2002).

The MOSS project has been used in preservice teacher courses at the University of Tennessee at the University Garden. As a result, students adopted and now sponsor a multicache (www.geocache.com) at the Gardens called *Bloomin' Gardens*. Figure 2 is an example of a map created by the preservice science teachers as a

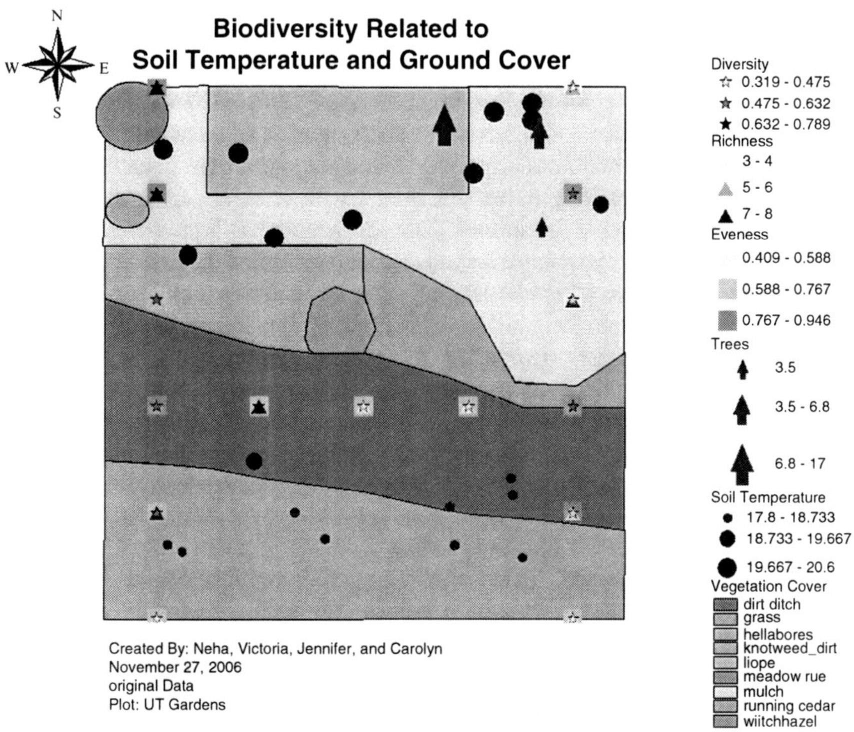

Fig. 2 Preservice teachers' MOSS analysis map

conclusion to their problem question, How is the abundance and variety of animals related to soil temperature and ground cover?

The MOSS project involves an outside data collection procedure (the study site) and an inside data analysis procedure on a computer. Students negotiate meaning as they discuss the data and results and compare it to past student data. They use multiple representations of data (the GIS maps, digital photographs, and graphs) and discourse throughout the collaborative learning activities to develop a shared understanding about the scientific processes that are represented in their study site. Preservice teachers use the MOSS unit to make meaning of how EE fits into the science curriculum by using the NAAEE *Guidelines for Excellence Workbook* (North American Association for Environmental Education 2000) and National and State Science Standards to analyze the curriculum. It is through these types of discourses that preservice teachers understand how environmental education can become an important part of their future science curriculum.

Incorporating GIT into Preservice Field Experience Courses

Field experience courses are another way to incorporate environmental education using GIT and nature study into the preservice science teacher curricula. For years, the authors have offered a variety of science education courses in the natural environment. These courses are taught at various locales including the North Carolina mountains at Grandfather Mountain, an International Biosphere Reserve; a wilderness barrier island, Ossabaw Island off the coast of Georgia; or in Brazil (Stubbs 2009). These courses usually include preservice and inservice teachers, nonformal educators, and others, studying with scientists and from across the United States and other countries.

In these courses, teaching duties are shared with scientists from the fields of ecology, biology, and other disciplines. This partnership has been extremely effective and demonstrates how science and education may bridge the gap between what Duggan-Hass (2004) has effectively described as two different cultures. During the courses, the students make extensive collections of specimens, use field guides, engage in nature journaling (Leslie 2003), and plan environmental-based lessons for their future K-12 students. Experiments are designed on site. It should be noted that most of the students who enroll in these courses have never conducted an extended scientific study outdoors or even thought about how the components of the environment relate to one another. However, preservice students quickly adapt to the course requirements and most report a positive experience that influences their life and their potential teaching careers.

The mountain course takes place at and around Grandfather Mountain, located in Linville, NC. Grandfather Mountain became North Carolina's 34th State Park in the Fall of 2008. At nearly 6,000 ft and approximately 2,600 acres of wilderness, it is the only privately owned United Nations International Biosphere Reserve. Seventy-three rare and endangered species, including certain salamanders, the Carolina northern flying squirrel and the Blue Ridge goldenrod inhabit Grandfather Mountain

(Mason September 29 2008). During the course, housing, dining, computer facilities for technology instruction, and evening activities are held at Lees-McRae College in Banner Elk, NC. Students visit Grandfather Mountain and other local sites such as Roan Mountain during the week-long experience.

Students enrolled in a graduate course at University of Tennessee, Knoxville (UTK), attend one pretrip meeting on campus, are in residence for 5 days, and finally share their experiences and projects at one posttrip meeting on campus after the trip. The nature experiment for this course is the MOSS project, which is completed in groups in a mountain meadow at a field site. Students use GPS units and create photo journals of their experiences incorporated along with GIS maps in their mountaintop nature projects. Nature journaling is an important part of the mountain experience. Examples of mountain top nature projects include "Formation of the Appalachian Mountains," "Adaptations of Living Things," and "Plate Tectonics, Mountains and Mapping."

Ossabaw Island, located near Savannah, Georgia, is home to maritime forests, hammocks, rivers, swamps, animals of these habitats, and an ocean beach of 9.5 miles. While on the island the UTK preservice teachers reside at the only camping site on the island. Ossabaw is a 26,000-acre barrier island, one of Georgia's "Golden Isles" and Georgia's first Heritage Preserve, the strongest level of protection offered by the state. Ossabaw Island is maintained as a wilderness preserve, to be used solely for natural, scientific, and cultural study, research and education, and environmentally sound preservation of the island's ecosystem. The former owner of the island, Mrs. Eleanor West, directed that primarily educational groups could visit the island. She feels that her island, accessible only by boat, is an ideal site for education, inspiration, and the development of creativity and community. Because the island is so remote, only drinking water, water for bathing (no bathrooms or electricity), a covered pavilion with picnic tables, and a primitive palm-surrounded "facility" are provided during the course. The focus of the course is the flora, fauna, and barrier island processes.

The UTK students in the course prepare for the 4-day camping trip through three pretrip classes. The course concludes with a sharing time during a posttrip class on campus. The students analyze relationships between the different environments on the island. GIS maps are used to visualize the physical features of the island and surrounding landscapes. Students complete the MOSS project while on the island. GPS units and digital photography are used together with nature journaling to gain a deeper appreciation of nature through an in-depth nature study. All students submit a final project called a "nature improvement" unit. These units focus on wetlands and include students' nature photographs, nature collections, and at least one mapping activity. Students can use GIT as a possible tool when designing, conducting, or displaying the results of their experiments completed on the island, but this is not a requirement. Some examples of experiments completed by students on the island include "Diversity of Organisms Caught by Seining in the Ossabaw South Beach Slough versus the Shoreline at Low and High Tide," "The Effects of Sun on the Activity of Cone Ants," and "Slough Discharge and its Effects on Crabs." Some examples of nature improvement units include "Waves and Their Properties," "Chemistry of Wetlands," and "Barrier Island Ecology."[1]

In Brazil, preservice and inservice teachers from the United States are joined by teachers from Brazil for 2 weeks. Through experiences such as visits to the Atlantic Rain Forest and the Pantanal, teachers learn about a few Brazilian ecosystems. They also visit Sao Paulo to study urban problems. In addition, teachers visit private and public schools in Brazil and discuss educational practices in the country. Brazilian teachers are using GIT with their students to study environmental problems such as water quality. A science education professor at the Federal University of Paraná in Curitiba, who has been involved in the Grandfather Mountain course and the Brazilian course, has been assisting teachers in Brazil with their GIT projects. She has recently written two books for inservice teacher education using GIS.

The first book, called *Primary School Science Evaluation* (Gioppo and Barra 2005), was developed by preservice science teachers in her courses and connects field, lab, and GIT. The lessons were validated in schools. An example of one of these lessons is analyzing animals at the beach. A 50 × 50 × 50 cm hole is dug every 5 m from the water to the sand dunes. The sand from each hole is put in a sieve, washed, and screened. All the animals in the holes are sorted and tabulated. Then an analysis is done using maps that compare animals that prefer wet to dry environments. The second book, called *Middle School Science Evaluation* (Gioppo et al. 2006), uses activities developed by a graduate student on dengue fever mosquitoes. First, students complete an ant activity using a site map. Then, GIS maps of municipalities, roads, climate, and population are used to decide how state money should be allocated to prevent the disease. Finally, maps illustrating the spread of the dengue fever in Brazil from 1993 to 2003 are used to guide students in creating and answering their own questions related to the disease.[2]

In summary, the field experience courses are based on the relating of living to nonliving components of the environment. The courses emphasize ways to observe nature using nature journaling, virtual or actual collections, and on-site investigations. In each course, there is an explicit connection of nature study to the science classroom. Over the years, teachers have stated that the benefits of these courses include learning together in a community in the out-of-doors, experiencing creating actual or virtual collections and learning how to use them in their classrooms, gaining an increased awareness through nature journaling, and learning how to use inquiry in the science classroom (Hagevik et al. in press).

These field experience courses have evolved over the past 12 years and GIT has been included for six of those years. GIT became an integral part of these courses because it is prevalent in our society, easy to use, and important to our understanding of the relationships in the environment. The instructors and students in these

[1]For more information about the Grandfather Mountain and Ossabaw Island course and to see examples of student projects go to http://web.utk.edu/~start8/506.

[2]For more information about Brazil and Grandfather Mountain workshops go to http://www.ncsu.edu/scilink.

courses form collaborative learning communities to investigate the natural world (Anderson and Stubbs 1993). As students become immersed in the natural world, they become inquisitive and want to investigate their own questions. GIT helps students learn about the natural world as they are able to visually see relationships in the environment as they investigate their questions. It is through discourse, collaboration, and shared experiences that the teachers apply what they have learned to their own science classrooms.

Summary

Many teacher educators become science teachers owing to a love of the subject resulting from an early exposure to outdoor settings. We, the authors, are prime examples. We learned to love and appreciate the out-of-doors because we were taken camping, or were raised on farms, or had at least one person who, as the great environmentalist Rachael Carson said, "…encouraged them to wonder about natural phenomena and introduced them to the love and importance of nature and natural settings."

Science education and environmental education must promote new partnerships and take advantage of curricula that use learning technologies such as GIT. We must provide teachers and students with awareness, knowledge, and skills to utilize the natural world that is literally "in their own backyards." We must enable educators to introduce a new generation of children to nature study and nature appreciation. It is imperative that our present and future generations become aware of and appreciate first their local environments, leading to a broader knowledge of the importance of ecosystem interactions on a global scale. Preparation of preservice and inservice teachers must include learning tools to help analyze the environment. With such tools, they are equipped to educate students who are both environmentally and scientifically literate.

References

Abraham, M. R. (1998). The learning cycle approach as a strategy for instruction in science. In B. J. Fraser & K. G. Tobin (Eds.), *International handbook of science education* (pp. 513–524). Dordrecht, Netherlands: Kluwer.

Alibrandi, M. (1998). GIS as a tool in interdisciplinary environmental studies: Student, teacher, and community perspectives. *Meridian, 1*(2), 1–10.

American Association for the Advancement of Science: Project 2061. (1990). *Science for all Americans*. New York: Oxford University Press.

Anderson, N., & Stubbs, H. S. (1993). SCI-LINK: An innovative project linking research scientists and science teachers. *Journal of Science Teacher Education, 4*(2), 44–50.

Archie, M., & McCrea, E. (1998). *NAAEE: Environmental education in the United States - past, present, and future*. Troy, OH: North American Association for Environmental Education.

Audet, R. H., & Paris, J. (1996). GIS implementation model for schools: Assessing the critical concerns. *Journal of Geography, 96*(6), 293–300.

Baker, T. R. (2002). *The effects of Geographic Information System (GIS) technologies on students' attitudes, self-efficacy, and achievement in middle school science classrooms*. PhD dissertation, University of Kansas, Lawrence, KA.

Bayer Facts of Education Survey V. (1999). *Nation's science teacher register concern over U.S. science education in new survey* [Electronic Version]. Retrieved August 13, 2008, from www.bayerus.com/msms/news/facts.cfm?mode=detail&id=survey99

Bodzin, A. M. (2008). Integrating instructional technologies in a local watershed investigation with urban elementary learners. *Journal of Environmental Education, 39*(2), 47–57.

Comstock, A. (1911). *Handbook of nature study*. Ithaca, NY: Comstock Publishing.

Coulter, B. (2003). *Mapping neighborhood trees to enhance math and science skills*. Paper presented at the National Urban Forest Conference. Retrieved April 2, 2009, from http://www.umsl.edu/~polmanj/papers/coulter-polman-aera04.pdf

Coulter, B., & Polman, J. L. (2004). *Enacting technology-supported inquiry learning through mapping our environment*. Paper presented at the American Educational Research Association. Retrieved April 2, 2009, from http://www.umsl.edu/~polmanj/papers/coulter-polman-aera04.pdf

Dewey, J. (1915). *The school and society*. Chicago: The University of Chicago Press.

Duggan-Hass, D. (2004). *From ecosphere to edusphere: Problems in (mis)managing ecosystems and edusystems*. Paper presented at the Complexity Science and Educational Research Caffey's Locks, Canada.

Engestrom, Y. (1993). Developmental studies on work as a testbench of activity theory. In S. Chaicklin & J. Lave (Eds.), *Understanding practice*. Cambridge, UK: Cambridge University Press.

ESRI. *ArcGIS* (Version 9.1). (2005). Redlands, CA: Environmental Systems Research Institute.

Gioppo, C., & Barra, V. M. (2005). *Primary school evaluation*. Curitiba, Brazil: UFPR/SEB-MEC.

Gioppo, C., da Silva, R. V., & Barra, V. M. (2006). *Middle school science evaluation*. Curitiba, Brazil: UFPR/SEB-MEC.

Hagevik, R. A. (1999). *Mapping Our School Site (MOSS)*. Retrieved 2003, from www.ncsu.edu/scilink/studysite

Hagevik, R. A. (2007). *GIS goes to school multiple paper set: Mapping our school site*. Paper presented at the National Urban and Regional Information Systems Association.

Hagevik, R. A. (2008). Facilitating scientific inquiry using CITYgreen and the problem-study framework. *Meridian Middle Schools Technology Journal, 11*(1).

Hagevik, R. A., Hales, D., & Harrell, J. (2007). GIS live and web problem-solving. *Meridian Middle Schools Technology Journal, 10*(2).

Harte, A. (2002). Taking a measure of community. *American Forests, Summer, 108*(2), 7–9.

Kerski, J. J. (2000). *The implementation and effectiveness of geographic information systems technology and methods in secondary education*. PhD dissertation, University of Colorado, Boulder, CO.

Kerski, J. J. (2003). The implementation and effectiveness of geographic information systems technology and methods in secondary education. *Journal of Geography, 102*(3), 128–137.

Kozma, R. (2003). The material features of multiple representations and their cognitive and social affordances for science understanding. *Learning and Instruction, 13*(2), 205–226.

Lave, J., & Wenger, E. (1991). *Situated learning: Legitimate peripheral participation*. New York: Cambridge University Press.

Leslie, C. W. (2003). *Nature journal: A guided journal for illustrating and recording your observations of the natural world*. North Adams, MA: Storey Publishing.

Liberman, G. A., & Hoody, L. L. (1998). *Closing the achievement gap: Using the environment as an integrated context for learning*. San Diego, CA: State Education and Environmental Roundtable (SEER).

Mason, S. (2008, September 29). *Grandfather Mountain becomes 34th state park*. WRAL.com.

McKowen-Ice, R. (2000). Environmental education in the United States: A survey of preservice teacher education programs. *The Journal of Environmental Education, 32*(1), 4–11.

McWilliams, H., & Rooney, P. (1997). *Mapping our city: Learning to use spatial data in the middle school science classroom.* Paper presented at the American Educational Research Association, Chicago.

National Research Council. (1996). *National science education standards.* Washington, DC: National Academy Press.

National Research Council. (2006). *Learning to think spatially.* Washington, DC: The National Academies Press.

North American Association for Environmental Education. (2000). *Environmental education materials: Guidelines for excellence workbook.* Rock Srping, GA: NAAEE.

Orr, D. W. (2004). *Earth in mind.* Washington, DC: First Island Press.

Robinson, M., & Crowther, D. (2001). Environmental science literacy in science education, biology, and chemistry majors. *The American Biology Teacher, 63*(1), 9–14.

Roth, C. (1992). *Environmental literacy: Its roots, evolution and direction in the 1990s.* Columbus, OH: ERIC/CMME, Ohio State University.

State Education and Environment Roundtable. (2008). The EIC model. Retrieved April 6, 2008, from http://www.seer.org/

Stubbs, H. S. (2009). Using technology to develop global teachers: An innovative model. *Meridian: A Middle School Computer Technologies Journal, 12*(1).

Stubbs, H., Devine, H., & Hagevik, R. (2002, July). *Thinking spatially: Geographic Information Systems (GIS) curricula K-16 and professional development for educators.* Paper presented at the 10th International Symposium, Sustainable Development in a Changing and Diverse World, IOSTE Conference, Iguasu Falls, Brazil.

Survey Research Center. (2000). *Environmental studies in the K-12 classroom: A teacher's view.* Washington, DC: North American Association for Environmental Education and Literacy Council.

Tryse, D. (2009). *Google earth files.* Retrieved April 13, 2009, from http://david.tryse.net/googleearth/

UCAR/CSU. (2004). *The GLOBE program.* Retrieved 2004, from http://www.globe.gov

Volk, T. L., & McBeth, W. (2005). Environmental literacy in the United States. In H. R. Hungerford, W. J. Bluhm, T. L. Volk, & J. M. Ramsey (Eds.), *Essential readings in environmental education* (3rd ed.). Champaign, IL: Stipes Publishing.

Vygotsky, L. (1978). *Mind in society: The development of higher psychological processes.* Cambridge, MA: Harvard University Press.

White, R. (2005). *Interaction with nature during the middle years: Its importance in children's development and nature's future* [Electronic Version]. *White Hutchinson Leisure and Learning Group* from http://www.whitehutchinson.com/children/articles/nature.shtml

Woodhouse, J. L., & Knapp, C. E. (2000). *Place-based curriculum and instruction: Outdoor and environmental education approaches* [Electronic Version]. ERIC Clearinghouse on Rural Education and Small Schools.

Using Podcasting to Address Nature-Deficit Disorder

Beth Shiner Klein and Starlin Weaver

Introduction

As teacher educators, we are faced with issues associated with the digital native student population and their disconnection or alienation from nature. To address this, we developed projects at both the elementary and secondary preservice levels aimed at using technology to encourage connections with nature.

Who is the Digital Native Student?

Who are the preservice teachers we are preparing in our methods courses and who are the students they will be teaching? According to Pattengale (2008), current college students have been dubbed the first wave of "digital native" students. They grew up in a technology-rich environment. Computers and other technologies have always been a part of their lives. They are fascinated by technology. This fascination entices them to be connected virtually all the time. They use email, instant messaging, text messaging, internet-linked gaming systems, social networking web sites such as *Facebook* and *MySpace,* and iPods to stay plugged in. It seems like a natural step to use these technologies, such as podcasting, as a tool for instruction in order to connect with the digital native student.

B.S. Klein (✉)
School of Education, SUNY Cortland, P.O. Box 2000, Cortland, NY 13045, USA
e-mail: Beth.Klein@cortland.edu

S.D. Weaver
Seidel School of Education and Professional Studies, Department of Education Specialties, Salisbury University, Salisbury, MD 21801, USA
e-mail: sdweaver@salisbury.edu

A.M. Bodzin et al. (eds.), *The Inclusion of Environmental Education in Science Teacher Education*, DOI 10.1007/978-90-481-9222-9_21,

What is Podcasting?

The term podcast is a combination of the words iPod and broadcast. In the purest sense of the definition, podcasts are a means of publishing audio, video, and graphics or a combination of these to the internet to allow individuals to "subscribe" and automatically receive updated content. There are three basic types of podcasts. An audio podcast contains audio only content. An enhanced podcast contains audio plus a photo/image slideshow, and a video podcast or vodcast contains a movie. Podcasts are delivered on the Internet via Really Simple Syndication Feed (RSS Feed) that allows the subscribers to receive new episodes of content. Apple, Inc created iTunes, which is a free digital media management application that allows for easy playing and sharing of podcasts. There are various podcast development tools available for both the Macintosh and Windows platforms. Two methods of developing podcasts are described later in this chapter.

A multitude of examples of podcast classroom usage at all instructional levels is available with even a cursory internet search; for example, Mrs. Brackett's third-grade class's enhanced podcast talks about owl pellet dissections at Walnut Park Elementary (2009). At Willowdale Elementary School in Omaha Nebraska, fourth-graders created audio podcasts where the students presented their research on the Endangered Species Act and discussed specific endangered animals (2006).

Why Should Teachers Use Podcasting in the Classroom?

A source of guidance for technology literacy for both teachers and students is provided by International Society for Technology in Education (ISTE). ISTE has two sets of technology standards: the National Educational Technology Standards for Teachers (NETS-T) (ISTE 2007b) and the National Educational Technology Standards for Students (NETS-S) (ISTE 2007a). The NETS-T includes several standards that are addressed by using podcast development in teacher preparation. *Standard 1: Facilitate and Inspire Student Learning and Creativity* states that *teachers use their knowledge of the subject matter, teaching and learning, and technology to facilitate experiences that advance students learning, creativity, and innovation in both face-to-face and virtual environments*. The construction of podcasts by preservice teachers clearly meets this standard. *Standard 2: Design and Develop Digital-Age Learning Experiences and Assessments* is also addressed by the creation of podcasts by preservice teachers. This standard states *that teachers design, develop, and evaluate authentic learning experiences and assessments incorporating contemporary tools and resources to maximize content learning in context*. By using iPods, GarageBand, Audacity, and iTunes in curricular contexts, teachers address this standard. *Standard 3: Model Digital-Age Work and Learning*. For this standard, teachers should be able *to model knowledge, skills, and work processes that are illustrative of an innovative professional in the twenty-first Century*. By developing their own podcasts to use with students, a preservice teacher models the use of current digital-age tools to communicate information. *Standard 4: Promote and Model Digital Citizenship and Responsibilities* states that *teachers understand local and global societal issues and responsibilities in an evolving digital culture and exhibit legal and ethical behavior in their professional practices.* When learning to construct the podcasts, preservice teachers should understand copyright laws. We advocate that preservice teachers use copyright free or original music and graphics in the podcasts they develop.

In addition to meeting literacy standards, podcasting is a learning tool that motivates digital age learners. According to Michael Rappa ("Podcasting" 2008), "when students create a podcast for class, they not only learn the content in a creative way, they learn twenty-first Century communication skills at the same time" (p. 8). Bob Sprankle (2006) writes that his elementary students remarked that they enjoy school more now that they use podcasts in the classroom. Student comments included, "It didn't feel like we're going to school" and "We're creating our own learning" (Sprankle 2006, p. 62). Incorporating technologies such as podcasting can provide necessary motivational contexts to support content area learning for elementary, secondary, and postsecondary students.

What Do Studies Say About Environmental Education?

As other chapters in this book have indicated, in order to have an environmentally literate citizenry equipped to meet sustainability challenges, we must find ways to integrate environmental education into the curriculum. As teacher educators, we also need to develop a cadre of teachers who can support children's inquiry and learning about these environmental issues and sustainability challenges.

Public support for environmental education (EE) in school curricula has been well established (NEETF/Roper 2001). Research has indicated that schools that incorporated EE have improved standardized test scores (Lieberman and Hoody 1998; NEETF/Roper 2001; Ernst and Monroe 2004; SEERS 2005). In addition, these studies reported that students had gains in science, math, literacy, problem-solving, and critical thinking skills and improved in overall grade point average. Despite these findings, there has been little progress in incorporating EE into K-12 school curricula (Ramsey, Hungerford, and Volk 2001).

What Do Studies Say About Teacher Preparation in Environmental Education?

One reason for the lack of progress incorporating environmental education into PreK-12 school curricula is the limited preparation of teachers in environmental education. Most preservice teacher education programs do not incorporate environmental education into their programs (Rakow 1985; McKeown-Ice 2000).

Observed barriers to including environmental education in preservice teacher education programs include the inflexible structure of preservice teacher education courses and the lack of time in the typical teacher education curriculum (Scott 1996; McKeown-Ice 2000).

Hart (2003) found that teachers committed to teaching environmental education have a "sense of nature" and "connection" to the natural world as well as "an intellectual foundation that is highly experiential in nature" (p. 199). However, the current preservice population is made up of a majority of students who are included in the digital native population. Unfortunately, research indicates that these individuals are increasingly disconnected from the natural world (Clements 2004; Hofferth and Curtin 2006; Louv 2005; Wridt 2004). This disconnection from the natural world has become known as Nature–Deficit Disorder (Louv 2005). We, as well as Louv, emphasize that this is not a clinical term. However, Louv used this term to bring attention to the issue. In addition, Louv's book provided information about the consequences of this disconnection with nature that may include health issues, psychological issues, and a lack of desire to address environmental issues.

Chawla (2006) discovered that individuals need a connection with the natural world in order to seek to protect it. Adults who had developed a relationship with the natural world as children are found to be more likely to care for the land they live on and near, disapprove of destructive environmental practices, and experience "simple pleasure at being out in nature and a fascination with the details of other living elements of the earth and sky" (Chawla, p.72). This effect translates into adult actions that range from making career choices in the sciences or environmental protection fields, to taking active political measures when aspects of the environment are thought to be at risk.

Hart (2003) found that many of the teachers committed to teaching EE had this connection with the natural world. Elements of their professional preparation that helped them develop this connection, the commitment, and skills to incorporate environmental education into their teaching included field trips or other similar activities that involved actually being in the "out-of-doors in natural or rural settings or in open urban spaces" (Hart, p. 199). Preservice teachers who have not had these experiences, or have not developed connections to the natural world, will be less likely to provide such experiences for their future students. This suggests the importance of involving preservice teachers in outdoor activities that help them explore EE in natural settings as a significant part of their professional preparation. Without direct outdoor experiences in natural settings, preservice teachers will likely have difficulty in creating meaningful outdoor learning experiences for their students.

Description of the Project

One issue this teaching implementation sought to address was that of curricular time constraints for preservice programs. To address this concern, the authors combined technology-enhanced instruction with science learning that encouraged elementary and middle-school students and teachers to engage and explore outdoor environments.

Preservice teachers at two institutions were introduced to podcasts developed by science experts and other reliable educational resources. For example, National Oceanic and Atmospheric Administration (NOAA), National Geographic Society (NGS), and National Aeronautics and Space Administration (NASA) all provide podcasted content the preservice teachers could use in their future classrooms. Preservice teachers were also provided instruction for the planning and production of podcasts.

To learn more about *Nature-Deficit Disorder*, preservice teachers researched and discussed the ideas presented in Richard Louv's book, *Last Child in the Woods: Saving our Children from Nature-Deficit Disorder* (2005). These preservice teachers then created a high-interest podcast to excite and encourage elementary or middle-school students to explore the outdoors using science inquiry techniques on a future field trip.

Implementation at SUNY Cortland: Elementary Preservice Level

For several years, I (first author) have been involved with teaching elementary preservice teachers in the Environmental Thematic Methods Block (evTMB) learning community at SUNY Cortland. These evTMB preservice teachers have been exposed to opportunities to learn and teach in the outdoors including a 3-day residential experience at an outdoor education facility. This experience exposes them, depending on the season, to different outdoor education activities including canoeing, kayaking, challenge course, hiking, snowshoeing, and cross-country skiing. The preservice teachers are also introduced to team-building activities, geocaching, natural history activities, and regional New York State history.

I have developed partnerships with environmental educators and state foresters from the New York State Department of Environmental Conservation to provide the evTMB preservice teachers with environmental education experiences through *Project* curricula workshops including *Project Learning Tree* (2008) and *Project Wild* (Council for Environmental Education 2008a). These workshops take advantage of the outdoor areas on the SUNY Cortland campus. For more details about these *Project* curricula, see Chapter 3.

As part of the podcast project, I began by introducing my preservice teachers to environmental education research and literature. As part of this introduction, they were oriented to the concepts and ideas presented in Richard Louv's book, *Last Child in the Woods: Saving Children from Nature Deficit Disorder* (2005). In addition, I indentify the health, academic, and spiritual benefits for children of having environmental education experiences in the outdoors (Chawla 2006; Clements 2004; Lieberman and Hoody 1998; Lord 2008; Wells 2000).

I also take my elementary preservice teachers on a field trip to Lime Hollow Center for Environment and Culture. Located in proximity to the SUNY Cortland campus, this facility provides EE opportunities for the community and the surrounding school districts. While at the nature center, preservice teachers explore trails, ponds, a unique bog area, and multiple cultural and geologic features of the 375-acre site. Preservice teachers are also exposed to the different teaching tools that are available at the center including bug boxes, sampling nets, hand lenses, and stream survey charts.

After providing my students with their own set of EE experiences, the preservice teachers have the background to plan and implement their own EE activities for local elementary school students. The preservice teachers are assigned a specific age group, either first-grade or sixth-grade. The teachers from collaborating school districts suggest content topics that fit within their curriculum. The preservice teachers work in small groups and use their *Project* curricula guides, additional environmental education curricula, and internet resources to develop their lessons. The minimal costs for the field trip to Lime Hollow include bus transportation and facility use fees. For the last several years, these have been covered by a partnership grant.

As noted, my preservice teachers create lessons on a variety of topics. For example, one group focused on the theme of ecosystems and developed a set of activities that examined the bog, forest, and field ecosystems. In addition to my own observation and written feedback, faculty from the participating schools and colleagues from the university also observe the preservice teachers and provide feedback on lesson development and implementation.

This outdoor teaching and learning activity had been ongoing for a few years after I attended a podcast professional development workshop (Zembal-Saul et al. 2006). I had been looking for some way to expand the field trip experience for my students so it was not a "one shot" experience for the elementary students. The approach presented at this workshop seemed to offer some promise.

This workshop occurred at the same time that my institution became an *Apple iTunes University* (iTunesU) partner and I was asked to be part of the pilot team of faculty to use podcasting with students. This *iTunesU* partnership enabled me to obtain resources at my institution through *Apple Computer* to carry out this project. At that point, podcasting, as an educational tool, was still relatively new.

To prepare the preservice teachers for the project, I presented them with background on podcasting and directions for using *iTunesU*. We used the *GarageBand* software, which is part of the *iLife* suite of software that comes on every Macintosh computer. *GarageBand* had just added a podcast creation feature that was a simple way to develop a podcast. I also used the tutorial videos and documentation available free from *Apple Computer* via their web site to assist my students in learning the technical skills needed to create the podcasts. The preservice teachers, who all had access to our class *iTunesU* site, uploaded their podcasts. After review (and revision if necessary), I moved the podcasts to the public community area of our *iTunesU* site to be accessible to the local school partners. The first podcasts were audio only; later podcasts have been *enhanced* podcasts, including photos and short video clips along with the music, sound effects, and voice-overs. The content for the podcasts included background information on the fieldtrip site facility, the various concepts of the lesson, field trip expectations, and a personal introduction of the preservice teachers to the elementary students. This last point is important to create a personal relationship between the preservice teachers and the elementary students.

A recent extension of this has been to use podcasting as a means for elementary student assessment. I went into the school with a set of laptops and worked with a small group of sixth-grade elementary students to teach them how to develop a podcast. These students were tasked with creating a podcast that presented what they had learned about the bog, forest, and field ecosystems from the preservice teachers. The sixth-graders developed an anonymous survey instrument and designed a set of questions that they used to interview their peers. These podcasts were then presented to the preservice teachers as a means for assessing their teaching.

The podcast development has had a most interesting effect on my preservice teachers. I have found, as have the classroom teachers from the partner districts, that since the inclusion of the podcast component, the preservice teachers are much better prepared for the teaching at the nature center. The development of the podcasts requires them to study and better understand the environmental science content. Because of their often limited experiences with teaching, and especially their unfamiliarity with learning experiences in an outdoor setting prior to this coursework, this assignment often creates a high level of anxiety. Since adding the podcast component, I have seen much of this anxiety replaced by more enthusiasm for this project. As an example of their enthusiasm, preservice teachers suggested that they would develop follow-up podcasts (on their own time) to send to the elementary students to reinforce the concepts they learned on the field trip and included photos of the students conducting the learning activities. This last point clearly indicates the enhanced personal connections among the project participants.

Implementation at Salisbury University: Middle/Secondary Preservice Level

Salisbury University's location offers numerous sites for exceptional field trips. The University is located between the Chesapeake Bay and the Atlantic Ocean on the eastern shore of Maryland. There are many wildlife preserves and refuges, state and national parks, and local wetlands available for the purpose of education. As a science teacher educator dedicated to environmental education, I (second author) was disappointed that few inservice teachers in the local schools utilize the wealth of environmental education sites that are available to them. This prompted me to create a field trip project to instill in secondary preservice teachers an awareness of the importance of taking students outside in order to help them understand, appreciate, and care for the world in which they live.

Like SUNY Cortland's elementary preservice teachers, Salisbury University's secondary preservice teachers are provided with inservice in four of the national EE PreK-12 *Project* curricula: Project Wild (Council for Environmental Education 2008a), Project Wild Aquatic (Council for Environmental Education 2008b), Project WET (1995), and Project Learning Tree (2008). These *Project* workshops are facilitated by university faculty, cooperative extension agents, and other wildlife management professionals from the local area. Workshops are held at local parks and other outdoor venues. For example, one venue is a demonstration forest, which also provides preservice teachers with future field trip locales. Richard Louv's *Last Child in the Woods: Saving Children from Nature Deficit Disorder* (2005) is also introduced and discussed as a rationale for doing a comprehensive field trip project as a major performance-based assessment in the course.

This comprehensive project was also designed to meet *Standard 7 – Science in the Community* of the National Science Teachers Association (NSTA) Standards for Science Teacher Preparation (2003). It involves planning a comprehensive field trip with introduction and summary lessons. This standard also involves planning all activities that middle/secondary students could do while visiting the field trip site. Preservice teachers select a local outdoor area, visit the site, and plan a full-day potential field trip for students. Unfortunately, a lack of local school funding sometimes prevents the preservice teachers from actually conducting the field trips with middle and secondary students during their preservice preparation.

As a requirement of the project, preservice teachers must visit the site and meet with educators who might be on the staff at the field trip location. They must also design a full day of activities and instruction using the site. Although the assessed project is only a plan, many preservice teachers have actually taken students on their field trip as a part of their student teaching internship. To address issues of nature-deficit disorder and the needs of digital native students, I have also incorporated podcasting into this project. Construction of the podcast supports *Standard 5- General Skills of Teaching* of the NSTA Standards for Science Teacher Preparation (2003) that includes using technological tools to facilitate the learning of science. In this assignment, preservice teachers decide where and how to use a podcast in the project. In preservice teacher projects, podcasts have been used to introduce the field trip site to students prior to visiting and to provide students important safety and background information once they arrive at the site.

To acquire the technology skills necessary to create the podcast, preservice teachers receive instruction in *Audacity*, a free downloadable open source audio editor and recorder. This introduction takes place as a part of the instructional technology course that is offered in conjunction with my methods course. Some of my more motivated students have also used the *GarageBand* software on personal Macintosh computers to produce enhanced and video podcasts. All computer labs on campus contain Windows PCs, not Apple Macintosh computers. Therefore, *GarageBand* is not readily available for preservice teachers to use for course assignments. Salisbury University is, as of Fall 2008, an iTunesU partner, so publishing the podcasts on campus is now available.

I have observed interns leading field trips and using their podcasts. On a field trip to Pemberton Historical Park, a diverse outdoor site, I watched as middle-school students eagerly received their iPod shuffles and listened intently to a podcast designed to introduce them to the history of the park and also provided them with important safety rules and guidelines. As students were listening, the teacher was able to leave the bus and speak with park staff and get ready for the rest of the day's activities. This example demonstrates the utility and success of the implementation. By having technology take care of routine "housekeeping" activities that provided important content background information, logistical planning time for the teachers was taken care of while the students were engaged with listening to the podcast.

Conclusion

As this chapter explains, there is evidence to support the incorporation of outdoor learning experiences in teacher preparation. While field trips for gaining science and social science content are encouraged when they engage the students in active learning (Knapp 2007), learning is only maximized when the content of the field trips are clearly connected to the curriculum (Noel 2007: Ramey-Gassert 1997). "Teachers can support a deeper level of learning from a field trip by implementing preparatory lessons that result in further development or 'construction' of knowledge in the content of the field trip, as well as assist students in organizing new and existing content" (Noel 2007, p. 43). Podcasts, developed by both elementary and secondary preservice teachers, can provide some of this preparatory knowledge in an engaging, and, to the students, a "native" format. In other words, this approach uses a medium that the digital native students, both preservice teachers and the preK-12 students, are comfortable with and interested in using.

Louv (2005) outlines in his book the potential problems with having a population mostly disconnected from nature. These include health issues, psychological issues, and a generation of individuals who do not see nature as something they are part of, and may see no need to be a caretaker of the natural world. This has dire consequences for our future. This project is one way to use something the students are familiar with (digital media) to reconnect them to the natural world. The hope is that once they've had some positive experiences interacting with the natural world, they will no longer need technology to be the motivator or the go-between to get them outside to experience inquiry learning.

References

Chawla, L. (2006). Learning to love the natural world enough to protect it. *The Journal Barn, 2*, 57–78.

Clements, R. (2004). An investigation of the state of outdoor play. *Contemporary Issues in Early Childhood, 5*(1), 68–80.

Council for Environmental Education. (2008a). *Project Wild K-12 Curriculum and Activity Guide*. Houston, TX: Council for Environmental Education.

Council for Environmental Education. (2008b). *Project Wild Aquatic Curriculum and Activity Guide*. Houston, TX: Council for Environmental Education.

Ernst, J., & Monroe, M. (2004). The effects of environment-based education on students' critical thinking skills and disposition toward critical thinking. *Environmental Education Research, 10*(4), 507–522.

Hart, P. (2003). *Teachers' thinking in environmental education: Consciousness and responsibility*. New York: Peter Lang.
Hofferth, S. L., & Curtin, S. C. (2006). *Changes in American children's time, 1997–2003*. Retrieved September 2, 2008 from: http://www.popcenter.umd.edu/people/hofferth_sandra/papers/Changesinkidstime97–03v2.pdf
International Society for Technology in Education. (2007a). The ISTE National Educational Technology Standards (NETS-S) and performance indicators for students. Available: http://www.iste.org/Content/NavigationMenu/NETS/ForStudents/NETS_for_Students.htm
International Society for Technology in Education. (2007b). The ISTE National Educational Technology Standards (NETS-T) and performance indicators for teachers. Available: http://www.iste.org/Content/NavigationMenu/NETS/ForTeachers/NETS_for_Teachers.htm
Knapp, D. (2007). A longitudinal analysis of an out-of-school science experience. *School Science and Mathematics, 107*(2), 410–417.
Lieberman, G., & Hoody, L. (1998). *Closing The Achievement Gap*. Pomay, CA: State Education and Environment Roundtable.
Lord, T. (2008). If you go down to the woods today. *Journal of College Science Teaching, 37*(4), 90–93.
Louv, R. (2005). *Last child in the woods: Saving our children from nature-deficit disorder*. Chapel Hill, NC: Algonquin Books.
McKeown-Ice, R. (2000). Environmental education in the United States: A survey of preservice teacher education programs. *Journal of Environmental Education, 32*(1), 4–11.
National Environmental Education and Training Foundation (NEETF) and Roper Starch Worldwide. (2001). *Lessons from the environment: Why 95% of adult Americans endorse environmental education / the ninth annual national report card on environmental attitudes, knowledge, and behaviors*. Washington, DC: Author.
National Science Teachers Association. (2003). *Standards for Science Teacher Preparation*. Reston, VA. Available: http://www.nsta.org/preservice?lid=tnavhp
Noel, A. M. (2007). Elements of a winning field trip. *Kappa Delta Pi Record, 44*(1), 42–44.
Pattengale, J. (2008). Millennial's characteristics: Implications for campus life. *Student Affairs Leader, 36*(16), 4–5.
Podcasting in the classroom. (2008). Retrieved October 1, 2008 from http://www.acteonline.org/uploadedFiles/Publications_and_Online_Media/files/files-techniques-2008/Classroom-Connection-March-2008.pdf
Project Learning Tree. (2008). *PreK-8 Environmental Education Activity Guide*. Washington, DC: American Forest Foundation.
Project WET: curriculum and activity guide. (1995). *Bozeman*. MT: Council for Environmental Education.
Rakow, S. J. (1985). A review of teacher inservice in environmental education: 1970–1980. *Journal of Environmental Education, 16*(4), 7–10.
Ramey-Gassert, L. (1997). Learning science beyond the classroom. *The Elementary School Journal, 97*(4), 433–450.
Ramsey, J. M., Hungerford, H. R., & Volk, T. L. (2001). Environmental education in the K-12 curriculum: Finding a niche. In H. R. Hungerford, W. J. Bluhm, T. L. Volk, & J. M. Ramsey (Eds.), *Essential readings in environmental education* (pp. 111–124). Champagne, IL: Stipes Publishing L.L.C.
Scott, W. A. H. (1996). Pre-service "environmental teacher education": A critique of recent arguments about constraints, approaches and course design. *International Journal of Environmental Education and Information, 15*(3), 307–318.
Sprankle, B. (2006). Podcasting with purpose. *Principal, 58*(4), 62–63.
State Education and Environment Roundtable (SEER). (2005). *California Student Assessment Project Phase Two: The effects of environment-based education on student achievement*. Poway, CA: SEER.
Walnut Park Elementary School. (2009). Mrs. Brackett's 3rd grade class's Owl Pellet podcast. Retrieved March 24, http://www1.gcs.k12.al.us/~podcast/

Wells, N. M. (2000). At home with nature: Effects of "greenness' on children's cognitive functioning. *Environment and Behavior, 32*(6), 775–795.

Willowdale Elementary School. (December 7, 2006). *Endangered animals, everywhere*. RadioWillowWeb. http://www.mpsomaha.org/willow/Radio/shows/Willowcast21.html

Wridt, P. J. (2004). A historical analysis of young people's use of public space, parks and playgrounds in New York City. *Children, Youth and Environments, 14*(1), 86–106.

Zembal-Saul, C., Hershberger, K., & Starr, M. (2006). Podcasts as part of the EDUCATE at Penn State Project: Producing and sharing educationally sound podcasts. In C. Crawford et al. (Eds.), *Proceedings of society for information technology and teacher education international conference 2006* (pp. 4057–4066). Chesapeake, VA: AACE.

Integrating Web-based Activities and Site-based Experiences to Investigate Environmental Issues

Alec M. Bodzin

> Important reserves of oil, gas, and minerals lie deep beneath the seafloor; however, prospecting and drilling for these poses a major threat to sensitive marine habitats and species. Rising energy prices coupled with growing concerns about global warming have sharpened the debate over government funding for offshore drilling versus investing in renewable energy.
>
> Dead zones where fish and most marine life can no longer survive are spreading across the continental shelves of the world's oceans at an alarming rate as oxygen vanishes from coastal waters. Scientists point to tons of nitrogen and phosphorus in fertilizers that run-off from farms and spill into rivers, streams, and bay as well as by fallout from power plants that burn fossil fuels as contributing factors to these dead zones (Diaz and Rosenberg 2008).

These are just two of many issues related to the environment that have risen to the top of the public agenda.

Environmental issues are quite complex, involve conflicting interests and values, and are often controversial. They frequently involve interrelationships between economic, social, cultural, scientific, and political factors. Most issues are open-ended problems in which there is rarely one correct solution to solve the problem and concern multiple stakeholders who view the issue from varying perspectives. Solving an environmental issue requires understanding the context, seeing the problem from varying perspectives, and exploring different possibilities (Environmental Literacy Council 2007). The process involves understanding the practicality of various solutions that are proposed, evaluating scientific evidence, and critically assessing arguments that may involve economic and environmental consequences. In many cases, both the subject matter and interpretations of that subject matter are influenced by value judgments.

Developing skills for understanding and addressing environmental issues is a key component of environmental literacy and is advocated by the North American Association for Environmental Education (NAAEE) as an essential component of preparing preservice teachers (NAAEE 2004). Studying environmental issues provides learners with meaningful contexts by connecting their daily lives and local

A.M. Bodzin (✉)
Lehigh University, College of Education, A113 Iacocca Hall, 111 Research Drive, Bethlehem, PA 18015, USA
e-mail: amb4@lehigh.edu

A.M. Bodzin et al. (eds.), *The Inclusion of Environmental Education in Science Teacher Education*, DOI 10.1007/978-90-481-9222-9_22,

community issues to content of study (Pennock and Bardwell 1994). In addition to understanding the underlying science, investigating environmental issues and their solutions actively involves learners in practicing and improving skills such as critical reflection, problem-solving, and decision making -- each important skill inherent in science teacher education programs. Providing teachers with learning experiences that challenge and enhance their conceptions of environmental issues by confronting them with alternative viewpoints can help them to better understand the viewpoints of others, and become aware of inadequacies and inconsistencies in their own conceptions of environmental issues (Ballantyne and Bain 1995).

This chapter describes how the *Environmental Education* (EE) course at Lehigh University uses a hybrid approach of instruction using web-based activities and face-to-face site-based experiences to primarily focus on the study of environmental issues in the Lehigh River watershed. A watershed is an ideal way to segment the environment for analysis. Watersheds are scalable, topographic, and hydrologic basins that lend themselves to systems analysis. The EE course is designed to meet Pennsylvania Department of Education program standards for EE certification and preparation competencies and is offered during the summer to accommodate schedules of both inservice and preservice teachers.

The Hybrid Approach

Studies have shown that participation in outdoor site-based learning experiences is a promising technique for improving students' environmental attitudes and knowledge (Bogner 1998; Crompton and Sellar 1981; Lisowski and Disinger 1991; Orion and Hofstein 1991, 1994). Learning activities within one's local environment can have a strong effect on the students' environmental learning, enhance environmental attitudes, promote a sense of environmental stewardship, and foster environmental behavior (Bodzin 2008; Fisman 2005; Sobel 2004). Outdoor field settings have also been successful in teaching awareness of environmental issues. Strategies such as field trips to selected sites with environmental problems, and case studies are among the experiences that have been most effective (Howe and Disinger 1988). Such findings support the use of site-based activities for learning about environment issues within a defined watershed area. The local watershed provides for a locale that is a geographically familiar setting for students and is easily accessible for daylong field trips.

Internet technology can be used to support and extend learning activities rooted in outdoor, site-based experiential learning (Moore and Huber 2001). Examples of the existing web-based learning resources that may be used to promote the teaching and learning of environmental issues include descriptive photojournals for virtual watershed explorations of locations that are both accessible and inaccessible by conventional transportation, environmental databases that can be used for analyzing pollution emissions in local areas, and web-based inquiry activities (Bodzin and Cates 2003). More recently, geospatial information technologies and their products such as web-based Geographic Information Systems (GIS) and Google Earth have become readily available. They are quite accessible as internet-based interactive mapping applications with point-and-click access to numerous environmental data sets. Both tools allow for visualizing, mapping, and analyzing multiple layers of georeferenced environmental data. Most web-based GIS require little time to learn, drastically reducing the time it takes for educators to incorporate them into their curricula when compared to desktop-based GIS. No special software is needed to view these maps other than a web browser with an internet connection. The Geospatial One Stop Web site at http://gos2.geodata.gov/wps/portal/gos contains links to many national, state, and local web-based GIS data sites that can be used by educators. Google Earth is a form of digital or virtual globe that allows users to examine satellite imagery and digital aerial photographs overlaid on a three-dimensional representation of the Earth. The interface is simple and intuitive and provides a basic set of navigation controls to allow the user to zoom and pan around the view, as well as the ability to locate a specific place of interest using the search function. Similar to a web-based GIS, Google Earth provides a means to overlay thematic data layers and allows users navigation, data retrieval, and visualization functionality. Using Keyhole Markup Language (KML) or by creating a *mashup* using Google Maps (for example, see Lucking et al. 2008), data information for any location can be customized to create a resource for environmental studies and exploration. Google Earth is available via a free internet download at http://earth.google.com. In educational settings, both web-based GIS and Google Earth have proven to be valuable tools in the process of understanding the environment and of making responsible environmental decisions (Bodzin and Anastasio 2006; Bodzin and Cirruci 2009; National Research Council 2006).

The hybrid approach combines online instructional supports with in-class and outdoor face-to-face interactivity to promote learning. There are many advantages to web-enhanced learning environments for instruction. When properly designed, students enrolled in web-enhanced courses have access to many resources otherwise not accessible in a traditional classroom setting. Web-enhanced classes make learning more accessible and more accommodating to a variety of learners (Grasha and Yangarber-Hicks 2000). In addition, a web-enhanced hybrid approach to EE learning better serves the needs of students over a web-based course since students experience the physical, sensory nature of a live classroom in outdoor field-based settings (Wright 2008).

The EE course at Lehigh University takes advantage of using web-based learning activities during the first face-to-face class session and also within a series of course modules that students complete asynchronously. Many of these curricular materials take advantage of using geospatial technology tools to promote EE learning. The modules include: teaching and learning about environmental issues; geospatial technologies in EE; designing and implementing water quality curricular projects; environmental laws and regulations; EE essentials; and activity selection for EE. The course modules take advantage of many instructional materials that have been developed at the Lehigh Environmental Initiative (EI) to promote the teaching and learning of environmental issues. These materials are primarily located on the

Lehigh Earth Observatory EnviroSci Inquiry Website (http://www.leo.lehigh.edu/envirosci) and also on other Lehigh EI project websites. Since these materials are both interdisciplinary and web-based, they are flexible and portable to use in other disciplines in select secondary- and college-level courses that include geology, environmental science, environmental studies, Earth system science, ecology, or geography.

I have taught the EE course as a lead instructor for the past 6 years. Each year, the course is modified to reflect current local environmental issues in the watershed. New web-based activities and materials are incorporated each year to keep teachers updated on emerging web-based tools and instructional resources that can be used to promote environmental learning with classroom learners. This chapter presents and discusses select course activities that were implemented in 2008. As with other chapters in this section, my goal is to illustrate how technology can be used effectively to support EE teaching and learning with prospective and current science teachers.

Course Activities

The class meets face-to-face for five full days. The first day is spent on-campus at the University and four other days are spent off-campus at site-based settings investigating environmental issues in the watershed. The students complete the web-based modules prior to the off-campus field sites.

To begin the on-campus day, I ask students to draw a sketch map of the Lehigh River watershed. The sketch map task provides me with an understanding of the students' mental maps in terms of how they spatially view the watershed. In one section of the course, seven of nine students' watershed sketch maps included only the southern portion of the Lehigh River, omitting more than half of the watershed area. Each student accurately labeled the three largest cities – Allentown, Bethlehem, and Easton – on their maps. Surprisingly, only two of the seven students who were born and raised in the watershed area had fairly detailed sketch maps that included many features and landmarks. These two students included the Delaware River, specific creek tributaries that flowed into the Lehigh River, prominent geographic features such as the Lehigh Gap and the Kittatinny Ridge/Blue Mountain, names and locations of many boroughs, towns, and smaller cities, farm areas, and industry locations that included a cement plant and Mack trucks. In addition, one student noted environmental issues on his sketch map that included: "farm-based fertilizer runoff, industry outflows, housing development runoff (too much pavement), and recreational usage (litter)." On the other students' sketch maps, only one or two such features were included in each sketch map: farms (1), the Kittatinny Ridge/Blue Mountain (1), the Lehigh Canal (1), Monocacy Creek (1), Bethlehem Steel area (2), the Lehigh Valley Mall (1), the locations of four different bridges over the Lehigh River (1), and the historic Bethlehem area (1).

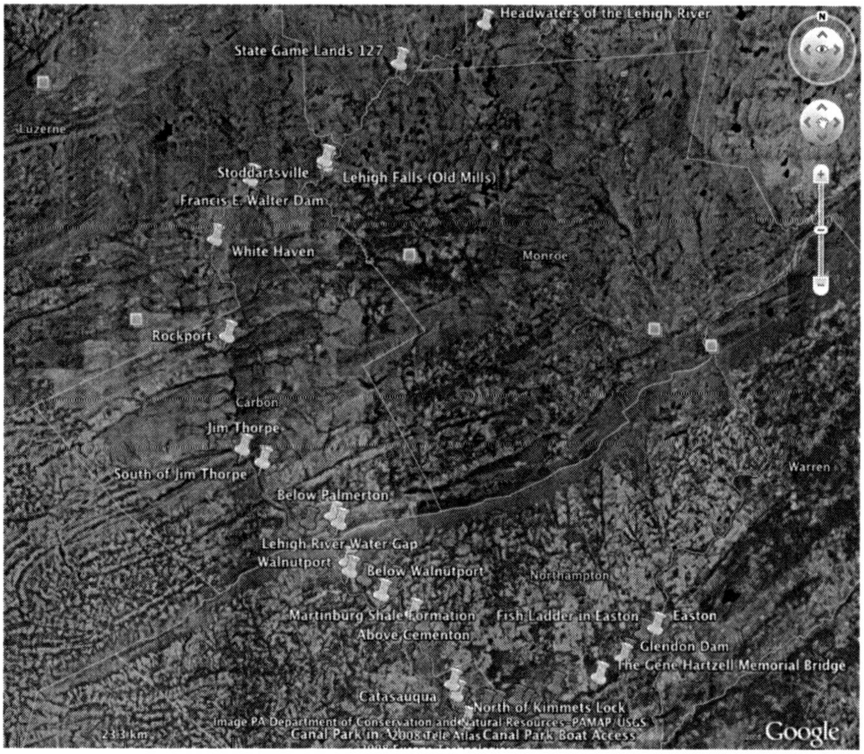

Fig. 1 Google Earth display showing the Lehigh River watershed. Yellow pushpin placemarks denote specific locations of interest in the watershed (Source web address: http://www.leo.lehigh.edu/envirosci/watershed/pjournal/kml/Lehigh River Watershed.kml)

To assist students in understanding the scope and size of the watershed, students use the Google Earth version of the *Lehigh River Photojournal* [http://www.leo.lehigh.edu/envirosci/watershed/pjournal/] (Fig. 1). This virtual photojournal contains yellow pushpin placemarks that denotes specific locations of interest in the Lehigh River watershed. Clicking on a placemark provides the user with a pop-up box containing a digital image of the location and text information about the historical and geological significance about that location (Fig. 2). Certain locations briefly describe environmental issues that are of interest at that location. For example, the *Rockport* placemark informs the user "Buck Mountain Creek (Indian Run) is one of four tributaries containing mine drainage entering into the Lehigh River." Students are shown how to use the drop-down menu in Google Earth's left frame to navigate from one location to the next (Fig. 2). Each placemark has been developed with preset altitudes in the visualization, making specific contextual details easy to observe. In class, I use Google Earth as a virtual fieldtrip to highlight specific areas that we will later visit during the course field trips and also highlight other significant locations pertaining to environmental issues that will not be visited on-site

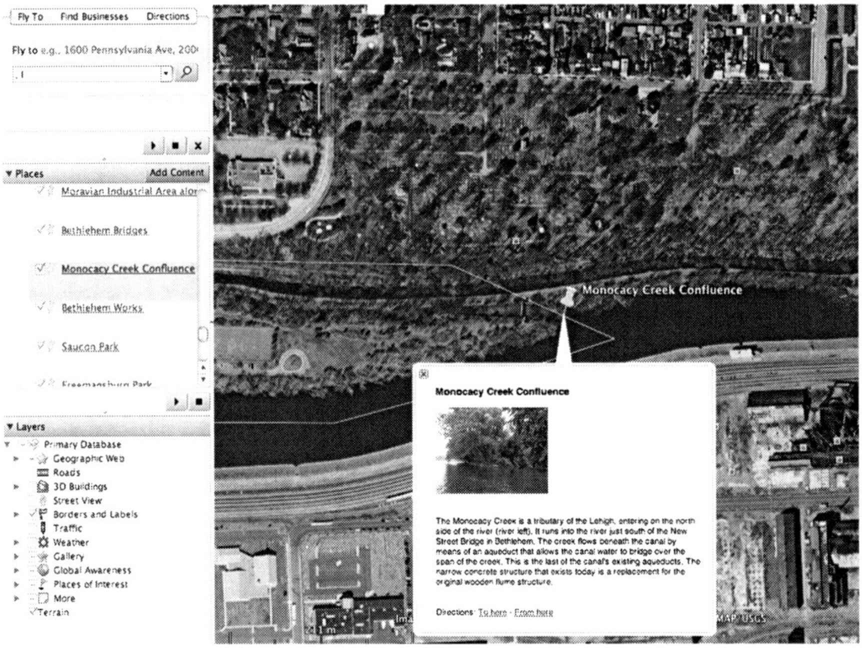

Fig. 2 Google Earth display highlighting the Monocacy Creek Confluence placemark. The drop-down menu in Google Earth's left frame can be used to navigate from one location to the next (Source web address: http://www.leo.lehigh.edu/envirosci/watershed/pjournal/kml/Lehigh River Watershed.kml)

during the course but relate to issues presented in other course materials (such as the location of sinkholes, sources of abandoned mine drainage, and the location of proposed wind turbines to be used for energy generation located in areas of raptor migratory paths). I provide students with exploration time to use the Google Earth visualizations to further develop their spatial concept of the watershed.

I build upon these initial explorations by using a series of web-based GIS maps of the Lehigh River watershed area to specific promote aspects of scientific inquiry and environmental literacy. The GIS maps are disseminated over the internet using a web server and are available at: http://www.leo.lehigh.edu/envirosci/watershed/gis/investigations.html. I use four main topic areas to help learners understand the complex networks of interactions and dependencies within watersheds: *underlying science, human resources, people centers,* and *human impacts*. *Underlying science* focuses on the interdisciplinary study of the complex and interconnected issues of natural watershed processes, natural resources, populations, and pollution. *Human resources* address materials consumed or reused by humans to meet their needs, including air, water, minerals, fuels, building materials, and open space. *People centers* refers to societal needs for human activities, including housing, transportation, agriculture, industry, and recreation; while *human impacts* attends to how human activities affect both biotic and abiotic conditions of the environment.

As a way of illuminating these interactions and complexities, each GIS map is organized to promote inquiry with driving investigative questions about a particular aspect of the Lehigh Valley watershed. The GIS maps are designed around driving investigative questions that incorporate two main properties: *scalability* and *portability*. Scalability refers not only to the need for the problems addressed by the learner to be small enough so that they can derive conclusions in a reasonable length of time but also of sufficient detail so that in completing them they will understand concepts that apply to larger and more complex environmental problems. Portability means the problems addressed in the activities should involve concepts and practices that apply to diverse locations and situations, allowing learners to extrapolate their derived understandings to problems other than those to which they were exposed.

One example that I use in the course focused on the question:

Which Part of the Lehigh River watershed is the best place to build your new home? This GIS map (see Fig. 3) provides learners with a variety of different data layers one may wish to examine when selecting a site to build a new home. Learners can display land use types to determine locations of urban, forested, and agricultural areas in the watershed. Map layers of major, state, and local roads can be shown to determine transportation patterns throughout the watershed. The map also contains data about sites that may be prone to natural hazards. A *limestone* data layer may be displayed to consider locations that may be prone to sinkhole occurrences, and a *flood plains* data layer may be viewed to identify areas where flooding may occur. Industries that release regulated toxic chemicals into the environment can also be located. The *toxic chemical release inventory* data layer provides the name, address, and location of specific industries, and a complete list of chemicals that each site discharges. Recreational and preserved land areas including County and PA State Parks and State Game Lands areas may also be displayed. Census data for each municipality in the watershed for the years 1990 and 2000 are included and can be explored to determine population growth trends in the area. Using this GIS activity, students learn that there are many factors one must consider when selecting a location to build a new home. Such factors involve natural hazards while others involve anthropogenic influences that have environmental consequences to a once natural landscape.

Another core on-campus activity involves student groups analyzing and discussing web-based EE curricular activities that focus on specific environmental issues in the Lehigh River watershed area. The activity involves the analysis of key characteristics of high-quality EE materials (see NAAEE 2000) with a primary focus on examining pedagogical features and supports to promote the learning of environmental issues. This activity is also used to discuss the complexities of particular environmental issues that will later be revisited during the course field trips and in the course modules. Summaries of the four web-based activities are described below.

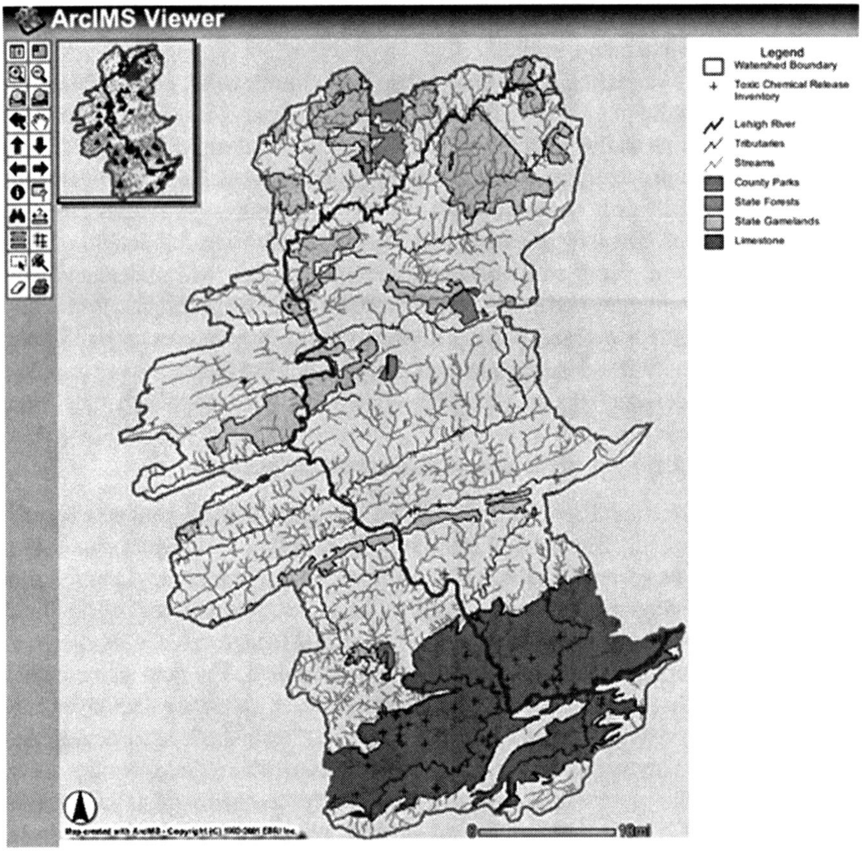

Fig. 3 GIS map of the Lehigh River watershed displaying the location of recreational and preserved lands, limestone areas, and industries discharging toxic chemicals (Source web address: http://128.180.10.97/website/activity3/viewer.htm)

Sprawl in the Lehigh River Watershed Activity

Land use and development in the form of urban or suburban sprawl has always been a problem in the minds of many people. This activity (http://www.leo.lehigh.edu/envirosci/enviroissue/sprawl/) uses web-based GIS maps to explore sprawl issues in the Lehigh River watershed. Learners are first introduced to historical population growth patterns in the Lehigh Valley watershed. Next, they are prompted to use a GIS map to explore trends in population change in the watershed area. The impacts of zoning laws created by multiple municipalities are then presented. Learners are prompted to use a GIS map to explore the effects of transportation infrastructure on land use. Information on the effects of sprawl on human and environmental health is then presented in the activity. Environmental issues that include pollution, effects of creating impervious surfaces, deforestation of riparian buffers, and the reduction of open spaces and farmlands are discussed. Learners are then guided to use GIS maps to examine patterns of land use and population centers. Best practices in land use including smart growth initiatives, brownfield redevelopment, and the creation of conservation easements are discussed. As a culminating activity, learners are presented with two differing viewpoints about creating a new highway extension in the area. They are prompted to select a viewpoint and write a position statement with supporting facts to either favor the highway extension construction or encourage land preservation.

The Land Use Change Unit

The Land Use Change (LUC) unit (http://www.ei.lehigh.edu/eli/luc) is designed to assist students in understanding land use change issues by investigating land use features and issues in the greater Lehigh Valley area. The LUC activities use Google Earth in conjunction with NASA and USGS images to assist learners with enhanced qualitative analysis of land use on the earth's surface. To understand concepts involved in the formation of urban heat islands, students investigate how shopping malls change natural environments. They learn how communities can use certain heat island reduction strategies to reduce the impact of an urban heat island effect. Students complete a case study of the greater Atlanta area to understand environmental issues that are typically associated with sprawl. Their investigations continue with a case study of the Lehigh Valley area in Pennsylvania to identify area land features. They then compare land use types around five different shopping mall areas using Google Earth to examine the significance of mall locations. Students then analyze and interpret satellite data images and aerial photographs to examine landscape changes over time in different locations around the world. In the culminating *Where should we build the new Wal-Mart Supercenter?* activity, students take on the role of a Lehigh Valley Planning Commission employee and recommend a plan for locating a new Wal-Mart Supercenter in the greater metropolitan Lehigh Valley area to have minimal impact on the environment. They use Google Earth to analyze and evaluate features of different land areas for proposed development sites and then develop a proposal to apply "smart growth" principles to their planning decisions and communicate their plan in a simulated planning commission meeting.

Stockertown Sinkhole Dilemma

In the Stockertown Sinkhole Dilemma activity (http://www.leo.lehigh.edu/envirosci/enviroissue/sinkholes), students learn about the sinkholes and decide who should be responsible for compensating property damage caused by a sinkhole. Students adopt different stakeholder roles and access a variety of resources that they will use to develop a position statement about who should be responsible for the investigation and remediation of the sinkholes. They decide what should be done to solve the sinkhole problem, what might be causing the sinkholes, and what new policies should be created to protect the interest of homeowners affected by sinkholes. In this activity, students are responsible for presenting a long-term action plan to prevent and/or remediate sinkhole destruction during a simulated town hall meeting.

Abandoned Mine Drainage in Pennsylvania

Abandoned Mine Drainage in Pennsylvania (http://www.leo.lehigh.edu/envirosci/enviroissue/amd/) is a science-technology-society role-playing debate simulation. In this activity, learners investigate the abandoned mine drainage (AMD) issue from differing perspectives. In their investigation, they identify AMD problems caused by Pennsylvania's long history of coal mining, search for a solution by learning about active and passive treatment systems, and prepare a statement indicating what they believe is the best course of action for treating abandoned mine drainage in Pennsylvania. In class, a debate is held in the form of a town meeting for the Commonwealth of Pennsylvania. Students evaluate active and passive treatment options and decide on a course of action to treat and clean up AMD in Pennsylvania.

Field Trip Site Visits

The course field trips consist of site visits to a variety of locations related to environmental issues in the watershed area. The first trip is spent with a local conservancy organization that works to preserve, protect, restore, and enhance the land, water, ecological, and recreational resources in the Lehigh Valley watershed area. Environmental issues pertaining to agricultural practices serve as case studies for this day. These include allowing unrestricted livestock access to streams and the placement of crop fields that extend out to a waterway making the banks more susceptible to erosion due to a lack of root structure, thus causing increased sedimentation, which is harmful to fish and aquatic macroinvertebrate habitats. Land use practices that seek education and compromise with landowners are highlighted. These include the importance of establishing riparian buffers, in-stream habitat improvement, and streambank stabilization efforts to minimize erosion and restore the stream to a more "natural" condition.

The second site visit is to the Lehigh Gap Nature Refuge (LGNF), the location of the largest EPA Superfund site east of the Mississippi River. At this site, students learn the historical significance and environmental consequences of two large zinc-smelting plants that emitted large amounts of sulfur dioxide for over 80 years. The sulfur dioxide emissions reacted with moisture and gasses in the atmosphere to produce sulfuric acid, which destroyed the plant and microbial life on the nearby Kittatinny Ridge and surrounding areas. When the vegetation died, approximately 2 ft of topsoil washed off about 2,000 acres of the mountain creating a barren environment with soil containing high concentrations of zinc, cadmium, and lead with a subsoil devoid of microbes and organic matter (Kunkle 2004). Since these metals are a potential human health threat, this area was designated a Superfund Site in 1983. At a tour of the site, students learn how the nonprofit LGNF was formed, and the political issues involved with working collaboratively with the Environmental Protection Agency to consider novel revegetation ideas to restore the ecologically degraded mountainside with mixtures of warm season grasses.

The third field trip focuses on recycling initiatives in manufacturing and energy production. The first stop is at a cement manufacturing plant that uses tire-derived fuel and plastic-derived fuel – the burning of plastic types 4–7 (that are currently landfilled) – to offset 50% of coal-burning produced energy and reduce nitrogen oxide emissions. At the plant, students learn about the many state legislative issues involved in obtaining permits for recycling used in a manufacturing process. The issues and concerns that were raised during public forums during the 3-year permitting process are discussed. The second stop is at a power-generating plant that uses culm (anthracite waste coal) in addition to other alternative fuel sources to generate electricity. The site itself is quite unique since it resides on land that once belonged to one of the largest operating cement plants in the world. Environmental permitting issues are highlighted as students become aware of the legislation involved for using the plant's ash by-product to fill abandoned mine strip pits in order to reclaim the land for other commercial purposes.

The fourth field trip is a canoe trip through ten miles of the Lehigh River to examine land use practices. Throughout the trip, we discuss environmental issues pertaining to zoning, industrial water discharges, invasive species, pollution and water quality related to urban development. During the trip, stops are made to gather water-quality data to assess the health of the river.

Discussion

The hybrid approach of using web-based modules and materials in conjunction with site-based experiences to investigate environmental issues in the watershed appears to be an effective course delivery design. The EE course activities provide teachers with an in-depth content understanding of local environmental issues as well as opportunities to explore pedagogical strategies to promote issues-based approaches to learning that hopefully will be adopted for later classroom implementation. The EE course modules are web-based, making learning quite accessible and accommodating for the students. A key feature of the EE course design is that it is "web-enhanced" and not entirely "web-based". An advantage of using a web-enhanced course is that learners experience many interactive dynamics with the course instructor in face-to-face settings at the university and field trip sites. These interactions are highly valuable and cannot be completely replicated in typical web-based learning courseware environments (such as Blackboard or WebCT). Furthermore, learning about an environmental issue at a site-based location with first-hand accounts from people who are intricately involved with an issue is a powerful learning experience. While video and web-base media can be used to learn about environmental issues, physically being at a location where an issue takes place provides for a more compelling setting to promote learning and understanding.

Previous chapters in this book emphasize the importance on incorporating inquiry teaching and learning to promote environmental literacy. It is important that classroom science teachers gain a theoretical and practical understanding on how to take advantage of both web-enhanced and web-based instructional materials to promote essential skill development for understanding and addressing environmental issues in classroom contexts. The EE course takes advantage of the existing web-based curricular materials that highlight the complexity of environmental issues that entail conflicting interests and values and involve understanding the interrelationships between economic, social, cultural, scientific, and political factors. The role-playing simulations described in this chapter highlight the open-endedness of environmental problems that concern multiple stakeholders who view the issue from varying perspectives. Well-designed web-based curricular materials can help learners view a problem or issue from varying perspectives, prompt learners to evaluate scientific evidence, and critically assess arguments that may involve economic consequences. When learners critically examine a local environmental issue with inquiry-based methods, they develop conceptual understandings and practices that can be transferred to related issues in different geographical areas. The web-based materials and activities described in this chapter have such geographical portability. For example, the concepts and understandings one learns from the *Stockertown Sinkhole Dilemma* can be transferred to understand geoenvironmental engineering and policy issues in other areas of the United States that contain limestone geology that might be prone to sinkhole occurrences.

Geospatial information technologies can be used to spatially support learners in EE learning activities. Web-based GIS and Google Earth visualizations are interactive images that are *information-rich* (they include layers representing various types of information) and *dynamic* (learners can explore them by observing spatial patterns and by selecting more or less detail). These applications support learners with the ability to make use of data visualizations for analysis and interpretation when examining environmental issues such as sprawl and land use decision-making.

The incorporation of geospatial information technologies in the course activities appears to be an effective strategy to assist teachers with enhancing their spatial concept of the watershed. Watershed sketch maps that were produced by the students at the end of the course were much more expansive and highly detailed compared to their initial sketch maps that were constructed at the beginning of the course. The end-of-the-course sketch maps included many geographical features and landmarks (such as sinkhole locations and the limestone belt), numerous tributaries of the Lehigh River including those with mine drainage in the northern portion of the watershed, the locations of all sites visited during the field trips, and the names of many boroughs, towns, and small cities throughout the watershed.

Conclusion

This chapter describes how a hybrid approach to EE instruction using web-based activities and site-based experiences can be used to promote the teaching and learning of environmental issues in a local watershed with preservice and inservice teachers. In the course activities presented, teachers learn essential skills for linking EE methods and interdisciplinary content of the natural and social sciences to understand the complexities of environmental issues based in a local watershed. Developing skills for understanding and addressing environmental issues is a key component of environmental literacy. Teacher professional development courses can take advantage of well-designed web-based curricular materials to have learners participate in real-world environmental problem-solving. Such materials present environmental issues in appropriate and engaging ways for learners. Instruction may also take advantage of easily available geospatial information technologies to foster spatial literacy in the curriculum. In a hybrid course, site visits to areas of environmental concern can support and extend the EE concepts and skills that are initially developed with web-based materials.

References

Ballantyne, R., & Bain, J. (1995). Enhancing environmental conceptions: an evaluation of structured controversy learning units. *Studies in Higher Education, 20*(3), 293–303.

Bodzin, A. (2008). Integrating instructional technologies in a local watershed investigation with urban elementary learners. *The Journal of Environmental Education, 39*(2), 47–57.

Bodzin, A., & Anastasio, D. (2006). Using web-based GIS for earth and environmental systems education. *The Journal of Geoscience Education, 54*(3), 295–300.

Bodzin, A., & Cates, W. (2003). Enhancing preservice teachers' understanding of Web-based scientific inquiry. *Journal of Science Teacher Education, 14*(4), 237–257.

Bodzin, A., & Cirucci, L. (2009). Integrating geospatial technologies to examine urban land use change: A design partnership. *Journal of Geography, 108*(4–5), 186–197.

Bogner, F. X. (1998). The influence of short-term outdoor ecology education on long-term variables of environmental perspective. *Journal of Environmental Education, 29*(4), 17–29.

Crompton, J. L., & Sellar, C. (1981). Do outdoor education experiences contribute to positive development in the affective domain? *The Journal of Environmental Education, 12*(4), 21–29.

Diaz, R. J., & Rosenberg, R. (2008 August). Spreading dead zones and consequences for marine ecosystems. *Science, 321*(5891), 926–929.

Environmental Literacy Council. (2007). *Resources for Environmental Literacy*. Arlington, VA: NSTA Press.

Fisman, L. (2005). The effects of local learning on environmental awareness in children: An empirical investigation. *The Journal of Environmental Education, 36*(3), 39–50.

Grasha, A., & Yangarber-Hicks, N. (2000). Integrating teaching styles and learning styles with instructional technology. *College Teaching, 48*(1), 2–10.

Howe, R. W., & Disinger, J. F. (1988). *Teaching environmental education using out-of-school settings and mass media.* (ERIC/SMEAC Environmental Education Digest No. 1). Columbus OH: ERIC Clearinghouse for Science Mathematics and Environmental Education.

Kunkle, D. R. (2004). *Lehigh Gap History and Restoration*. Slatington, PA: Wildlife Information Center.

Lisowski, M., & Disinger, J. F. (1991). The effect of field-based instruction on student understandings of ecological concepts. *The Journal of Environmental Education, 23*, 19–23.

Lucking, R. A., Christmann, E. P., & Whitting, M. J. (2008). Make your own mashup maps. *Science Scope, 31*(8), 58–61.

Moore, C. J., & Huber, R. A. (2001). Support for EE from the National Science Education Standards and the Internet. *The Journal of Environmental Education, 32*(3), 21–25.

National Research Council. (2006). *Learning to think spatially: GIS as a support system in K-12 education*. Washington, DC: National Academy Press.

North American Association for Environmental Education. (2000). *Environmental education materials: Guidelines for excellence workbook; Bridging theory and practice*. Rock Springs, GA: NAAEE.

North American Association for Environmental Education. (2004). *Guidelines for the initial preparation of environmental educators*. Rock Springs, GA: NAAEE.

Orion, N., & Hofstein, A. (1991). The measurement of students' attitudes towards scientific field trips. *Science Education, 75*, 513–523.

Orion, N., & Hofstein, A. (1994). Factors that influence learning during a scientific field trip in a natural environment. *Journal of Research in Science Teaching, 31*, 1097–1119.

Pennock, M. T., & Bardwell, L. V. (1994). *Approaching environmental issues in the classroom*. Ann Arbor, MI: University of Michigan.

Sobel, D. (2004). *Place-based education: Connecting classrooms and communities*. Great Barrington, MA: The Orion Society.

Wright, J. M. (2008). Web-based versus in-class: An exploration of how instructional methods influence postsecondary students' environmental literacy. *The Journal of Environmental Education, 39*(2), 33–45.

Author Index

Subject Index

T

U

Lightning Source UK Ltd.
Milton Keynes UK
20 October 2010

161628UK00006B/1/P

9 789048 192212